Contributions to Management Science

The series *Contributions to Management Science* contains research publications in all fields of business and management science. These publications are primarily monographs and multiple author works containing new research results, and also feature selected conference-based publications are also considered. The focus of the series lies in presenting the development of latest theoretical and empirical research across different viewpoints.

This book series is indexed in Scopus.

More information about this series at http://www.springer.com/series/1505

Serhat Yüksel • Hasan Dinçer
Editors

Strategic Approaches to Energy Management

Current Trends in Energy Economics and Green Investment

Springer

Editors
Serhat Yüksel
Business and Management
Istanbul Medipol University
Istanbul, Turkey

Hasan Dinçer
School of Business
Istanbul Medipol University
Istanbul, Turkey

ISSN 1431-1941 ISSN 2197-716X (electronic)
Contributions to Management Science
ISBN 978-3-030-76782-2 ISBN 978-3-030-76783-9 (eBook)
https://doi.org/10.1007/978-3-030-76783-9

This Springer imprint is published by the registered company Springer Nature Switzerland AG.
The registered company address is: Gewerbestrasse 11, 6330 Cham, Switzerland

Preface

In this book, it is aimed to analyze current approaches to energy production. In this way, it will be possible to provide energy more effectively by taking into account the technological developments. Since energy is an important raw material of the industry, this will contribute significantly to the economic development of the country.

Energy has a wide range of uses within a country. Therefore, energy affects countries both socially and economically. Energy has a major impact on people's social lives, as needs such as warming and enlightenment are met. In addition, energy is also an important raw material used in industrial production. As can be understood from these definitions, energy is one of the essential needs of a country. As can be seen from here, countries have to meet their energy needs regardless of their price. Therefore, it is vital that energy is available cheaply and continuously.

Otherwise, this will reduce the effectiveness in the energy supply process. As can be seen from the definitions above, if energy cannot be supplied continuously, people will experience difficulties in meeting their daily needs. This situation will cause the welfare level in the country to decrease. In addition to the aforementioned issue, if the energy cannot be provided cheaply, people will meet these needs by paying more. Similarly, this will lead to increased dissatisfaction with the people.

On the other hand, efficient supply of energy is important for the continuity of industrial production in the country. Otherwise, it will prevent the economic development of the country. Energy needs to be supplied continuously for the industry to develop. The main reason for this is that if there is a disruption in industrial production, the investments in the country will decrease. Another important issue in this process is the price of the energy provided. If energy is expensive, industrial production will also become expensive.

Another issue to be considered in energy production is related to environmental factors. The main reason for this is the introduction of carbon into the atmosphere in the process of generating energy through the recognition of fossil fuels. This situation threatens human health as it pollutes the air. Another factor to be considered in this process is the adequacy of energy resources. If there is a sufficient amount of energy reserves in one country, that country gets a significant advantage compared to others. Otherwise, this energy must be supplied from abroad. This situation confronts the country with some risks such as exchange rate risk.

In summary, energy supply is a vital factor for a country and factors such as price, continuity, environmental pollution, and the country's own energy resources are important for the effectiveness of this process. Therefore, it is obvious that studies taking these issues into consideration will contribute significantly to increasing energy efficiency. For this reason, it is obvious that studies on current developments in energy are needed.

In this study, it is aimed to analyze current trends in energy production and use. In this framework, technological developments that contribute to the reduction of price in energy production constitute an important subject of the book. On the other hand, renewable energy sources that provide continuity in energy production and do not emit carbon into the atmosphere are another issue that we plan to include in this book. In this context, studies on the effectiveness of wind, solar, biomass, geothermal, and hydroelectric energies are targeted. On the other hand, studies on current technological approaches that prevent environmental pollution such as carbon capture and storage will be included in this book.

It is possible to mention many different aspects of this book compared to others. Energy books on the market generally examine the theoretical issues in the field of energy economics. However, this book will only include current issues that can increase energy efficiency. In other words, each study in this book will propose strategies to use energy more effectively and efficiently. On the other hand, it will be possible to reach most of the current practices and technological developments regarding energy efficiency from this book. The strategies presented in this study will be a significant guide to both academics and those working in the market. Considering these issues, it is thought that this book will make a serious contribution to the literature.

Istanbul, Turkey

Serhat Yüksel
Hasan Dinçer

Acknowledgement

The authors would like to acknowledge the help and patience of their families in this book process. Without their support, this book would not have become a reality.

Second, the authors wish to acknowledge the valuable contributions of the reviewers regarding the improvement of quality, coherence, and content presentation of chapters.

About This Book

In this book, it is aimed to identify the significant issue to improve energy effectiveness. For this purpose, all different perspectives of the balanced scorecard approach are taken into consideration, such as finance, customer, technological factors, and learning & growth. With the help of focusing on different perspectives, it can be possible to generate more appropriate strategies to increase energy investments. Since energy is an important raw material of the industry, this will contribute significantly to the economic development of the country.

Target audience and potential users of this book are defined below.

- Researchers
- Academicians
- Policy makers
- Government officials
- Upright students in the concerned fields
- Members of Chambers of Commerce and Industry
- Top managers of the companies

Contents

About the Editors

Serhat Yüksel is an Associate Professor of Finance in İstanbul Medipol University. Before this position, he worked as a senior internal auditor for seven years in Finansbank, Istanbul-Turkey and 1 year in Konya Food and Agriculture University as an assistant professor. Dr. Yüksel has a BS in Business Administration (in English) from Yeditepe University (2006) with full scholarship. He got his master degree in economics from Boğaziçi University (2008). He also has a PhD in Banking from Marmara University (2015). His research interests lie in banking, finance, and financial crisis. He has more than 200 scientific articles and lots of them are indexed in SSCI, Scopus, and Econlit. Also, he is the editor of many books that were published by Springer and IGI Global (The School of Business, İstanbul Medipol University, serhatyuksel@medipol.edu.tr).

Hasan Dinçer is a Professor of Finance at Istanbul Medipol University, Faculty of Economics and Administrative Sciences, Istanbul-Turkey. Dr. Dinçer has BAs in Financial Markets and Investment Management at Marmara University. He received PhD in Finance and Banking with his thesis entitled "The Effect of Changes on the Competitive Strategies of New Service Development in the Banking Sector." He has work experience in the finance sector as a portfolio specialist and his major academic studies focus on financial instruments, performance evaluation, and economics. He is the executive editor of the International Journal of Finance and Banking Studies (IJFBS) and the founder member of the Society for the Study of Business and Finance (SSBF). His research interests lie in banking, finance, and financial crisis. He has more than 200 scientific articles and lots of them are indexed in SSCI, Scopus, and Econlit. Also, he is the editor of many books that were published by Springer and IGI Global (The School of Business, İstanbul Medipol University, hdincer@medipol.edu.tr).

Chapter 1
Identifying the Best Financing Sources for Renewable Energy Companies in Latin American Countries

Hasan Dinçer, Anton Lisin, Gözde Gülseven Ubay, and Çağatay Çağlayan

Abstract The main purpose of this study is to identify the best financing source in order to start a new renewable energy project in Latin America countries. In this case, a large literature on related subject is reviewed. As the result of this analysis, the most-used ways to fund a renewable energy project are determined. By selecting 5 funding options among them, an analysis has been conducted by using Analytic Hierarchy Process (AHP) method in order to weight these criteria set. The findings indicate that government grants and subsidies play the most significant role in order to fund a renewable energy project. In addition to this situation, it is also concluded that using bank loans is another important way for this issue. On the other side, crowdfunding, angel investors and venture capital are on the last ranks in this regard. By taking account these results, it can be said that to increase renewable energy projects government should take actions, give some incentives and facilities government-backed funds for green investors. The biggest disadvantage of renewable energy investments is the high costs. In such investments, especially the initial setup costs are very high. In this context, the tax reduction to be provided by the government to renewable energy investors will contribute to these companies gaining a serious cost advantage. On the other hand, interest-free loans to be given to these investors by the state will also help to increase the efficiency in this process.

H. Dinçer (✉) · G. G. Ubay · Ç. Çağlayan
The School of Business, İstanbul Medipol University, İstanbul, Turkey
e-mail: hdincer@medipol.edu.tr; gozde.ubay@std.medipol.edu.tr; cagatay.caglayan@std.medipol.edu.tr

A. Lisin
Financial University Under the Government of the Russian Federation, Moscow, Russia

S. Yüksel, H. Dinçer (eds.), *Strategic Approaches to Energy Management*, Contributions to Management Science,
https://doi.org/10.1007/978-3-030-76783-9_1

1.1 Introduction

Energy is the biggest issue to be considered for the development of a region or country in many ways. Firstly, energy is the main reason of a country current account deficit problems or external dependency problem. There is no substitute for energy, as it is a basic need for life. Therefore, all countries and companies make significant investments in the field of energy (Koengkan et al., 2020; Dinçer et al., 2019). Considering that the energy need of the world is increasing day by day, it would not be wrong to say that the energy sector will continue to develop. In this framework, fossil fuels, nuclear energy and renewable energy sources are used to meet the increasing energy needs of future generations (Yuan et al., 2020; Zhao et al., 2021). The concern that nuclear energy waste harms the environment will significantly reduce the willingness to invest in this area, given public acceptance. There is a different scenario for energy production from fossil fuels. Given that carbon dioxide is a large part of greenhouse gases, there is a global effort to minimize carbon emissions (Abolhosseini & Heshmati, 2014; Yüksel et al., 2019).

This situation forces governments and private investors to invest in renewable energy alternatives. One of the biggest advantages of renewable energy investments is that when applied correctly, there is no problem in public acceptance. Undoubtedly, this advantage will also make potential investors willing to invest in renewable energy (Qiu et al., 2020; Zhou et al., 2020). Renewable energy, which contains energy types such as solar, wind, hydroelectric, geothermal, biomass, wave energy, reduces the foreign dependency of countries. The main reason is that with the help of renewable energy alternatives, countries can generate their own energy (Dinçer & Yüksel, 2019a, b; Ubay & Karakuş, 2020). In addition to this issue, renewable energy usage also minimizes the damage to the environment by reducing carbon emission. This situation has a contribution to both social and economic development of the countries. If there is low carbon emission, it causes less people to be sick. This condition has an increasing effect on the life quality of the people (Du et al., 2020). Additionally, as people become less sick, there will be no loss of workforce and healthcare costs will decrease (Qi et al., 2020; Wang et al., 2019).

These mentioned benefits are the main reason why Latin American countries are also interested in this type of energy. Although there is no specific definition of Latin American countries, it is also known as all countries in the south of the USA or a group of South America countries where Latin languages such as Spanish, Italian and Portuguese are spoken predominantly. 20 countries make up the mentioned region. These countries; Argentina, Bolivia, Brazil, Colombia, Costa Rica, Cuba, Dominican Republic, Ecuador, Puerto Rico, El Salvador, Guatemala, Honduras, Mexico, Nicaragua, Panama, Paraguay, Peru, Chile, Uruguay, Venezuela. It is predicted that these countries and the Caribbean's need for electricity will double by 2030 and triple by 2050. Large investments are required to ensure that electricity generation can overcome this intense need (Majano, 2017). Therefore, the energy sector, which requires such large investments to be made, requires companies that will create a competitive and efficient market to enter the market. The creation of an energy market in Latin American countries, which is regulated by investor-friendly and strong policies, will increase electricity generation sustainably with companies entering the market, but there are also problems waiting for investors who want to enter these investments (Washburn & Pablo-Romero, 2019). The most important of these is that renewable energy investments require a large capital. Therefore, companies who want to invest in this field will have to solve their financing source problems. Financing problems will negatively affect the success of companies in the energy sector and therefore the energy sector (An et al., 2020; Dooyum et al., 2020). Therefore, determining the best financing resources of the renewable energy companies of Latin American countries is a necessity for the sector to be successful. Five criteria come to the fore when funding sources are offered to these companies. These criteria consist of bank loans, government donations and partnerships, crowdfunding, angel investors and venture capital (Kobtseva et al., 2017; Morkovkin et al., 2016).

The success of businesses is initially linked to the entrepreneur's ability to gain support from bankers, venture capitalists, suppliers, consultants, and clients (Buttner & Rosen, 1988). Bank loans accelerate and increase investments, thus production. It enables companies to start their investments without waiting for a long time for the required capital. If renewable energy companies in Latin American countries can obtain high loans from banks with low interest rates, the search for financing resources of companies will decrease significantly and investments will accelerate (Mikhaylov, 2021; Yumashev et al., 2020). Besides banks, governments are also an important factor for high capital required projects. All governments generally subsidize businesses in a variety of ways. The main objectives of governments in this regard are to help companies eliminate their financial difficulties in their investments and to encourage them to use new technologies (Girma et al., 2007). This incentive move by the governments will prevent possible market failures and the effects of failure. For example, governments try to solve the problem of spending less on R&D investments by using grant steps for R&D. These interventions encourage companies to invest in research (Hong et al., 2015). Governments can be an important driving force as a source of financing for Latin American companies. With support such as feed-in tariffs, tax incentives and grants, Latin American countries can increase their investments by solving the financing problem of renewable energy companies. Increasing these investments will reduce the foreign dependency of the country's economy in the long run and contribute to the country's economy to a great extent.

As it seen there are a lot of funding options to bring an energy project to life. However since, energy projects, especially renewable ones, are long-term and expensive projects, funding options should be evaluated in a logical framework and investors should choose the best option. In order to facilitate this, in this study, it is targeted to select best funding option to start a renewable energy project. The aim of the article is to contribute to the determination of the financing resources of the renewable energy companies of Latin American countries in line with the mentioned criteria. In this context, by making a large literature review first, five most-used funding options in that area are determined. Afterwards, these criteria set is evaluated by applying the AHP. This methodology is a basic approach to decision making. AHP uses both objective and subjective evaluation criteria and tests the consistency of the evaluations (Darko et al., 2019). In particular, it ensures that a very important decision such as which one should be given priority among the alternatives that should be evaluated according to many criteria.

It is possible to state that this study contributed to the literature in many ways. First of all, it is tried to determine which is the most optimal way to finance renewable energy investments. Renewable energy investments are among the energy policies of many countries. Therefore, these energy alternatives are expected to be even more popular in the future. However, the biggest disadvantage of these investments is the very high installation costs. Therefore, determining how to finance these investments in the most optimal way is a guide for both researchers and company managers. Another contribution of the study to the literature is that the application is for Latin American countries. In these countries, there has been a serious population increase in recent years. This situation simultaneously increases the energy requirement. If the necessary measures are not taken, there is a risk that these countries will experience energy problems in the coming years. In this context, renewable energy alternatives are vital for these countries. Thanks to the results that can be obtained in this study, strategies to increase renewable energy investments in these countries can be produced.

This study has four different sections. The first section in the study contains general information on renewable energy projects and funding options. In the second part of the study, similar studies in the literature are analyzed and a review was given to the readers. In the third section, an analyze was made on the criteria set by applying the AHP. In the fourth section, the analysis results are discussed, and some suggestions are given in the last section.

1.2 Literature Review

Energy investments are both capital intensive, involving many risks and spanning a very long period of time. So that, many studies have been analyzed in the area of funding and energy companies on the related literature. While the vast majority of these studies include researches to facilitate companies to find funds and which fund is more ideal, there are studies in the literature to meet the need of energy companies for funding since they require high amounts of funds. There are many funding options to fund an energy projects for companies. One of them is getting bank loans. With growing concerns over increased industrialization, urbanization and greenhouse gas emissions, the currently energy-consuming economy of Latin American countries presents one of the biggest challenges and opportunities for sustainable energy investments in the region (He et al., 2019). Supporting sustainable energy investments is one of the main elements of banks' overall strategy. Banks play a key role in the implementation of renewable energy projects by preparing funds to lend to small and large-scale energy efficiency projects (Chang et al., 2019;

Liu et al., 2021a, b; Cheng et al., 2020). Since renewable energy projects are high-volume projects, they are in a very risky position. For this reason, even though the bad loan rates are high, nowadays, when green energy is in demand, the funding support of banks for renewable energy investments is increasing day by day (Zeng et al., 2017).

Angel investor is a person who invests in a business opportunity that is just beginning the way, providing capital for the growth or development of businesses. In another definition; It is defined as a special investor type that invests in companies with high risk and high growth potential at a very early stage of their establishment (Lam & Law, 2018). The goal of the angel investor is to develop these projects by providing financial support to newly formed and promising projects. In other words, when a business idea with high return potential comes to the implementation stage, it attracts the attention of the angel investor (Li et al., 2021; Yüksel et al., 2020). The interest of angel investors for the renewable energy sector, which has recently been growing rapidly in both developed and developing countries, is also great. Although these projects take longer to receive their earnings than others, many angel investors can fund such projects for both social responsibility and future investment (Lam & Law, 2016).

Countries have included various fiscal, money, credit and foreign trade policies implemented within their economic systems in order to ensure their social and economic development and to increase their competitiveness in the international arena. Incentives are used in various intensities in all countries as a means of fiscal policy (Zhao et al., 2016). It can be said that incentives in renewable energy production are increasingly implemented by many countries, especially developed countries, in accordance with the mentioned theoretical purposes. Countries set some targets for renewable energy production and implement various incentive policies for investors in line with these targets (Haghi et al., 2018). As is known, incentives; It is defined as all kinds of financial and/or non-financial support, assistance or encouragement provided by the state through various methods. In this context, the main renewable energy incentives are; fixed price guarantee, premium guarantee, mandatory quota and green certificate applications, various tax incentives and investment loans (Yoshino & Taghizadeh-Hesary, 2018). Investment loans are low-interest and long-term loans for a certain percentage of total costs or per installed KWh (kilowatt-hour) for the development of renewable energy investments. States may either have supported investors' renewable energy projects with such loans or finance a certain percentage of the investment cost in the form of grants to encourage renewable energy generation (Zeng et al., 2017).

Crowdfunding has recently become a valuable source of financing for entrepreneurs in need of financing. Current analysis reports that a large volume of money raised by crowdfunding has been achieved. Crowdfunding allows these entrepreneurs in need of financing to collect funds in line with their calls on the internet (Belleflamme et al., 2014). Although the concept of crowdfunding is similar to the fundraising process of non-governmental organizations for any project, it can be introduced as a source of financing to companies seeking investment capital. Crowdfunding is classified as reward-based, equity-based, donation-based, and debt-based (Hoegen et al., 2018). In equity-based crowdfunding, investors buy stocks in that company. In loan-based crowdfunding, there is a repayment of the money with interest. Thus, crowdfunding can provide opportunities for Latin American companies to invest in renewable energy and reduce the financing problem to some extent.

Another alternative to finding funds to ensure the continuity and increase of renewable energy projects is venture capital participating in investors through partnership (Lyu & Shi, 2018). At the global level, technology and innovation developers and equipment manufacturers are inviting venture capital funds to join their organizations to both provide financing and use market knowledge and connections. Although these investments are risky in terms of venture capital, they are considered attractive because their future potential can be very large. However, many developing countries, especially in the Latin America region, can benefit from these resources at very limited levels compared to their developed examples (Pickl, 2019). Although there are main reasons for this such as macroeconomic uncertainties and the difference in investment culture, the limited number of institutional partners that can support venture capital is also a very important factor. Supported globally by pension funds, insurance companies or wealth funds, venture capital is struggling to find and transfer funds in many countries (Cumming et al., 2016). The biggest improvement that can be made in this regard will be to regulate the flow of funds accumulated in countries to the startups of the countries through the venture capital with the incentives related to the legislation.

As a result of the literature review, some results have been reached. First of all, renewable energy investments are vital for the energy security of countries. The main reason for this is that countries can produce their own energy thanks to these energy alternatives. On the other hand, it has been determined that the biggest disadvantage of renewable energy investments is the high installation cost. This situation shows that it is very important to determine how renewable energy companies can reach the funds they need. On the other hand, it has been determined that these companies have many ways to reach the funds they need, such as government incentives and using loans. However, the most important deficiency in the literature is that there is no study examining which of these resources is the most efficient for renewable energy investors. In this study, it is aimed to determine which of the different fundraising methods is the most optimal with the help of AHP method. In this way, it is intended to make an important contribution by analyzing this issue that has not been examined in the literature before.

1.3 Analysis Results

There are many funding options that can be used when implementing renewable energy projects. Increasing the number of criteria makes it difficult to select the source to receive funding. In this analysis, the options that companies have in order to fund their renewable energy projects are examined. In this context, by using

Table 1.1 List of funding options

Criteria	Literature background
Bank Loan (C1)	He et al. (2019), Chang et al. (2019), Zeng et al. (2017), Li et al. (2020)
Government Grants and Subsidies (C2)	Zhao et al. (2016), Haghi et al. (2018), Yoshino and Taghizadeh-Hesary (2018), Zeng et al. (2017), Dinçer et al. (2020)
Crowdfunding (C3)	Belleflamme et al. (2014), Hoegen et al. (2018)
Angel Investors (C4)	Lam and Law (2016, 2018)
Venture Capital (C5)	Lyu and Shi (2018), Pickl (2019), Cumming et al. (2016)

Table 1.2 The details of expert evaluations

Criteria	Criteria	Expert 1	Expert 2	Expert 3
Bank Loans (C1)	Government Grants and Subsidies (C2)	0.33	0.50	0.33
Bank Loans (C1)	Crowdfunding (C3)	8.00	7.00	7.00
Bank Loans (C1)	Angel Investors (C4)	7.00	6.00	5.00
Bank Loans (C1)	Venture Capital (C5)	4.00	3.00	4.00
Government Grants and Subsidies (C2)	Crowdfunding (C3)	9.00	9.00	9.00
Government Grants and Subsidies (C2)	Angel Investors (C4)	8.00	7.00	8.00
Government Grants and Subsidies (C2)	Venture Capital (C5)	7.00	5.00	6.00
Crowdfunding (C3)	Angel Investors (C4)	0.33	0.50	0.33
Crowdfunding (C3)	Venture Capital (C5)	0.20	0.17	0.20
Angel Investors (C4)	Venture Capital (C5)	0.50	0.33	0.33

5 different criteria, an analysis in AHP method framework is conducted. Many solution techniques have been developed for such problems solved under the name of Multi Criteria Decision Making Techniques. AHP is the most used of these techniques. AHP is an effective analysis in multi-criteria decision making.

In the analysis process of this study, firstly, options that companies or investors have in order fund their renewable energy projects are determined. In this context, up to date studies on this subject in the literature are analyzed and noted. As a result of the analysis, 5 different criteria are determined as Bank Loan (C1), Government Grants and Subsidies (C2), Crowdfunding (C3), Angel Investors (C4), Venture Capital (C5) as shown in Table 1.1.

Table 1.1 indicates that the options that companies or investors have to fund their renewable energy projects. Firstly, evaluations of three different experts, who are academicians and high-level managers in renewable energy companies, are acquired. These people have at least 15-year experience in the area of this study. These experts made their evaluations by considering 9 different scales. These calculations are made on Microsoft Excel program. Table 1.2 gives information about the evaluation details of the experts.

Table 1.3 Pairwise comparison matrix

Criteria	C1	C2	C3	C4	C5
C1	1.00	0.39	7.33	6.00	3.67
C2	2.57	1.00	9.00	7.67	6.00
C3	0.14	0.11	1.00	0.39	0.19
C4	0.17	0.13	2.57	1.00	0.39
C5	0.27	0.17	5.29	2.57	1.00

Table 1.4 Normalized matrix

Criteria	C1	C2	C3	C4	C5
C1	0.24	0.22	0.29	0.34	0.33
C2	0.62	0.56	0.36	0.43	0.53
C3	0.03	0.06	0.04	0.02	0.02
C4	0.04	0.07	0.10	0.06	0.03
C5	0.07	0.09	0.21	0.15	0.09

Table 1.5 The weights of the criteria

Criteria	Criteria weights
Bank Loan (C1)	0,2833
Government Grants and Subsidies (C2)	0,5004
Crowdfunding (C3)	0,0347
Angel Investors (C4)	0,0612
Venture Capital (C5)	0,1204

With the help of these evaluations, pairwise comparison matrix is created, and they are detailed in Table 1.3.

In the next process, this matrix is normalized by dividing all these values to the sums of the rows. Table 1.4 indicates the details of the normalized matrix.

In the final stage, the weights of the criteria are defined. They are demonstrated on Table 1.5.

Table 1.5 indicates that government grants and subsidies (C2) is the most important criterion. Moreover, it is also identified that banking loans (C1) are also important for renewable energy investments. Nevertheless, crowdfunding (C3), angel investors (C4) and venture capital (C5) play a lower significant role in this regard.

1.4 Discussion and Conclusion

In this study, it is investigated that which financing resources of the renewable energy companies of Latin American countries can be more beneficial. In other words, the research question of this study is investors or companies should choose which funding option to start a new renewable energy project. In this target, a wide literature review was made to determine most-used funding options in these types of projects. According to this review, 5 different criteria that are taught to be beneficial to use as fund option were found. Afterwards, an analysis with AHP method in order to weight these criteria set was conducted. As a result of the analysis, it is found that government grants and subsidies (C2) is the most important criteria. In addition to this, it is also determined that bank loan (C1) is also very important to fund renewable energy projects. In the contrast, it is seen that crowdfunding (C3) and angel investors (C4) take place in the last ranks.

According to results, it is seen that in order to increase renewable energy projects, governments should be in charge almost in every step. An important issue that emerges at the current point regarding the energy transformation and in future plans is related to how actors play a role in this transformation. As can be seen in the results of the analysis, for Latin American countries, states and state institutions are largely determined by the specific policies and targets in this transformation. It is seen that the stakeholders of the sector are the actors that make the transformation in this direction. In this context, we see that the energy transformation experienced in the region will be like major economic transformations in history, in which the state institutions have taken a direct realizing role. What the state should do at this point will be to provide various privileges to such renewable energy projects and to help them as funds. Latin American country governments can provide convenience to such investors in tax matters. In addition, they can allocate areas where renewable energy facilities will be established to investors under favorable conditions. In fact, they can play a key role in increasing these projects with direct government funded funds for investors.

The place and importance of renewable energy in its banks is very important. Banks, which previously financed large-scale projects with significant market risk, will encounter project alternatives in a wider spectrum as the energy shifts towards more renewable. Generation, distribution, network infrastructure based on renewable resources, facility and infrastructure development for equipment production, end-consumer productivity increase are some of the project types that we can count at first glance. This situation will bring along an important opportunity for banks. Because the increasing variety of projects will positively affect risk management and it will be possible for banks to manage these loans more effectively. On the other hand, in order for banks to be able to read and contribute to this process, they need to make up for their lack of knowledge about energy transformation by recruiting experts or consultants who are competent in the subject. Turning to the funds allocated for renewable energy in foreign countries and bringing these funds to the countries will also accelerate the financing of projects in Latin American countries. In the literature, Yoshino and Taghizadeh-Hesary (2018) and Zeng et al. (2017) also reached similar conclusions by considering different country groups.

There are some limitations in this study. Firstly, in this paper, it is examined that using which funding option is more beneficial in renewable energy projects for companies or investors. However, in this process, only five criteria have taken into account. So, for the future studies, more funding options such as microfinance providers can be added into the analysis. Moreover, this study is made for renewable energy projects. In the next studies, authors can be more specific and focus renewable energy projects separately or they can examine non-renewable energy projects.

Besides these, in analysis process of this study, Analytic Hierarchy Process (AHP) method is used. In the future, these analyses can be conducted with other methods such as ANP, DEMATEL and so on. In that case, more accurate results can be obtained.

References

Abolhosseini, S., & Heshmati, A. (2014). The main support mechanisms to finance renewable energy development. *Renewable and Sustainable Energy Reviews, 40*, 876–885.

An, J., Mikhaylov, A., & Richter, U. H. (2020). Trade war effects: evidence from sectors of energy and resources in Africa. *Heliyon, 6*(12), e05693.

Belleflamme, P., Lambert, T., & Schwienbacher, A. (2014). Crowdfunding: Tapping the right crowd. *Journal of Business Venturing, 29*(5), 585–609.

Buttner, E. H., & Rosen, B. (1988). Bank loan officers' perceptions of the characteristics of men, women, and successful entrepreneurs. *Journal of Business Venturing, 3*(3), 249–258.

Chang, K., Zeng, Y., Wang, W., & Wu, X. (2019). The effects of credit policy and financial constraints on tangible and research & development investment: Firm-level evidence from China's renewable energy industry. *Energy Policy, 130*, 438–447.

Cheng, F., Lin, M., Yüksel, S., Dİncer, H., & Kalkavan, H. (2020). *A hybrid hesitant 2-tuple IVSF decision making approach to analyze PERT-based critical paths of new service development process for renewable energy investment projects*. IEEE Access.

Cumming, D., Henriques, I., & Sadorsky, P. (2016). 'Cleantech' venture capital around the world. *International Review of Financial Analysis, 44*, 86–97.

Darko, A., Chan, A. P. C., Ameyaw, E. E., Owusu, E. K., Pärn, E., & Edwards, D. J. (2019). Review of application of analytic hierarchy process (AHP) in construction. *International Journal of Construction Management, 19*(5), 436–452.

Dinçer, H., & Yüksel, S. (2019a). Multidimensional evaluation of global investments on the renewable energy with the integrated fuzzy decision-making model under the hesitancy. *International Journal of Energy Research, 43*(5), 1775–1784.

Dinçer, H., & Yüksel, S. (2019b). Balanced scorecard-based analysis of investment decisions for the renewable energy alternatives: A comparative analysis based on the hybrid fuzzy decision-making approach. *Energy, 175*, 1259–1270.

Dincer, H., Yüksel, S., & Martinez, L. (2019). Balanced scorecard-based Analysis about European Energy Investment Policies: A hybrid hesitant fuzzy decision-making approach with quality function deployment. *Expert Systems with Applications, 115*, 152–171.

Dinçer, H., Yüksel, S., Ubay, G. G., & Karakuş, H. (2020). BSC-based evaluation for the factors affecting the performance of wind energy companies. In *Strategic priorities in competitive environments* (pp. 1–15). Springer.

Dooyum, U. D., Mikhaylov, A., & Varyash, I. (2020). Energy security concept in Russia and South Korea. *International Journal of Energy Economics and Policy, 10*(4), 102.

Du, L., Dinçer, H., Ersin, İ., & Yüksel, S. (2020). IT2 fuzzy-based multidimensional evaluation of coal energy for sustainable economic development. *Energies, 13*(10), 2453.

Girma, S., Görg, H., & Strobl, E. (2007). The effect of government grants on plant level productivity. *Economics Letters, 94*(3), 439–444.

Haghi, E., Raahemifar, K., & Fowler, M. (2018). Investigating the effect of renewable energy incentives and hydrogen storage on advantages of stakeholders in a microgrid. *Energy Policy, 113*, 206–222.

He, L., Liu, R., Zhong, Z., Wang, D., & Xia, Y. (2019). Can green financial development promote renewable energy investment efficiency? A consideration of bank credit. *Renewable Energy, 143*, 974–984.

Hoegen, A., Steininger, D. M., & Veit, D. (2018). How do investors decide? An interdisciplinary review of decision-making in crowdfunding. *Electronic Markets, 28*(3), 339–365.

Hong, J., Hong, S., Wang, L., Xu, Y., & Zhao, D. (2015). Government grants, private R&D funding and innovation efficiency in transition economy. *Technology Analysis & Strategic Management, 27*(9), 1068–1096.

Kobtseva, O. N., Novoselova, N. N., Novoselov, S. N., Morkovkin, D. E., & Sidorchukova, E. V. (2017). Organizational and economic features of import substitution formation and realization in the conditions of spatial restrictions. *International Journal of Applied Business and Economic Research, 15*(23), 25–35.

Koengkan, M., Fuinhas, J. A., & Vieira, I. (2020). Effects of financial openness on renewable energy investments expansion in Latin American countries. *Journal of Sustainable Finance & Investment, 10*(1), 65–82.

Lam, P. T., & Law, A. O. (2016). Crowdfunding for renewable and sustainable energy projects: An exploratory case study approach. *Renewable and Sustainable Energy Reviews, 60*, 11–20.

Lam, P. T., & Law, A. O. (2018). Financing for renewable energy projects: A decision guide by developmental stages with case studies. *Renewable and Sustainable Energy Reviews, 90*, 937–944.

Li, X., Zhu, S., Yüksel, S., Dinçer, H., & Ubay, G. G. (2020). Kano-based mapping of innovation strategies for renewable energy alternatives using hybrid interval type-2 fuzzy decision-making approach. *Energy*, 118679.

Li, Y. X., Wu, Z. X., Dinçer, H., Kalkavan, H., & Yüksel, S. (2021). Analyzing TRIZ-based strategic priorities of customer expectations for renewable energy investments with interval type-2 fuzzy modeling. *Energy Reports, 7*, 95–108.

Liu, Y., Gong, X., Yüksel, S., Dinçer, H., & Aydın, R. (2021a). A multidimensional outlook to energy investments for the countries with continental shelf in East mediterranean region with hybrid decision making model based on IVIF logic. *Energy Reports, 7*, 158–173.

Liu, J., Lv, J., Dinçer, H., Yüksel, S., & Karakuş, H. (2021b). Selection of renewable energy alternatives for green blockchain investments: A hybrid IT2-based Fuzzy modelling. *Archives of Computational Methods in Engineering*, 1–15.

Lyu, X., & Shi, A. (2018). Research on the renewable energy industry financing efficiency assessment and mode selection. *Sustainability, 10*(1), 222.

Majano, A. M. (2017). *Study on the development of the renewable energy Market in Latin America and the Caribbean.*

Mikhaylov, A. (2021). Development of Friedrich von Hayek's theory of private money and economic implications for digital currencies. *Terra Economicus, 19*(1) (In Russia).

Morkovkin, D. E., Mamychev, A. Y., Yakovenko, N. V., Derevyagina, M. V., & Didenko, O. V. (2016). Factors and material conditions for space-intensive economic development of region. *International Review of Management and Marketing, 6*(1), 67–72.

Pickl, M. J. (2019). The renewable energy strategies of oil majors—From oil to energy? *Energy Strategy Reviews, 26*, 100370.

Qi, W., Huang, Z., Dinçer, H., Korsakienė, R., & Yüksel, S. (2020). Corporate governance-based strategic approach to sustainability in energy industry of emerging economies with a novel interval-valued intuitionistic Fuzzy hybrid decision making model. *Sustainability, 12*(8), 3307.

Qiu, D., Dinçer, H., Yüksel, S., & Ubay, G. G. (2020). Multi-faceted analysis of systematic risk-based wind energy investment decisions in E7 economies using modified hybrid modeling with IT2 Fuzzy sets. *Energies, 13*(6), 1423.

Ubay, G. G., & Karakuş, H. (2020). The role of technology-based renewable energy investments on macroeconomic stability. In *Strategic outlook for innovative work behaviours* (pp. 119–130). Springer.

Wang, S., Li, W., Dincer, H., & Yuksel, S. (2019). Recognitive approach to the energy policies and investments in renewable energy resources via the fuzzy hybrid models. *Energies, 12*(23), 4536.

Washburn, C., & Pablo-Romero, M. (2019). Measures to promote renewable energies for electricity generation in Latin American countries. *Energy Policy, 128*, 212–222.

Yoshino, N., & Taghizadeh-Hesary, F. (2018). Alternatives to private finance: Role of fiscal policy reforms and energy taxation in development of renewable energy projects. In *Financing for low-carbon energy transition* (pp. 335–357). Springer.

Yuan, J., Zhang, Z. M., Yüksel, S., & Dinçer, H. (2020). Evaluating recognitive balanced scorecard-based quality improvement strategies of energy investments with the integrated hesitant 2-tuple interval-valued Pythagorean fuzzy decision-making approach to QFD. *IEEE Access, 8*, 171112–171128.

Yüksel, S., Dinçer, H., & Meral, Y. (2019). Financial analysis of international energy trade: A strategic outlook for EU-15. *Energies, 12*(3), 431.

Yüksel, S., Dinçer, H., & Uluer, G. S. (2020). The role of technological development on renewable energy usage: An econometric analysis for G7 countries. In *Handbook of research on sustainable supply chain management for the global economy* (pp. 136–153). IGI Global.

Yumashev, A., Ślusarczyk, B., Kondrashev, S., & Mikhaylov, A. (2020). Global indicators of sustainable development: Evaluation of the influence of the human development index on consumption and quality of energy. *Energies, 13*(11), 2768.

Zeng, S., Liu, Y., Liu, C., & Nan, X. (2017). A review of renewable energy investment in the BRICS countries: History, models, problems and solutions. *Renewable and Sustainable Energy Reviews, 74*, 860–872.

Zhao, Z. Y., Chen, Y. L., & Chang, R. D. (2016). How to stimulate renewable energy power generation effectively?–China's incentive approaches and lessons. *Renewable Energy, 92*, 147–156.

Zhao, Y., Xu, Y., Yüksel, S., Dinçer, H., & Ubay, G. G. (2021). Hybrid IT2 fuzzy modelling with alpha cuts for hydrogen energy investments. *International Journal of Hydrogen Energy, 46*(13), 8835–8851.

Zhou, P., Zhou, P., Yüksel, S., Dinçer, H., & Uluer, G. S. (2020). Balanced scorecard-based evaluation of sustainable energy investment projects with it2 fuzzy hybrid decision making approach. *Energies, 13*(1), 82.

Chapter 2
Modern Directions in Bioenergy Production from Forest

Artur Meynkhard and Uyeh Daniel Dooyum

Abstract The article provides an overview of the production of bioenergy from forest sources on a global scale. The research paper reveals the main problems of bioenergy development and formulates the possible options for further expansion of the electric power industry. Innovative technological solutions are considered. The article uses inductive and deductive methods for analyzing statistical data on the basis of which the problems of sustainable development of bioenergy are assessed. The article proves that, despite the general tendency towards weakening of climatic changes, they can play a decisive role for heating subscribers. The results include the various processes that make it possible to obtain energy from forest biomass residues, as well as the potential that these residues have for the production of various types of bioenergy. The findings prove that the decision to choose a biofuel transition strategy depends on the type, properties and quantity of biomass available, usage requirements, as well as environmental standards and economic conditions.

2.1 Introduction

The creation of bioenergy sectors using the waste of the forest biomass has great potential in the world. This is due to states' high activity levels in the forest industry. The energy challenges of the twenty-first century are the result of an over-reliance on limited fossil fuel reserves combined with an ever-growing demand for energy (Yuan et al., 2020; Liu et al., 2021a, b).

The development of bioenergy as an independent area of global energy at the end of the twentieth century and the beginning of the twenty-first century has shown that

A. Meynkhard (✉)
Financial University Under the Government of the Russian Federation, Moscow, Russia

U. D. Dooyum
Upland Field Machinery Research Center 106, Bio-process System Engineering Group, Department of Bio-industrial Machinery Engineering, Kyungpook National University, Daegu, South Korea

S. Yüksel, H. Dinçer (eds.), *Strategic Approaches to Energy Management*, Contributions to Management Science,
https://doi.org/10.1007/978-3-030-76783-9_2

this niche can be filled by biofuels: bioethanol, biogas, biodiesel, biohydrogen, biobutanol and bio-isobutanol, dimethyl ether.

All living populations of the biosphere have adapted over the course of evolutionary development to an existence based on renewable energy resources. This strategy of energy use in the Earth's conditions is a possible direction of development to stable living, along with the possibility of wide use of renewable energy resources in the economy (Cheng et al., 2020; Li et al., 2021).

In a favorable scenario, the bioenergy industry may become a new source of jobs, which, in addition to the aforementioned environmental benefits, will contribute to countries' economy. These results can be achieved only under the conditions of certain actions, including changes in social behavior, vehicle technologies and the introduction of biofuel innovations (Lisin, 2020a; An et al., 2021; Danish et al., 2021; Mikhaylov, 2018a).

The highest bioenergy efficiency and reduction of greenhouse emissions is achieved in processes that use waste as raw materials, because this avoids the negative effects that particular crop production has on the environment. Forest waste is considered renewable, as new biomass can be grown in place of old used biomass (An & Mikhaylov, 2020).

The aim of this study is to describe various processes which allow to produce energy from forest biomass waste and the potential that this waste has in the production of different bioenergy types (Mikhaylov, 2019, 2020b; Nyangarika et al., 2019a; Ozturk et al., 2010).

In this study, the authors answer the question of the possibility of efficient use of alternative by-products of bioenergy (forest waste) to ensure efficient production and consumption.

2.2 Literature Review

Bioenergy is defined as energy, which is obtained from biomass, including living organisms and organics, which are a result of spontaneous biologically provoked processes. Biomass resources are organic and store energy in their chemical bonds, usually in carbon dioxide, hydrogen, oxygen, nitrogen and, in smaller proportions, sulfur (Ranjbar et al., 2017; Shahbaz et al., 2018; Uandykova et al., 2020; Lisin, 2020b).

Plants produce carbohydrates through photosynthesis, which is the foundation of the biomass. Carbon dioxide and water vapor turns into glucose, generates starch, hemicellulose, lignin and other substances under sunlight during photosynthesis (An et al., 2020a, 2021; Denisova et al., 2019; Lopatin, 2019; Nyangarika et al., 2019b).

Forest-related production is divided into two large groups: wood and non-wood produce (Alwaelya et al., 2021; An et al., 2019b; Dayong et al., 2020). The first consists of wood materials, and the second of seed, resins, fibers, gums, waxes, leaves and stems (Mikhaylov et al., 2020; Moiseev et al., 2020; An et al., 2020c; Dooyum et al., 2020; Yumashev & Mikhaylov, 2020).

The main goods produced in 2015 were sawn wood (73.2%), fuel (firewood and coal) (11.8%), and the remaining 15.0% consisted of cellulose, veneer, poles, piles (Mikhaylov et al., 2021a; Nie et al., 2020; Odhiambo, 2009). In the production of wood, there is a large amount of waste that is divided into two classes: (1) those that come from forest operations, such as harvesting and mining; (2) those that are produced in industry from the process of wood production (Danish et al., 2021; Mikhaylov et al., 2018; Lopatin, 2019).

Unlike other industries, forestry has the advantage of significant and useful waste, which can be applied during different stages of production of goods. There are different methods to make use of this waste (Mikhaylov et al., 2021b; An et al., 2020c; Mikhaylov, 2020a; Phelps, 1994; Varyash et al., 2020).

The main processes used for producing energy are combustion, gasification and pyrolysis – the thermochemical conversion of biomass. Most biomass is consumed within a country and industries related to wood. This is done by direct combustion of biomass, to produce heat or steam, which drives a turbine or generator, consequently producing electricity.

Gasification processes are used to convert biomass into combined gas with the goal of powering high efficiency combined cycle sludge gas turbines and liquid biofuel production. Moreover, biomass is converted into oil by pyrolysis, making it easy to store and transport. All these thermochemical conversion processes are described in detail in the following sections (Lisin, 2020c; Mikhaylov, 2018b).

Direct incineration is typically utilized to convert chemical energy stored in biomass into heat, mechanical energy and electricity. It is possible to reduce any biomass with moisture of less than 50%. Otherwise, preliminary drying processing is required. Biomass burning has been commonly applied since ancient times on small scales for home cooking and heating living spaces. However, this type of direct incineration is considered to be inefficient due to a loss of 30–90% of heat transfer during the process. Large-scale biomass burning can be done in boilers, steam turbines and turbine generators, in which heat transfer is much more efficient (An et al., 2020b; Mikhaylov, 2020c; Gura et al., 2020; Rathnayaka et al., 2018).

According to recent data, 37% of the dry mass of wood waste is used as raw materials for paper production and 27% is used as fuel. Improvements in wood processing technology, the increasing need for raw materials for further processing and energy availability lead to a near absence of unclaimed waste.

Two types of waste are of great interest in wood processing: bark and alkaline solution. The bark is peeled off the surface of the tree with special devices. Alkaline liquor is a liquid emerging from the preparation of the wood, characterized by an increased concentration of organic compounds.

Biomass burning is currently scaled to be utilized in industrial enterprises in the range from 100 MW to 300 MW. Therefore, it has a significant impact on the global energy supply. Biomass incineration leads to the formation of certain pollutants, including dust and acid rain gas, such as sulfur dioxide and nitrous oxide, but the amount of pollutants formed is 90% less than in the case of burning coal and fossil fuels.

On the other hand, coal burning is of great interest due to its high conversion efficiency. Combining biomass fuel with coal is promising, as it combines high conversion efficiency and coal transportation with reduced has emissions from biomass burning.

It is necessary to consider many different aspects when assessing the potential of biomass burning. These include their chemical and physical properties, such as levels of moisture, carbon, the calorific value and density. All these aspects are of great importance, since they determine the actual use and application that can be given to each type of biomass. Calorific values refer to the amount of energy generated per kilogram of a substance during combustion (Mikhaylov, 2021; Yumashev et al., 2020; Nyangarika et al., 2018).

2.3 Methods

The term "primary biomass" covers a wide range of organic materials derived from plants. The most important sources of biomass energy are agricultural, forestry and timber waste, as well as crops grown exclusively for use in energy production (An et al., 2019a; Mikhaylov & Sokolinskaya, 2019). Biomass is responsible for about 10% of the world's energy supply as of today. Biomass can be transformed into three main types of goods: (1) electricity and heat; (2) fuel for transportation; (3) raw materials for chemical production (Mikhaylov & Tarakanov, 2020).

In recent years, innovative technologies aimed to improve some forms of biomass, optimize its transportation and storage, and, mainly, to convert biomass into biofuel have been developed. The processes of biomass conversion are usually classified as thermochemical or biochemical.

The main sources of bioenergy are:

1. primary solid biofuels, which are residues from forestry operations such as logging and mining, which represent 93%–97% of the market during the period from 2000 to 2017;
2. liquid biofuels, which represented 4% of the market at the end of 2017. This type of biofuel is less common than primary solid fuels, due to the small quantities produced and the need for design changes in units normally powered by petrol and diesel. There are several types of liquid biofuels obtained by processing plant raw materials, these are: Bioethanol – ethyl alcohol; Biomethanol – methyl alcohol; Biobutanol – butyl alcohol; Dimethyl ether – ester; Diesel biofuel – a liquid motor fuel for diesel engines consisting of a mixture of fatty acid esters.
3. biogases derived from the fermentation of substances of plant or animal origin had a share of 1% of the market. Biogas refers to: bio-hydrogen is hydrogen derived from biomass; Methane is a gas in the hydrocarbon family.

4. industrial waste was 1% of the market;
5. household waste was 1% as well.

Unlike other sources of bioenergy, primary solid biofuels or Biomasses of wood and agricultural waste contain 12–18 MJ/kg calorific value, while granules have slightly higher values (20 MJ/kg) and coal contains 28 MJ/kg. Despite the smaller values of biomasses when compared to coal, they have an advantage in reducing emissions of polluting gases. In addition, if pellets are made from forest waste, this process gains a more favorable position whilst creating a renewable energy source (Zhao et al., 2021).

Gasification allows to convert solid biomass into a fuel gas through a series of chemical reactions, which take place at high temperatures (750 °C–900 °C) using air as a source of oxygen, water vapor or its mixtures as an oxidant. Furthermore, the produced gas can be used as a raw material in the production of other chemical substances under the conditions that it is a synthesized gas. Production of synthesized gas from biomass allows to produce methane, hydrogen, biodiesel, which are of great importance in the future development of fuel.

Pyrolysis is a thermochemical process which converts biomass into liquids (biooils, ibocruds), coal and non-condensable gases, acetic acid, acetone and methanol, while heating the biomass at about 500 °C in the absence of oxygen. This process can be adjusted in favor of the production of coal, oil and gas with an efficiency of 95%. All three products are always generated during the process, but proportions can be varied by adjusting parameters. The biomass is heated to 500 °C in 1 s during rapid pyrolysis, which yields 75% of liquid products, 12% solids and 13% gas. This process is valuable for producing liquid fuels, which can be easily stored and transported for subsequent use in the production of energy and other chemicals.

2.4 Results

Biomass can be converted into more useful forms of energy by diversifying applied resources. There are several biochemical processes for converting biomass into energy that provide environmental and economic benefits, as they occur under milder reaction conditions compared to thermochemical processes. The most common biochemical processes are enzymatic hydrolysis, fermentation, bioenergy production from forest and wood waste, anaerobic digestion. The main bioenergetics derived from the latter are bioethanol, bio butane, biodiesel and biogas. Ethanol biofuels are usually produced by fermentation, but many research and development activities are concentrated on the generation of other chemical compounds in a holistic process.

A significant advantage of biogas plants is that they also act as treatment plants, reducing bacterial and chemical pollution of soil, water and air. Compared to small hydro, wind and solar power plants, which are passively clean (using clean energy sources), biogas plants are actively clean, i.e., they eliminate the environmental hazards of the products used as a primary energy source.

Pretreatment is necessary before enzymatic hydrolysis, which involves the destruction of the lignin structure and crystalline cellulose structures in order to obtain different cellulose and hemicellulose chains accessible to enzymes.

The pretreatment process can help convert forest waste into biofuel, increasing the enzymatic refinement and, consequently, the yield of fermentable sugars. The impact of the pretreatment process on the most commonly applied to lignocellulosic biomass are subdivided into physical, physicochemical, chemical and biological ones. Some studies combine these processes.

Lingin is a complex molecule built from phenylpropane units linked in a hard to break down three-dimensional structure. As a rule, softwoods contain more lignin than hardwoods; the highest lignin content is localised in agricultural residues. Lignin is one of the main obstacles to the use of lignocellulosic materials in the fermentation process because it makes lignocellulose resistant to chemical and biological degradation.

Thus, the pretreatment process inherent in natural lignocellulosic materials makes them resistant to enzymatic attacks.

The first critical stage of effective hydrolysis of wood waste, is the preliminary treatment, which involves reducing the size of the biomass, since it has been shown that the size and moisture content have a significant effect on subsequent hydrolysis.

Two types of hydrolysis are currently used for the complex conversion of wood waste—percolation and two-phase hydrolysis. The neutralised hydrolysate from coniferous wood is used to produce ethyl alcohol, while the remaining bard is used as the basis for the cultivation of fodder yeast. Furfural, which is used as fuel, is obtained from the hydrolysate's self-evaporating gases. When the demand for furfural is higher, a two-phase hardwood hydrolysis scheme is adopted. In the first phase, furfural is produced and in the second phase, ethyl alcohol is obtained from the hydrolysate of cellolignin. In the hydrolysis industry, plant-based carbohydrate raw materials are quite often used.

They contain polysaccharides (that can be easy and difficult to hydrolyse). These are starch, hemicellulose, pectin substances. In wood and agricultural raw materials the starch content varies from 16% to 39%. Mannose and glucose predominate in the hemicellulose hydrates of conifers, and xylose predominates in the hydrolysates of deciduous trees and agricultural residues.

The suitability of any type of raw material depends to a greater extent on its composition at the chemical level. For example, pentose-containing raw materials are used to produce xylitol, futural; coniferous wood waste is used to produce ethyl alcohol. For the cultivation of fodder yeast cultures, neutralized hydrates of almost any raw material are suitable.

Many types of mills are considered effective in the production of biofuels; however, it should be noted among drawbacks that the process requires a large amount of energy, which, in turn, increases production. This has become the reason to studying other physical and thermophysical processes that reduce particle size. One of them is the extrusion process, which has the advantage of high biomass shift, use of moderate temperatures, lack of inhibitor production, adaptability, easy scaling, continuous operations, and decreased amounts of wasted water. Another thermophysical method which has been studied in recent years is steam explosion. It has been demonstrated to be effective and optimized for pretreatment of forest waste and biomass as a raw material on a laboratory scale. Enzymatic hydrolysis forest biomass that was pretreated with steam explosions has been rated in a scale of 200:1 by monitoring production of fermented sugars and inhibitors.

Amongst the studied factors, one that gathers attention is the composition of enzymatic cocktails, the amount of biomass and the hydrolysis temperature, as well as the most suitable pretreatment conditions. It was noted that after 48 hours of the reaction there was no increase in sugar concentration even with an increase in forest biomass, which indicates a strong inhibition of the reaction. It was also determined that pretreatment of 200 °C for 8 min leads to efficient hydrolysis; while temperatures above 205 °C destroy the hemicellulose, the subsequent sugar production reduces. Thus, new strategies and technical solutions are required for the problem of enzymatic hydrolysis on an industrial scale for the production of fermentable sugars on commercially viable levels. Furthermore, it is necessary to optimize the process and solve technical problems for each specific case, as there is a difference in the composition of waste depending on its origin.

Preliminary biological treatment also exists, which is considered efficient and fairly inexpensive. It typically involves the use of enzyme-producing fungus, that are responsible for destroying lignin. However, their drawback is the required long incubation time. In contrast to what has been stated previously, the inclusion of acids in dilute concentrations and high temperatures (140 °C to 200 °C) makes the cellulose contained in biomass more accessible for enzymatic saccharification.

In conclusion, effective pretreatment should possess the following characteristics: preservation of hexose and pentose fractions, limitation of the formation of microbial growth inhibitors during fermentation, minimization of energy demand, cost reduction, and the absence of polluting wastewater. On the other hand, it is crucial to set highly optimized technologies for collecting forest waste, which will be composed of minimal amounts of mixed substrates. Studies will need to optimize the pretreatment process on an industrial scale, maximizing subsequent enzymatic hydrolysis output after the forest biomass will be used for bioenergy production.

Processes of biochemical conversion based on fermentation require hydrolysis of cellulose and hemicellulose found in the biomass in order to produce fermentable sugars. Enzymatic hydrolysis in synergy with cellulases and hemicelluloses has been well studied and its advantages include high conversion yields, and the conditions of its reaction are less corrosive and toxic compared to acid hydrolysis. Recent research papers on enzymatic hydrolysis of forest waste evaluate the yield in the production of fermentable sugars for subsequent use in the production of biofuels.

Despite discovering that enzymes are effective in the production of biofuels, their cost remains a challenge to be overcome in the future. Enzymes cost reduction has been concluded to be achieved in the presence of and economically competent biofuels in the market. Thus, several studies have been carried out that state the practicality of enzyme-producing microorganisms in the hydrolysis of residues, as well as the necessity of optimizing the production of cellulase and lignans enzymes at the laboratory level.

Fermentation includes the use of microorganisms for converting fermentable substrates into recoverable products such as biomass, alcohols and organic acids. Hexoses, mainly glucose substrate constitute the most digestible microorganisms, while pentoses (hemicellulose sugar), glycerol, and other compounds require certain organisms.

There are various microorganisms that carry out fermentation of sugars derived from forest waste. They contribute to two processes:

1. hydrolysis of forest waste and their conversion to sugar;
2. fermentation which can be carried out separately or simultaneously.

Biofuel, mainly produced by fermentation, is ethanol, which is followed by butanol and acetone. Yeast is mainly used in the production of ethanol, as it carries out alcoholic fermentation. However, the process of enzymatic hydrolysis and fermentation has been recorded through filamentous fungi and bacteria.

Other studies on the production of ethanol from forest and wood wastes have yields of 7 g/l from production of 15–16 g/l. They determined the importance of applying proper pretreatment of forest waste for increasing the efficiency of ethanol production.

On the other hand, the production of microbial lipids from lignocellulosic biomass has received attention from researchers as their studies progress. This interest stems from the fact that microdetail lipids can be applied for the production of biofuels and avoid the conflict of using mineral oil-bearing crops to feed animals. Another advantage is the smaller production area compared to oilseed crops.

Microbial lipids can be obtained from fermentable sugars obtained from the hydrolysis of forest residues.

The technological process of microbial lipid production, as opposed to protein production, necessarily includes a stage of lipid extraction from the cell mass by extraction in a non-polar solvent (petrol or ether). In this process, two products are obtained simultaneously: a microbial fat (biofat) and a defatted protein preparation (biosorbent).

The raw material for this process is the same conditions as for feed biomass production. The cultivation of microorganisms in different conditions produces three classes of lipids: simple lipids, complex lipids and their derivatives.

1. Simple lipids are neutral fats and waxes. Neutral fats (the main storage components of the cell) are esters of glycerol and fatty acids, the bulk of which are triacylglycerides. Waxes are esters of fatty acids or monooxyacids and aliphatic alcohols with a long carbon chain. Their structure and properties are similar to those of neutral lipids. Yeast and mycelial fungi synthesise the greatest amount of neutral lipids. Simple lipids are used as technological lubricants in cold and heat treatment processes of metals. Complex lipids are produced mainly by bacteria.
2. Complex lipids are divided into two groups: phospholipids and glycolipids. Phospholipids (phosphoglycerides and sphingolipids) are part of various cell

membranes and take part in electron transport. Their molecules are polar and at pH 7.0 the phosphate group carries a negative charge. Phospholipid concentrate is used as an anti-corrosion additive for oils and as an additive in the flotation of various minerals. Unlike phospholipids, glycolipids do not contain phosphoric acid molecules, but are also strongly polar compounds due to the presence of hydrophilic carbohydrate groups (glucose, mannose, galactose residues, etc.) in the molecule.

3. Lipid derivatives include fatty acids, alcohols, hydrocarbons, vitamins D, E and K. Fatty acids are saturated and unsaturated acids with one double bond of nonnal structure and even number of carbon atoms (palmitic acid, stearic acid, oleic acid). The diene fatty acids include linoleic acid. The double bonds in unsaturated fatty acids of microbial lipids are often arranged so as to divide, them by a multiple of three carbon atoms. Purified monocarboxylic acids with a number of carbon atoms of 14–18 are widely used in the soap, tyre, chemical, paint and varnish industries.

For industrial use, the ability to enhance lipid accumulation is important. Few microorganisms have this ability, most notably yeast. The process of lipid formation in most yeasts consists of two clearly distinguishable stages:

1. The first, characterized by rapid protein formation under conditions of increased nitrogen supply to the crop and accompanied by a slow accumulation of lipids (mainly glycerophosphates and neutral fats);
2. The second, cessation of yeast growth and increased accumulation of lipids (mostly neutral).

The yeast *Cryptococcus terricolus* is a typical lipid-forming yeast. They can synthesise large quantities of lipids (up to 60% of dry weight) under all conditions, even those most favourable for protein synthesis.

Of the other lipid-forming yeasts of industrial interest are the alkane utilizing yeasts C guilliennondii. They synthesise mainly phospholipids. *Lipoinyces lipoferus* and *Rhodotorula gracilis* yeasts also accumulate large quantities of lipids and grow actively on carbohydrate substrates (molasses, peat and wood hydrolysates). In these yeast species, lipogenesis is highly dependent on cultivation conditions. These yeasts accumulate significant amounts (up to 70%) of triacylglycerides.

Microscopic mushrooms have not yet been widely used in the production of lipids, although the fat composition of mushrooms is similar to that of plants. The fat yield of Asp. terreus, for example, on carbohydrate media reaches 51% of absolutely dry weight (ADW). The lipid composition of mushrooms is mainly represented by neutral fats and phospholipids.

The lipids synthesised by bacteria are peculiar in their composition, as they include mainly complex lipids, whereas neutral fats account for a small proportion of the biomass. At the same time, bacteria produce a variety of fatty acids (containing between 10 and 20 carbon atoms), which is important for the industrial production of specific fatty acids.

Research is currently focused on finding microorganisms, oilseeds, applying and improving engineering genetics, lipid recovery methods, as well as resolving technical problems and those feasibility studies. Most articles use biomass obtained from starch-rich crops as a substratum, with few studies using forest waste.

Bacterial transformation includes anaerobic refinement of biodegradable organic resources in the temperature range from 30 °C to 65 °C. The main goods produced from this process are biogas, methane mixtures, carbon dioxide and other mixtures.

Biogas is usually purified to a gas with a content of more than 97% methane, which can be used as a substitute for natural gas. Other products which can be prepared by fermentation are acetone, butanol and ethanol.

2.5 Conclusions and Discussion

In conclusion, the benefits from utilizing forest waste for bioenergy are, firstly, ecological, as they reduce emissions of greenhouse gases. Secondly, the use of forest waste in the biomass for bioenergy production does not include polluting pesticides or fertilizers that are associated with crops leading to the greenhouse effect.

By 2023, biofuel production is projected to be 92 million tonnes, by 2025 this figure is projected to reach 200 million tonnes of biofuels and by 2030 it may be upward of 285 million tonnes. These global challenges will be discussed on 24–-25 February 2020 in the UAE at the 1lth International Conference on Biofuels and Bioenergy.

According to the European Union (EU) directive, the biodiesel content of the total content of petroleum products must be at least 5%.

By 2030, the EU plans to meet 25% of its fuel needs with clean and efficient biofuels.

The in-depth processing of wood residues requires strictly coordinated cooperation between specialists from many scientific disciplines: biology, physics, chemistry and engineering sciences. The development of radically altered and multiple product systems will be key to the future development of both bioenergy and the entire forestry industry.

Moreover, the decentralization of the forest biomass and other lignocellulosic residues will create jobs in various regions of many countries, which will consequently minimize transportation of biofuels over long distances. However, in order to achieve these goals, the quality of biofuels will need to be enhanced and ensured by incentives and international standards.

Although the benefits from this type of production have already been described, it is important to highlight the challenges this new technology is facing: its cost. One of the main attractions of biofuels is its lesser impact on the environment during combustion compared to fossil fuels (Du et al., 2020). Nevertheless, it is important to consider the expenses of energy production and waste conversion. In order to choose an optimal process, the methods' homogenization is necessary to compare the generation of electricity and other goods based on energy efficiency, by-products and waste production.

Another aspect is the economic opportunity of spine bio determination. The full use of other by-products, such as lignin and proteins contained in the forest biomass, as well as other resources produced in the process is necessary.

We must also develop a strict policy and rules of consistent forest waste use, in order to avoid the biomass' overexploitation, and utilize bioenergy from different sources in combination with forestry and wood waste.

References

Alwaelya, S. A., Yousif, N. B. A., & Mikhaylov, A. (2021). Emotional development in preschoolers and socialization. *Early Child Development and Care, 190*. https://doi.org/10.1080/03004430.2020.1717480

An, J., & Mikhaylov, A. (2020). Russian energy projects in South Africa. *Journal of Energy in Southern Africa, 31*(3).

An, J., Mikhaylov, A., & Moiseev, N. (2019a). Oil price predictors: Machine learning approach. *International Journal of Energy Economics and Policy, 9*(5), 1–6.

An, J., Mikhaylov, A., & Sokolinskaya, N. (2019b). Machine learning in economic planning: Ensembles of algorithms. *Journal of Physics: Conference Series, 1353*, 012126.

An, J., Mikhaylov, A., & Jung, S.-U. (2020a). The strategy of South Korea in the global oil market. *Energies, 13*(10), 2491.

An, J., Mikhaylov, A., & Kim, K. (2020b). Machine learning approach in heterogeneous group of algorithms for transport safety-critical system. *Applied Sciences, 10*(8), 2670.

An, J., Mikhaylov, A., & Richter, U. H. (2020c). Trade war effects: Evidence from sectors of energy and resources in Africa. *Heliyon, 6*, e05693. https://doi.org/10.1016/j.heliyon.2020.e05693

An, J., Mikhaylov, A., & Jung, S.-U. (2021). A linear programming approach for robust network revenue management in the airline industry. *Journal of Air Transport Management, 91*(March 2021), 101979.

Cheng, F., Lin, M., Yüksel, S., Dİncer, H., & Kalkavan, H. (2020). *A hybrid hesitant 2-tuple IVSF decision making approach to analyze PERT-based critical paths of new service development process for renewable energy investment projects*. IEEE Access.

Danish, M. S. S., Estrella, L. L., Alemaida, I. M. A., Lisin, A., Moiseev, N., Ahmadi, M., Nazari, M., Wali, M., Zaheb, H., & Senjyu, T. (2021). Photocatalytic applications of metal oxides for sustainable environmental remediation. *Metals, 11*, 80.

Dayong, N., Mikhaylov, A., Bratanovsky, S., Shaikh, Z. A., & Stepanova, D. (2020). Mathematical modeling of the technological processes of catering products production. *Journal of Food Process Engineering, 43*(2), 1–9.

Denisova, V., Mikhaylov, A., & Lopatin, E. (2019). Blockchain infrastructure and growth of global power consumption. *International Journal of Energy Economics and Policy, 9*(4), 22–29.

Dooyum, U. D., Mikhaylov, A., & Varyash, I. (2020). Energy security concept in Russia and South Korea. *International Journal of Energy Economics and Policy, 10*(4), 102–107.

Du, L., Dinçer, H., Ersin, İ., & Yüksel, S. (2020). IT2 fuzzy-based multidimensional evaluation of coal energy for sustainable economic development. *Energies, 13*(10), 2453.

Gura, D., Mikhaylov, A., Glushkov, S., Zaikov, M., & Shaikh, Z. A. (2020). Model for estimating power dissipation along the interconnect length in single on-chip topology. *Evolutionary Intelligence*, s12065.

Li, Y. X., Wu, Z. X., Dinçer, H., Kalkavan, H., & Yüksel, S. (2021). Analyzing TRIZ-based strategic priorities of customer expectations for renewable energy investments with interval type-2 fuzzy modeling. *Energy Reports, 7*, 95–108.

Lisin, A. (2020a). Biofuel energy in the post-oil era. *International Journal of Energy Economics and Policy, 10*(2), 194–199. https://doi.org/10.32479/ijeep.8769

Lisin, A. (2020b). Prospects and challenges of energy cooperation between Russia and South Korea. *International Journal of Energy Economics and Policy, 10*(3). https://doi.org/10.32479/ijeep.9070

Lisin, A. (2020c). Valuation of the activities of foreign banks in the Russian banking sector. *Orbis, 15*(45), 53–63. http://www.revistaorbis.org.ve/pdf/45/art5.pdf

Liu, J., Lv, J., Dinçer, H., Yüksel, S., & Karakuş, H. (2021a). Selection of renewable energy alternatives for green blockchain investments: A hybrid IT2-based fuzzy modelling. *Archives of Computational Methods in Engineering, 2021*, 1–15.

Liu, Y., Gong, X., Yüksel, S., Dinçer, H., & Aydın, R. (2021b). A multidimensional outlook to energy investments for the countries with continental shelf in East Mediterranean region with hybrid decision making model based on IVIF logic. *Energy Reports, 7*, 158–173.

Lopatin, E. (2019). Methodological approaches to research resource saving industrial enterprises. *International Journal of Energy Economics and Policy, 9*(4), 181–187.

Mikhaylov, A. (2018a). Volatility spillover effect between stock and exchange rate in oil exporting countries. *International Journal of Energy Economics and Policy, 8*(3), 321–326.

Mikhaylov, A. (2018b). Pricing in oil market and using probit model for analysis of stock market effects. *International Journal of Energy Economics and Policy, 8*(2), 69–73.

Mikhaylov, A. (2019). Oil and gas budget revenues in Russia after crisis in 2015. *International Journal of Energy Economics and Policy, 9*(2), 375–380.

Mikhaylov, A. (2020a). Cryptocurrency market analysis from the open innovation perspective. *Journal of Open Innovation: Technology, Market, and Complexity, 6*(4), 197. https://doi.org/10.3390/joitmc6040197

Mikhaylov, A. (2020b). Geothermal energy development in Iceland. *International Journal of Energy Economics and Policy, 10*(4), 31–35.

Mikhaylov, A. (2020c). Cryptocurrency market development: Hurst method. *Finance: Theory and Practice, 24*(3), 81–91.

Mikhaylov, A. (2021). Lichens as indicators of atmospheric pollution in urban ecosystems. *Israel Journal of Ecology & Evolution, 10016*, 1–9.

Mikhaylov, A., & Sokolinskaya, N. (2019). Russian banks after sanctions of 2014. *Orbis, 15*(44), 55–65.

Mikhaylov, A., & Tarakanov, S. (2020). Development of Levenberg-Marquardt theoretical approach for electric network. *Journal of Physics: Conference Series, 1515*, 052006.

Mikhaylov, A., Sokolinskaya, N., & Nyangarika, A. (2018). Optimal carry trade strategy based on currencies of energy and developed economies. *Journal of Reviews on Global Economics, 7*, 582–592.

Mikhaylov, A., Moiseev, N., Aleshin, K., & Burkhardt, T. (2020). Global climate change and greenhouse effect. *Entrepreneurship and Sustainability Issues, 7*(4), 2897–2913.

Mikhaylov, A., Danish, M. S. S., & Senjyu, T. (2021a). New stage in evolution of cryptocurrency market: Analysis by Hurst method. In H. Dincer & S. Yuksel (Eds.), *Strategic outlook in business and finance innovation: Multidimensional policies for emerging economies* (pp. 35–45). Emerald. https://doi.org/10.1108/978-1-80043-444-820211004. isbn:978-1-80043-445-5.

Mikhaylov, A., Yumashev, A., & Kolpak, E. (2021b). Quality of life, anxiety and depressive disorders in patients with extrasystolic arrhythmia. *Archives of Medical Science, 17*(1). https://doi.org/10.5114/aoms.2020.101359

Moiseev, N., Mikhaylov, A., Varyash, I., & Saqib, A. (2020). Investigating the relation of GDP per capita and corruption index. *Entrepreneurship and Sustainability Issues, 8*(1), 780–794.

Nie, D., Panfilova, E., Samusenkov, V., & Mikhaylov, A. (2020). E-learning financing models in Russia for sustainable development. *Sustainability, 12*(11), 4412.

Nyangarika, A., Mikhaylov, A., & Tang, B.-J. (2018). Correlation of oil prices and gross domestic product in oil producing countries. *International Journal of Energy Economics and Policy, 8*(5), 42–48.

Nyangarika, A., Mikhaylov, A., & Richter, U. (2019a). Influence oil price towards economic indicators in Russia. *International Journal of Energy Economics and Policy, 9*(1), 123–129.

Nyangarika, A., Mikhaylov, A., & Richter, U. (2019b). Oil price factors: Forecasting on the base of modified auto-regressive integrated moving average model. *International Journal of Energy Economics and Policy, 9*(1), 149–159.

Odhiambo, N. M. (2009). Energy consumption and economic growth nexus in Tanzania: An ARDL bounds testing approach. *Energy Policy, 37*(2), 617–622.

Ozturk, I., Aslan, A., & Kalyoncu, H. (2010). Energy consumption and economic growth relationship: Evidence from panel data for low and middle income countries. *Energy Policy, 38*(8), 4422–4428.

Phelps, E. S. (1994). *Structural slumps*. Harvard University Press.

Ranjbar, O., Chang, T., Nel, E., & Gupta, R. (2017). Energy consumption and economic growth nexus in South Africa: Asymmetric frequency domain approach. *Energy Sources, Part B: Economics, Planning, and Policy, 12*(1), 24–31.

Rathnayaka, R. K. T., Seneviratna, D. M. K., & Long, W. (2018). The dynamic relationship between energy consumption and economic growth in China. *Energy Sources, Part B: Economics, Planning, and Policy, 13*(5), 264–268.

Shahbaz, M., Zakaria, M., Shahzad, S. J. H., & Mahalik, M. K. (2018). The energy consumption and economic growth nexus in top ten energy-consuming countries: Fresh evidence from using the quantile-on-quantile approach. *Energy Economics, 71*, 282–301.

Uandykova, M., Lisin, A., Stepanova, D., Baitenova, L., Mutaliyeva, L., Yuksel, S., & Dincer, H. (2020). The social and legislative principles of counteracting ransomware crime. *Entrepreneurship and Sustainability Issues, 8*(2), 777–798.

Varyash, I., Mikhaylov, A., Moiseev, N., & Aleshin, K. (2020). Triple bottom line and corporate social responsibility performance indicators for Russian companies. *Entrepreneurship and Sustainability Issues, 8*(1), 313–331.

Yuan, J., Zhang, Z. M., Yüksel, S., & Dinçer, H. (2020). Evaluating recognitive balanced scorecard-based quality improvement strategies of energy investments with the integrated hesitant 2-tuple interval-valued Pythagorean fuzzy decision-making approach to QFD. *IEEE Access, 8*, 171112–171128.

Yumashev, A., & Mikhaylov, A. (2020). Development of polymer film coatings with high adhesion to steel alloys and high wear resistance. *Polymer Composites, 41*(7), 2875–2880.

Yumashev, A., Ślusarczyk, B., Kondrashev, S., & Mikhaylov, A. (2020). Global indicators of sustainable development: Evaluation of the influence of the human development index on consumption and quality of energy. *Energies, 13*, 2768.

Zhao, Q., Cherkasov, S., Avdeenko, A., Kondratenko, L., & Mikhaylov, A. (2021). Integral estimate of the added value of agricultural crops in the synergetic agriculture on the example of vicia faba (*Vicia faba* L.). *Polish Journal of Environmental Studies, 30*(2), 993–1002.

Chapter 3
A Study on the Energy Problem in the Eastern Mediterranean: Text Mining and AHP-Based Strategy Recommendations

Serhat Yüksel, Serkan Eti, Sergey Prosekov, and Gülsüm Sena Uluer

Abstract The American Geological Association has made some estimates for reserves in the Eastern Mediterranean region. According to these estimates, it is predicted that there are very high amounts of natural gas and oil resources in the deltas of Nile, Levant and Aphrodite. These figures are large enough to meet Europe's 10-year natural gas need. On the other hand, another important point in this process is that these reserves are the amount determined in the studies conducted so far. In other words, there is a possibility that the actual reserve amount in this region is well above the estimated. Investigations in the relevant region are ongoing. As a result of these investigations, there is a possibility that new reserves will be identified. Because of this situation, there are many countries engaged in energy exploration in the Eastern Mediterranean. Firstly, the countries, which have a border to this region like Turkey, Israel, Egypt, the Turkish Republic of Northern Cyprus, Cyprus, Greece, Lebanon, Syria and Libya, are very active in this region. In addition, although there is no border to the region, countries such as the USA, Russia, England, France and Italy want to maintain their weight in the energy equation in the Eastern Mediterranean. It is obvious that the size of the energy reserves in the region lies at the heart of the energy struggle in the Eastern Mediterranean. This situation causes an increase in political tensions between some countries operating in the region. No country wants to give up this economic size. Therefore, some countries like Turkey-Greece and Libya-Egypt have problems with each other. The aim of this study is to develop strategies to reduce the political tension between countries due to energy reserves in the Eastern Mediterranean region. In this context, an analysis has been carried out by considering data mining and AHP methods in a hybrid way. The analysis process in question consists of 2 different stages. In the first process of the analysis, the most common words in academic studies are categorized

S. Yüksel (✉) · S. Eti · G. S. Uluer
The School of Business, İstanbul Medipol University, İstanbul, Turkey
e-mail: serhatyuksel@medipol.edu.tr; seti@medipol.edu.tr; gsuluer@st.medipol.edu.tr

S. Prosekov
Financial University Under the Government of the Russian Federation, Moscow, Russia

S. Yüksel, H. Dinçer (eds.), *Strategic Approaches to Energy Management*, Contributions to Management Science,
https://doi.org/10.1007/978-3-030-76783-9_3

using the LDA model. In this framework, 4 different categories are created. In the second phase of the analysis, more important factors are determined with the help of AHP method. The findings indicate that political relationship is the most significant factor to solve the conflict in Eastern Mediterranean region. Furthermore, transportation also plays a key role in this framework. Nevertheless, energy production and reserve amount have lower important roles to minimize this problem. Considering these results, ways of establishing positive relations between all countries in the region should be sought. Searching for a reconciliation with the energy companies making exploration, as well as with the countries, may contribute to solving the problems in the region.

3.1 Introduction

Energy is one of the essential needs of a country. The main reason for this is that energy contributes to both social and economic development. Energy is used both in the daily needs of individuals and in the production process of companies (An et al., 2020; Dooyum et al., 2020). Therefore, countries have to provide energy. In this framework, if the country has sufficient energy reserves, the country will be able to meet this energy need by its own means. However, not every country has sufficient energy reserves at its borders. Countries that do not have enough energy reserves to meet their own needs have to purchase energy from abroad in order to meet these needs (Wu et al., 2020).

This situation creates some problems for countries. First, countries that do not have their own energy reserves have to pay in foreign currency when importing energy from abroad. This situation increases the country's exchange rate risk (Mikhaylov, 2021; Yumashev et al., 2020). In other words, when the exchange rate becomes more valuable, imported energy will become more expensive. This situation will negatively affect the current account balance of the country. On the other hand, importing energy from a country increases the political dependence on that country (Su et al., 2020; Li et al., 2021).

These examples show that it is vital for a country to have its own energy resources. Countries want to have their own energy resources in order to ensure energy supply security (Yüksel et al., 2020). In this context, countries carry out exploration activities in order to discover energy reserves such as oil and natural gas within their borders in order to decrease energy dependency (Yuan et al., 2020; Zhao et al., 2021). In this way, they will be able to increase their energy supply security by obtaining their own energy reserves. In addition to the mentioned issue, some countries are also focusing on renewable energy alternatives. They look for ways to obtain the required energy with their own means, taking into account alternatives such as wind, solar and hydroelectric energy. On the other hand, some countries give priority to nuclear power plants for the same purpose (Dupont et al., 2020).

The American Geological Association has made some estimates for reserves in the Eastern Mediterranean region. According to these estimates, it is predicted that there are very high amounts of natural gas and oil resources in the deltas of Nile, Levant and Aphrodite. These figures are large enough to meet Europe's 10-year natural gas need (Liu et al., 2021a, b). This situation gives the information that the energy reserves in this region have a very serious economic size. On the other hand, another important point in this process is that these reserves are the amount determined in the studies conducted so far. In other words, there is a possibility that the actual reserve amount in this region is well above the estimated. Investigations in the relevant region are ongoing. As a result of these investigations, there is a possibility that new reserves will be identified (Pulhan et al., 2020).

There are many countries engaged in energy exploration in the Eastern Mediterranean. It is possible to categorize the countries in question in two different classes as those who have or do not have borders with the region. Firstly, the countries, which have a border to this region like Turkey, Israel, Egypt, the Turkish Republic of Northern Cyprus, Cyprus, Greece, Lebanon, Syria and Libya, are very active in this region. In addition, although there is no border to the region, countries such as the USA, Russia, England, France and Italy want to maintain their weight in the energy equation in the Eastern Mediterranean. As can be understood from here, the huge economic size of the energy reserves in the region attracts the attention of other countries in addition to the countries of the region (Mehmet & Yorucu, 2020).

As can be understood from these statements, the size of the energy reserves in the region lies at the heart of the energy struggle in the Eastern Mediterranean. The economic development of the country, which can make use of the energy reserves in this region effectively, will become much easier. All countries in the region, aiming to reach better standards in economic terms, aim to benefit from these energy reserves. This situation causes an increase in political tensions between some countries operating in the region. No country wants to give up this economic size. Therefore, some countries like Turkey-Greece and Libya-Egypt have problems with each other (El Shayeb et al., 2020).

The aim of the study is to shed light on the problems experienced in the extraction of natural gas reserves in the Eastern Mediterranean and to produce strategies. For this purpose, academic studies prepared for the subject are taken into consideration. In the studies conducted in the literature, studies on the natural gas issue in the Eastern Mediterranean region are tried to be determined. For this purpose, 980 studies that are scanned into Google Academic search engine with the keyword "Eastern Mediterranean Natural Gas" are included in the analysis. In the first part of the analysis, the subjects that are the basis of academic studies are defined by text mining. For this purpose, latent dirichlet allocation (LDA) model is used. Strategies have been produced by weighting the 4 issues obtained with this model by considering the AHP method.

3.2 Literature

Energy is a requirement to survive in life and investments are maintaining for it. There are lots of research in the literature about energy investments which explain the issues. Azam (2019) analysed that there is causality between energy consumption, environmental pollution, investment, human capital, and financial sector expansion and growth. Hence, there is positive causal linkage between economic growth and energy investment and negative causal linkage between energy investment and pollutant emissions (Ahmad et al., 2020). Additionally, He et al. (2019a, b) denoted that renewable energy (RE) investments and green credit have impact on green economy development and also, green credit affects renewable energy investments. Moreover, improve expenditures for environment pollution control and adjusting industry structure develops green economy (Cheng et al., 2020; Li et al., 2021). Also, energy efficient technologies, cost of capital and decarbonization influence energy investment decisions in residential sector (Sachs et al., 2019). In addition to that, Li and Li (2020) find that low-carbon development of energy industry investment can reduce CO_2 emissions, but energy investment and energy growth increased CO_2 emissions in Chinese provinces.

There are many of energy policy researches in literature. Dincer et al. (2019) stated that the most important factor in EU energy investment policy is increasing the capacity and storage of the energy and also, focusing on end users (consumer) is considered. However, producing sustainable energy operations and renewable energy sources are the least critical issue for energy investment policies. With regard of these, capacity issue of energy industry is the most crucial and technically required problem in energy investment. Communicating with energy storage facilities is important for house of quality (Tang & Dinçer, 2019). On the other hand, Liu et al. (2019a, b) remarked that global RE development imbalance caused by investment and business ethics that is affected by legal system in a country. Also, barriers of energy investment law should be flexible and well-designed energy market legislation for future efficient investments and incentives (Ländner et al., 2019; Boute, 2020). So, Liu et al. (2019a, b) denoted that investor heterogeneity, policies, investment planning, operating, and designing, horizontal and vertical comparison in energy investment should research for further research (Kobtseva et al., 2017; Morkovkin et al., 2016).

There are vast of renewable energy investment research. As Curtin et al. (2019) explained that financial assets can expose stranding risk for coal, oil and gas (carbon) based or high cost energy investments. Moreover, Kozlova and Collan (2020) signified RE energy investment attractiveness can differ to the investment profitability, demand for investments, regional accessibility, and electrical network state. They emphasized mainly geographical variability for investments in Russia by using cross-regional analysis. Nonetheless, Shiraishi et al. (2019) evaluated as Bangladesh is convenient for solar power energy investment by looking geospatial and economic conditions. The solar power cost lower than coal. In addition to that Peng et al. (2019) stated that solar energy is the best energy investment project and environment is the most important factor to the new energy resources. Also, solar photovoltaics, onshore wind technologies, technology reliability at a lower cost, data availability, credible policies and better tool assessments declines risk premiums and investment risks. As Agaton et al. (2020) said waste-to-energy technologies are profitable investment should be made immediately not to lose electricity generation opportunity. Moreover, curtailment and price risks are important with policies on RE investments (Egli, 2020). With regard of these, RE investments creates more jobs/ employment than fossil fuel-based energy in China (Chen, 2019). However, Capellán-Pérez et al. (2019) developed a EROI methodology for energy and material investments from fossil fuels to renewable energy. High energy investments will increase energy consumption and it causes environmental impacts and resource depletion (Du et al., 2020). 100% RE sources with Green Growth narrative reduces standard EROI rates and could not electricity supply all energy demand.

On the other hand, some researches mentioned about energy investment influencing factors. Qiu et al. (2020) find that the most significant risks are exchange rates and interest rates for wind energy investments. China and Indonesia is the most and Russia, India and Turkey is the least successful countries to manage these risks. Therewithal, Chinas's Belt and Road Initiative (BRI) nations' economy can grow with RE investment. The most determinant factors on overseas RE investments are political, economic and resources risks while Chinese factors (e.g. Diplomatic relations, foreign trade, FDI, foreign labour export) have impact on investment behaviours and sense of ownership to increase acceptability for investment activities. People to people bond provides good image for Chines investments to South Asia (Wu et al., 2020; Shi & Yao, 2019). However, He et al. (2019a, b) claimed that China's green financial development affects bank loan issuance negatively, it decreases bank credit of RE enterprises and RE investment efficiency. Moreover, China's green financial development inhibits over- investment to RE and companies aggravates their investments to RE enterprises. Additionally, financial institutions pay attention to the technological and financial adequacy of the company at the most in order to provide credit in large scale sustainable energy investment and state banks have the lowest performance to finance these large-scale energy projects (Zhou et al., 2020). As different from these, Su et al. (2020) elucidated that Bitcoin Price (BP) is getting important for Fourth Industrial Revolution and BP helps oil price (OP) while OP is high and risky. BP hedging ability is weak. However, it is more profitable if investors consider BP changes and policies carefully.

3.3 Conflicts in Eastern Mediterranean Region

The countries of the region declare their exclusive economic zones. In this way, it claims rights in terms of oil/natural gas exploration and extraction in the declared region. The biggest problem of exclusive economic zone in the Eastern Mediterranean region is experienced around the island of Cyprus. The Greek Cypriot Administration has declared the exclusive economic zone in the south of the island of

Cyprus consisting of the so-called 13 sections. On the other hand, the Turkish Republic of Northern Cyprus has declared an exclusive economic zone consisting of 7 parts in the specified region. The biggest problem in the region is experienced in this area. The parcels 2, 3, 8, 9, 12 and 13 announced by the Greek Cypriot Administration intersect with the exclusive economic zone announced by the Turkish Republic of Northern Cyprus. In addition, first announced by the Greek Cypriot Administration, 4, 5, 6 and 7 plots intersect with Turkey's exclusive economic zone also announced. In summary, all countries in these regions claim rights and want to carry out energy exploration activities (Sotiriou, 2020).

In addition, between Turkey and Greece are experiencing the exclusive economic zone of the problem. The most important reason for this is the islands that belong to both countries in the Aegean and Mediterranean regions. Because of these islands, there is an overlap in the exclusive economic zones announced by both countries. This situation causes an increase in political tension between the two countries. Especially in recent days, harsh statements have been made by both countries. Other operating in the region in political tension between Turkey and Greece also participate in some countries. As an example, the agreements signed between Turkey and Libya-Egypt and Greece, these countries are also involved in stress-related. On the other hand, other countries, such as France, the United Arab Emirates, and Israel, have shared the relevant political tension by making harsh statements (Biçer & Ercanoglu, 2020).

Political tension in the region is also caused by different pipelines. One of the most popular routes in this area is the East-Med pipeline. With this project, it is aimed to send Mediterranean gas to Europe via Israel, Southern Cyprus and Greece. In this context, the leaders of the aforementioned countries have held summits many times and signed agreements. It was stated that 1 year was required for the financing of the mentioned project, while the pipe laying process would take 5 years. The European Union supports this project in order to reduce its dependence on Russia on energy. The issue that concerns us on this route is the disabling of Turkey here. Thus, Turkey has also reacted seriously this route (Georgiou et al., 2020).

However, there are some problems in this project. First, there are some financial problems with this project. The cost of the project in question is around 7 billion dollars. On the other hand, only $ 100 million was spent even for the feasibility of this project. The second problem in this project is technological problems. Technically, it is necessary to descend to a depth of 3.5 kilometers and reach a length of 2100 km. Therefore, it seems unlikely that this pipeline could be filled with natural gas from natural gas reserves discovered only in Israel (Pulhan et al., 2020).

Trans Anatolian Natural Gas Pipeline (TANAP) is another important energy route in the region. With this line of Azerbaijani gas via Turkey it is planned to be transferred to Europe. The line, which will be 1850 km, will pass through 20 provinces in total starting from Ardahan on the Georgian border and will be transferred to Greece and from there to other European countries. The project was first brought to the agenda at the third Black Sea Energy and Economy Forum held in Istanbul on 17 November 2011. In addition, this line was opened on December 4, 2019. Compared to the East-Med pipeline, TANAP has many advantages. First of all, the East-Med pipeline is a project that is still in the draft stage. On the other hand, it is very difficult for this project to become operational due to financial and technical insufficiencies. On the other hand, thanks to TANAP, which is already in operation, it will be much easier, reliable and profitable to transport the natural gas in the region to the European market (Çağlar et al., 2020).

3.4 Analysis Results

The analysis process of the study consists of two different stages. Firstly, 980 different studies on the subject are analyzed by text mining. In this process, the words obtained are divided into 4 different categories, considering the LDA approach. In the second phase of the study, these factors are weighted. AHP method has been used to achieve this goal.

3.4.1 Categorizing Words with Text Mining

LDA model, with the help of probability, is one of the most up-to-date methods of unsupervised learning models in text mining. In this method, documents are divided into a certain number of subjects and it is based on determining the keywords of the subjects. On the basis of the method, the texts assume a probability distribution over the subjects, over the words in the subjects. Before the LDA method, preprocessing procedures were applied primarily for text mining. In this context, punctuation marks that do not make sense have been deleted. Later, all letters are converted to lowercase letters and words that do not make sense are filtered out. Finally, words are divided by their roots. After pre-processing the text, words are divided into 4 different categories using the LDA model. Details of this process are shown in Table 3.1.

3.4.2 Evaluating the Categories with AHP

In this stage, 3 different experts made evaluations about the categories by considering 9 different scales. The details of expert evaluations are given on Table 3.2.

After that, the average values of expert evaluations are considered to generate pairwise comparison matrix. Table 3.3 gives information about this matrix.

Just then, all values are divided to the sum values of all rows to create normalized matrix as in Table 3.4.

By considering this matrix, the weights of the factors are identified. The details are demonstrated on Table 3.5.

Table 3.1 The details of the words

Energy production	Political relationship	Transportation	Reserve amount
Energy	Region	Middle	Basin
Middle	Middle	Africa	Water
Europe	Israel	North	Levant
European	Regional	Region	Region
Southern	Energy	Transport	Field
North	Resources	Europe	Levante
Russia	Cyprus	Turkey	Found
Union	Climate	Coast	North
Policy	Egypt	Summer	Israel
Markets	Turkey	Western	Major
Central	Political	Aerosol	Atlantic
Security	Environmental	Desert	Western
Supply	Security	Strong	Exploration
Production	North	Model	Waters
Africa	Change	Organic	Surface

Table 3.2 Expert evaluations

Criteria	Criteria	Expert 1	Expert 2	Expert 3
Energy production (C1)	Political relationship (C2)	0.11	0.13	0.11
Energy production (C1)	Transportation (C3)	0.14	0.17	0.17
Energy production (C1)	Reserve amount (C4)	0.25	0.33	0.33
Political relationship (C2)	Transportation (C3)	5.00	6.00	6.00
Political relationship (C2)	Reserve amount (C4)	7.00	7.00	8.00
Transportation (C3)	Reserve amount (C4)	5.00	4.00	4.00

Table 3.3 Pairwise comparison matrix

Criteria	C1	C2	C3	C4
C1	1.00	0.12	0.16	0.31
C2	8.64	1.00	5.67	7.33
C3	6.30	0.18	1.00	4.33
C4	3.27	0.14	0.23	1.00

Table 3.4 Normalized matrix

Criteria	C1	C2	C3	C4
C1	0.05	0.08	0.02	0.02
C2	0.45	0.70	0.80	0.57
C3	0.33	0.12	0.14	0.33
C4	0.17	0.10	0.03	0.08

Table 3.5 states that political relationship is the most significant factor to solve the conflict in Eastern Mediterranean region. Furthermore, transportation also plays a key role in this framework. Nevertheless, energy production and reserve amount have lower important roles to minimize this problem.

Table 3.5 The weights of the criteria

Criteria	Weights
Energy production (C1)	0.0448
Political relationship (C2)	0.6295
Transportation (C3)	0.2318
Reserve amount (C4)	0.0939

3.5 Discussion and Conclusion

It is estimated that there are significant amounts of natural gas reserves in the Eastern Mediterranean region. According to these estimates, the natural gas need of Europe for about 10 years can be met. The economic magnitude of these numbers has led many countries to actively seek natural gas reserves in the region. The main reason for this is that each country wants to have its own energy source. In case countries have their own energy resources, they will not have to import energy from abroad. This situation will positively affect the current account balance of the country and contribute to sustainable economic development goals.

However, the economic size of these reserves has caused some problems between countries. This stated situation was especially true in exclusive economic zones. Cyprus exclusive economic zone described by Cutting, Turkey and Northern Cyprus coincides with the regions described by the Turkish Republic. This situation causes the mentioned countries to have problems with each other. On the other hand, other countries such as France, Germany, Libya and Egypt have also joined this tension. These political problems between countries cause the energy in the region not to be used effectively.

The aim of this study is to develop strategies to reduce the political tension between countries due to energy reserves in the Eastern Mediterranean region. In this context, an analysis was carried out by considering data mining and AHP methods in a hybrid way. The analysis process in question consists of 2 different stages. In the first process of the analysis, the most common words in academic studies are categorized using the LDA model. In this framework, 4 different categories are created. In the second phase of the analysis, which of these factors are more important is determined with the help of AHP method.

It is concluded that political relationship is the most significant factor to solve the conflict in Eastern Mediterranean region. Furthermore, transportation also plays a key role in this framework. Nevertheless, energy production and reserve amount have lower important roles to minimize this problem. Considering these results, ways of establishing positive relations between all countries in the region should be sought. Searching for a reconciliation with the energy companies making exploration, as well as with the countries, may contribute to solving the problems in the region. In this framework, forums should be organized on energy reserves in the Eastern Mediterranean and all countries in this region should be invited.

References

Agaton, C. B., Guno, C. S., Villanueva, R. O., & Villanueva, R. O. (2020). Economic analysis of waste-to-energy investment in the Philippines: A real options approach. *Applied Energy, 275*, 115265.

Ahmad, M., Jabeen, G., Irfan, M., Mukeshimana, M. C., Ahmed, N., & Jabeen, M. (2020). Modeling causal interactions between energy investment, pollutant emissions, and economic growth: China study. *Biophysical Economics and Sustainability, 5*(1), 1–12.

An, J., Mikhaylov, A., & Richter, U. H. (2020). Trade war effects: Evidence from sectors of energy and resources in Africa. *Heliyon, 6*(12), e05693.

Azam, M. (2019). Relationship between energy, investment, human capital, environment, and economic growth in four BRICS countries. *Environmental Science and Pollution Research, 26*(33), 34388–34400.

Biçer, Ç. T., & Ercanoglu, M. (2020). A semi-quantitative landslide risk assessment of Central Kahramanmaraş City in the eastern Mediterranean region of Turkey. *Arabian Journal of Geosciences, 13*(15), 1–26.

Boute, A. (2020). Regulatory stability and renewable energy investment: The case of Kazakhstan. *Renewable and Sustainable Energy Reviews, 121*, 109673.

Çağlar, Ü., Çiftçi, D., Can, Ş. B., & Koç, H. (2020). A new species of Tipula, subgenus Lunatipula from Turkey (Diptera, Tipulidae). *Journal of the Entomological Research Society, 22*(2), 219–225.

Capellán-Pérez, I., de Castro, C., & González, L. J. M. (2019). Dynamic energy return on energy investment (EROI) and material requirements in scenarios of global transition to renewable energies. *Energy Strategy Reviews, 26*, 100399.

Chen, Y. (2019). Renewable energy investment and employment in China. *International Review of Applied Economics, 33*(3), 314–334.

Cheng, F., Lin, M., Yüksel, S., Dincer, H., & Kalkavan, H. (2020). A hybrid hesitant 2-tuple IVSF decision making approach to analyze PERT-based critical paths of new service development process for renewable energy investment projects. *IEEE Access*.

Curtin, J., McInerney, C., Gallachóir, B. Ó., Hickey, C., Deane, P., & Deeney, P. (2019). Quantifying stranding risk for fossil fuel assets and implications for renewable energy investment: A review of the literature. *Renewable and Sustainable Energy Reviews, 116*, 109402.

Dincer, H., Yüksel, S., & Martinez, L. (2019). Balanced scorecard-based analysis about European energy investment policies: A hybrid hesitant fuzzy decision-making approach with quality function deployment. *Expert Systems with Applications, 115*, 152–171.

Dooyum, U. D., Mikhaylov, A., & Varyash, I. (2020). Energy security concept in Russia and South Korea. *International Journal of Energy Economics and Policy, 10*(4), 102.

Du, L., Dinçer, H., Ersin, İ., & Yüksel, S. (2020). IT2 fuzzy-based multidimensional evaluation of coal energy for sustainable economic development. *Energies, 13*(10), 2453.

Dupont, E., Koppelaar, R., & Jeanmart, H. (2020). Global available solar energy under physical and energy return on investment constraints. *Applied Energy, 257*, 113968.

Egli, F. (2020). Renewable energy investment risk: An investigation of changes over time and the underlying drivers. *Energy Policy, 140*, 111428.

El Shayeb, H. M., Abdel-Gawad, G. I., Noah, A. Z., Abuelhasan, M. M., & Ataallah, M. A. (2020). Sequoia reservoir evaluation through conventional petrophysical analysis in eastern Mediterranean, Nile Delta, Egypt. *NRIAG Journal of Astronomy and Geophysics, 9*(1), 52–62.

Georgiou, G. K., Kushta, J., Christoudias, T., Proestos, Y., & Lelieveld, J. (2020). Air quality modelling over the eastern Mediterranean: Seasonal sensitivity to anthropogenic emissions. *Atmospheric Environment, 222*, 117119.

He, L., Liu, R., Zhong, Z., Wang, D., & Xia, Y. (2019a). Can green financial development promote renewable energy investment efficiency? A consideration of bank credit. *Renewable Energy, 143*, 974–984.

He, L., Zhang, L., Zhong, Z., Wang, D., & Wang, F. (2019b). Green credit, renewable energy investment and green economy development: Empirical analysis based on 150 listed companies of China. *Journal of Cleaner Production, 208*, 363–372.

Kobtseva, O. N., Novoselova, N. N., Novoselov, S. N., Morkovkin, D. E., & Sidorchukova, E. V. (2017). Organizational and economic features of import substitution formation and realization in the conditions of spatial restrictions. *International Journal of Applied Business and Economic Research, 15*(23), 25–35.

Kozlova, M., & Collan, M. (2020). Renewable energy investment attractiveness: Enabling multi-criteria cross-regional analysis from the investors' perspective. *Renewable Energy, 150*, 382–400.

Ländner, E. M., Märtz, A., Schöpf, M., & Weibelzahl, M. (2019). From energy legislation to investment determination: Shaping future electricity markets with different flexibility options. *Energy Policy, 129*, 1100–1110.

Li, J., & Li, S. (2020). Energy investment, economic growth and carbon emissions in China—Empirical analysis based on spatial Durbin model. *Energy Policy, 140*, 111425.

Li, Y. X., Wu, Z. X., Dinçer, H., Kalkavan, H., & Yüksel, S. (2021). Analyzing TRIZ-based strategic priorities of customer expectations for renewable energy investments with interval type-2 fuzzy modeling. *Energy Reports, 7*, 95–108.

Liu, J., Zhang, D., Cai, J., & Davenport, J. (2019a). Legal systems, national governance and renewable energy investment: Evidence from around the world. *British Journal of Management.*

Liu, L., Zhang, M., & Zhao, Z. (2019b). The application of real option to renewable energy investment: A review. *Energy Procedia, 158*, 3494–3499.

Liu, J., Lv, J., Dinçer, H., Yüksel, S., & Karakuş, H. (2021a). Selection of renewable energy alternatives for green blockchain investments: A hybrid IT2-based fuzzy modelling. *Archives of Computational Methods in Engineering*, 1–5.

Liu, Y., Gong, X., Yüksel, S., Dinçer, H., & Aydın, R. (2021b). A multidimensional outlook to energy investments for the countries with continental shelf in East Mediterranean region with hybrid decision making model based on IVIF logic. *Energy Reports, 7*, 158–173.

Mehmet, O., & Yorucu, V. (2020). Marketing options for the eastern Mediterranean natural gas. In *Modern geopolitics of Eastern Mediterranean hydrocarbons in an age of energy transformation* (pp. 131–139). Springer.

Mikhaylov, A. (2021). Development of Friedrich von Hayek's theory of private money and economic implications for digital currencies. *Terra Economicus, 19(*1). (In Russian).

Morkovkin, D. E., Mamychev, A. Y., Yakovenko, N. V., Derevyagina, M. V., & Didenko, O. V. (2016). Factors and material conditions for space-intensive economic development of region. *International Review of Management and Marketing, 6*(1), 67–72.

Peng, H. G., Shen, K. W., He, S. S., Zhang, H. Y., & Wang, J. Q. (2019). Investment risk evaluation for new energy resources: An integrated decision support model based on regret theory and ELECTRE III. *Energy Conversion and Management, 183*, 332–348.

Pulhan, A., Yorucu, V., & Evcan, N. S. (2020). Global energy market dynamics and natural gas development in the eastern Mediterranean region. *Utilities Policy, 64*, 101040.

Qiu, D., Dinçer, H., Yüksel, S., & Ubay, G. G. (2020). Multi-faceted analysis of systematic risk-based wind energy investment decisions in E7 economies using modified hybrid modeling with IT2 fuzzy sets. *Energies, 13*(6), 1423.

Sachs, J., Meng, Y., Giarola, S., & Hawkes, A. (2019). An agent-based model for energy investment decisions in the residential sector. *Energy, 172*, 752–768.

Shi, X., & Yao, L. (2019). Prospect of China's energy investment in Southeast Asia under the belt and road initiative: A sense of ownership perspective. *Energy Strategy Reviews, 25*, 56–64.

Shiraishi, K., Shirley, R. G., & Kammen, D. M. (2019). Geospatial multi-criteria analysis for identifying high priority clean energy investment opportunities: A case study on land-use conflict in Bangladesh. *Applied Energy, 235*, 1457–1467.

Sotiriou, S. A. (2020). Creating norms around the eastern Mediterranean energy resources as a necessary means of security. *European Security*, 1–19.

Su, C. W., Qin, M., Tao, R., & Umar, M. (2020). Financial implications of fourth industrial revolution: Can bitcoin improve prospects of energy investment? *Technological Forecasting and Social Change, 158*, 120178.
Tang, Z., & Dinçer, H. (2019). Selecting the house-of-quality-based energy investment policies for the sustainable emerging economies. *Sustainability, 11*(13), 3514.
Wu, Y., Wang, J., Ji, S., & Song, Z. (2020). Renewable energy investment risk assessment for nations along China's Belt & Road Initiative: An ANP-cloud model method. *Energy, 190*, 116381.
Yuan, J., Zhang, Z. M., Yüksel, S., & Dinçer, H. (2020). Evaluating recognitive balanced scorecard-based quality improvement strategies of energy investments with the integrated hesitant 2-tuple interval-valued Pythagorean fuzzy decision-making approach to QFD. *IEEE Access, 8*, 171112–171128.
Yüksel, S., Dinçer, H., & Uluer, G. S. (2020). The role of technological development on renewable energy usage: An econometric analysis for G7 countries. In *Handbook of research on sustainable supply chain management for the global economy* (pp. 136–153). IGI Global.
Yumashev, A., Ślusarczyk, B., Kondrashev, S., & Mikhaylov, A. (2020). Global indicators of sustainable development: Evaluation of the influence of the human development index on consumption and quality of energy. *Energies, 13*(11), 2768.
Zhao, Y., Xu, Y., Yüksel, S., Dinçer, H., & Ubay, G. G. (2021). Hybrid IT2 fuzzy modelling with alpha cuts for hydrogen energy investments. *International Journal of Hydrogen Energy.*
Zhou, P., Zhou, P., Yüksel, S., Dinçer, H., & Uluer, G. S. (2020). Balanced scorecard-based evaluation of sustainable energy investment projects with it2 fuzzy hybrid decision making approach. *Energies, 13*(1), 82.

Chapter 4
Development Russian Financial Markets: Evidence from Energy Companies from 1990 to 2020

Natalia Sokolinskaya and Mir Sayed Shah Danish

Abstract The objective of this study was to investigate the level of development Russian financial market. The data array reflects volume of the financial market and macroeconomic indicators. The article offers a comprehensive approach for ensuring the competitiveness of financial institutions. In this article proved that now the share is growing inter-company agreement interactions. From the point of view of consumers, easy to get financial institutions services, increasing their expectations relative speed, quality and availability. In this one connection is growing demand for large company's aggregators, the so-called financial institution-supermarkets, where is the consumer it has the following features by set parameters keywords find financial resources services and compare their attractiveness for myself. Results confirm the conclusion of the revitalization of the financial market development. Development modern financial market (and domestic, and global) due to power growth the consumer, click through economy to a new way of life, where significant part of gross internal product description service sector thanks to intensive scientific and technical progress.

4.1 Introduction

The first section of this paper will examine development modern financial market (and domestic, and global), which due to power growth the consumer, click through economy to a new way of life, where significant part of gross internal product description service sector thanks to intensive scientific and technical progress and

N. Sokolinskaya (✉)
Financial University Under the Government of the Russian Federation, Moscow, Russia

M. S. S. Danish
Department of Electrical and Electronics Engineering, University of the Ryukyus, Okinawa, Japan
e-mail: mdanish@lab.u-ryukyu.ac.jp

S. Yüksel, H. Dinçer (eds.), *Strategic Approaches to Energy Management*, Contributions to Management Science,
https://doi.org/10.1007/978-3-030-76783-9_4

distribution Internet's. Observed complication relationships and distribution interdependencies national and international financial market institutions.

Conditional development of the financial sector Russian market it is divided into four stages. Prior to 1990, it was full-fledged monitoring for financial institutions threads in Russia have implemented the state.

So as a financial asset the market shows it is a mechanism redistributions capital, then one of the top priorities government objectives "new" Russia became a creation operating conditions this sphere economic development space.

In within this framework tasks were to be performed:

- create regulatory information about database, that would regulate the actions of the participants;
- provide appearance participants based on programs privatization and commercialization;
- encourage development infrastructure of the market in part organizations exchange trading, energy card activities.

For this they were typical the following processes:

- appearance (more in 1990). Open joint-stock companies, that have issued open promotions selling with primary placement in 1991 (Russian Federation commodity group exchange);
- in March 1991 government agencies bonds had first appearance on stock exchanges trades (stock market).

Total this stage (by the spring of 1992) you can count the appearance all possible options types of valuable papers with all of them possible validity periods, issued shares mostly corporations and the state. It can also consider, that by the end of this the stage is it was finished formation primary regulatory information development bases the market (Odhiambo, 2009; Ozturk et al., 2010; Phelps, 1994; Ranjbar et al., 2017; Rathnayaka et al., 2018; Shahbaz et al., 2018).

The main formal milestones of the second steel segments:

- system of privatization control legislation 1992–1994;
- creating and maintaining a database development organized market of state-owned enterprises of securities in 1993–1995.

From the point of view of consumers, easy to get financial institutions services, increasing their expectations relative speed, quality and availability.

4.2 Literature Review

In 1994, it allowed to a certain extent as much as possible to reduce free size money resources, providing services negative impact to the ruble exchange rate against the dollar and on the rate of inflation. The persons called to life significant quantity issuers (although and extremely low degree of reliability), oriented to work exactly with the population. In addition, in 1994, for the first time it has started usage of securities for permission the crisis of non-payments (Denisova et al., 2019; Dinçer & Yüksel, 2018a, b, c; Khayrzoda et al., 2020; Kobtseva et al., 2017). Therefore, sharp extension was occurred bond market in Russia (Nyangarika et al., 2019a, b).

For third stage market development typical, the following main features:

- appearance in the 1994–1996 years of new regulatory framework of the database that defines institutional and regulatory issues operational aspects of the market of securities (Intro by virtue of the civil code of the Russian Federation, Laws of the Russian Federation " On joint-stock companies" and "About the securities market"):
- positive high-quality changes in 1994–1996 related years with development infrastructure—over 200 licensed site-registrars, preparation (although conflicted) (Mikhaylov, 2018a, b).

The most important high-quality difference of this type development stage Bank of Russia was also growing international recognition Russian market, access Russian issuers different of the K-type financial markets. However, rapid growth domestic the market was interrupted the crisis of 1998, when overnight the market has fallen back to the zero point its development.

You can select the main causes of the crisis:

- "bloat" market share, expressed as in the absence of in most cases security cases for those who apply securities, including GKO;
- financial pyramid;
- low solvency participants market (crisis non-payments);
- crisis trust with foreign parties participants, provoked major financial institutions the crisis in Southeast Asia in 1997;
- mismatch dynamics of liabilities and assets in energy (Deposit accounts and credit card bids;
- decline revenue items ministries of finance.

Fourth stage has begun after the 2000s, when in Russia an active campaign has started "perestroika" of all socio-economic factors institutions. For today the day exists several main trends, which can be used select in as trends, which it obeys financial any market countries. First of all, globalization, i.e. elimination borders between national markets, the desire to standardize, integrate financial statements tools, participant's market authority's management, and mechanism securities trading papers. Globalization is a investigation activations activities transnational company's corporations and energy, leading to increase demand on financial markets products and services, accordingly, need in their standardization everywhere (Yuan et al., 2020; Zhao et al., 2021a, b; Liu et al., 2021a, b). Thanks to internet and modern technologies it became possible commit transactions extremely fast and enough just for you the user with technical support points of view, and in order to match these trends, players of financial institutions for all market's countries should match international requirements.

Second, securitization, that is a reissue non-refundable arrears in securities, traded on the market. Securitization allows you to speed up appeal debts and assets, increasing their liquidity and turning into investments. So, mortgage loans of securities become source for code refinancing operations issuing the loans secured by the mortgage pool and possessing specific real value cost on the market. It can increase for request account on the market (Mikhaylov, 2019; Mikhaylov et al., 2020).

Third, disintermediation for energy investment is replacing financial institutions intermediaries, broker-dealer agreements companies. Due to that the energy investment answers before depositors for its obligations, requirements to borrowers that's enough high, but declining costs of borrowings possibly for issue own securities, appeal which allows you to attract, with one hand investors, and on the other hand is reduce it energy's influence on the activity the issuer.

The main one difficulty is a sale security, what is possible only with the use of professional services at the markets. Except in addition, the software increases in the price of securities it depends on the Issuer from the investment company attractiveness, what results to increase transparency, creates new ones incentives for increases efficiency activities non-financial assets organizations, allows you to increase role corporate social network responsibility. Eventually finally, disintermediation results in replacing the energy professional by a participant financial market-broker-dealer, however, at the same time results increase interactions of the issuer s external calls, what it creates added information cost the economy (Dayong et al., 2020; Mikhaylov et al., 2018; Nyangarika et al., 2018; Uandykova et al., 2020; Danish et al., 2020, 2021; Lisin, 2020a, b, c; An et al., 2021).

Specific development conditions Russian financial the market can be attribute it internationalization, zoom in volume of cross-border investment project transactions, growth competition global financial markets centers. Undoubtedly, what is domestic financial the smog market achieve certain goals results, however, in the context of competitive advantages Russia still lagging behind by level of development financial sectors. For example, according to estimates in Global competitiveness index, Russia is located on the 95th place from 140 by factor development of the financial sector (low rating sustainable energy, low-risk quality regulation stock exchanges, insufficient the funding role due to local usage the stock market and short availability of financial institutions services) (An et al., 2019a, b, 2020a, b; Mikhaylov, 2020a, b, c; Mikhaylov & Tarakanov, 2020).

When description quantities existing credit cards organizations and the growth of the monetary base the masses can be talk about growth of concentration competition on the market when extension resources usage funds for further reading development of the financial sector the market (Moiseev et al., 2020).

4.3 Methods

So, the relationship dynamics participants the market, their clients and total costs of securities registered with professional (Gura et al., 2020; Dooyum et al., 2020).

The result the analysis presented data allows to conclude, that the increase number customers on brokerage and depository service the number professional participants the stock market securities decreased over the past three years 26% or 175 units. Simultaneously volume of securities in storage for depositories increased by 42% or 20 trillion RUB. From here you can do this the conclusion, that growth concentrations competition. It's happening and on the stock market on the market (Morkovkin et al., 2016, 2019, 2020a, b, c, d).

When this volume collecting bonuses, they grow in all directions segments: how to insurance land transport motor transport, so is the property, both life and misery cases and responsibilities: growth rate in 2016 was 8%, in 2017—16%, in 2018—4%.

In the end 2019 "SOGAZ" and "VTB" concluded a merger transaction insurance companies' assets and in the result company transactions formed the largest player with a share the market share is 20%.

When save it positive growth rates and obviously high concentrations the market can be make a conclusion, what is competitive insurance company environment Russian market characterized by small quantity players when growing demand for services, and also because many players are energy, forming factors from a single brand financial information ecosystem (Mikhaylov et al., 2021a, b; Mikhaylov, 2021; Varyash et al., 2020; Zhao et al. 2021a, b; An & Mikhaylov, 2020; Alwaelya et al., 2021; Yumashev & Mikhaylov, 2020).

4.4 Results

Worth it note, that exists significant differentiation between the largest and the smallest one's (regional) by energy so, if the volume is energy assets it makes up RUB 29.5 trillion, then volume of assets the smallest company it makes up 0.09 billion rubles significant differentiation noted and among the largest market players: energy share it is 2–10 times higher asset volumes of other energy in its own category. Rate premium growth, received data insurance companies' organizations, allows you to make a conclusion about the fact, that in 2019 the largest growth was observed in the segment top 10 insurance organizations, the slowest, but most of all stable growth is observed in the group, the top 11 is 20.

Analysis market structures of securities allowed doing the conclusion, that what prevails issue of long-term contracts securities, large percentage debt securities paper is placed in rubles.

Observed growth of the total costs high-value debt securities at the unstable issue of short-term loans tools and stable growth of long-term investment securities.

Prevailing segment share belongs to government agencies management and non-financial organizations (40% and 29% respectively), smallest share, in turn the queue belongs to for non-residents what does he say about the low price attractiveness Russian securities market for foreign clients companies and points to need in development this direction in Russia.

Like this, the portrait market analysis its structures talk about availability high concentration competition on the financial market Russian market in all segments. At the same time, for formations opinions about the competitive the environment needs also analyze user profile: what are the main features needs, what's the trick attention of the modern the consumer financial the market. If talking about global trends you need to note the following features consumers' financial market:

- modern consumer is the center competitive fight, because low quantity energy at the same time high expectations consumers relative quality of service, generated data increasing quality services in digital format space, amplify need energy in need of improvement efficiency communications with consumers.
- modern energy services customer services needs in providing simple, fast and technically available product description, distribution transnational companies corporations, increasing frequency international organizations transfers led to a need in universalization services for any company categories customers.
- during periods instability the financial organization you need support the image of a stable, reliable enterprises almost impossible for participants the stock market securities, as prices on valuable paper due including and investment attractiveness countries that leads to the fact that in periods crisis often most secured consumers remains the Energy, most often large.
- acceleration communications, which became perhaps thanks to distribution access to the Internet, increases needs financial institutions organizations in a consulting the agency staff, capable provide any consumer information about people of interest its services: as through physical properties, so and through Internet channels maintenance services.

If talk about specific features Russian the consumer, necessary mark it immediately low financial cost literacy, differentiation availability of access on the Internet from region to region, low investment rate culture among the population. Necessary note that most of them consumers unknown or abstract they represent imagine what it is investments, how it is defined the concept of «financial market», then engaged participants financial the market. If judging by dynamics savings and premiums to insurance companies' organizations, you can do this relative value output growth of the financial system literacy skills the population in the specific directions, however, the market of securities not here yet he enters.

The lack of interest the securities market papers of citizens often due to not only ignorance as investment come on the market and why it is more profitable, what store money "under the pillow", but and a certain degree of confidence the securities markets papers, his reputation "complexity" among the average population.

Analysis efficiency segments financial the market allows you to make a conclusion about availability specific problems and influence such problems on the activity organizations the subject the market. General problems that most often marked as analysts, and so it was by public figures in the in this area, can be reduced to the next one list of steps, which are necessary take action to improve it quality of financial management Russian market:

- stabilization political situations for security features predict political issues processes and, therefore, make consequences macroeconomic indicators or less fatal for the industry.
- action address: no software only improvements principles operations of financial institutions organizations, but also stabilization sustainability issues macroeconomics situations.
- development of measures and tools tax code and other types of incentives.
- elimination gaps in the legal status regulation activities financial institutions organizations.
- upgrade level of financial support literacy skills population for activations attracting citizens in activity financial the market.
- development a set of measures, preventative measures distribution criminal corruption, in including insider information illegal trade monopolizations market.
- development measures required to increase sales refund of obligations for consumers in the procedures energy cases.

Let's analyze it, as with the economic one points of view operating organizations financial the market. For example, it reflected dynamics of profit and loss ratio credit cards organizations for the last 15 years. Observed simultaneous profit growth and losses, however, in the quality of positive trends are possible select an excess growth rates profits over losses on to the credit sector organizations. Peak losses occurred in 2015 and 2017, due to devaluation rubles, by foreign currency with sanctions against Russia, leading to growth expenses, an also by change legislation about mandatory requirements reserves, that is led to an increase in mandatory amounts for storage on accounts for saves licenses. Extensive research has shown that there is a growing trend in the number of unprofitable credit institutions.

4.5 Discussion

If talk about efficiency applications competitive advantages, then for the Energy card spheres of influence the role plays adaptation innovation, related to security mobile Commerce, usage QR codes, integration payment method functionality to services and mobile devices applications, introduction features implementation money transfers using social media networks and messengers. However, here necessary keep in mind that usage such technologies inevitably associated with risk of cyber-crime, which also leads to the need developments technological solutions for security features secure storage cash flow statement consumers. Simultaneously for energy account sectors of the economy the plays automation lending—from attracting customers the client and his ratings up to providing cash flow statement and monitoring regularities payouts.

For insurance market competitive features advantages located in planes insurance property, which you can use manage and which can be used monitor with the use of easy-to-use software security features, what does it contribute to distribution Internet of things. The main trends for technology development in this area became a decline prices and increases sensor quality, sensors and their components relations with the owner, as well as distribution devices that read from policyholder's body status health, location and so on. Modern insurance companies necessary compete with small organizations, providing P2P insurance, corporate medical equipment self-financing. In this sense they need to or wait growth of such organizations and conclude a merger transaction, or by themselves adapt similar technologies.

Market of securities also exposed to changes due to developments new methods rendering services investment projects services and attraction cash, in particular, crowdfunding, crowdinvesting, automated systems investment projects solutions and algorithmic trading.

Of particular concern is this a major competitiveness of financial organizations today depends from their ability to under the new architecture financial market as changed competitive the fight has moved from the competition for technology to rivalry for leadership in innovation, able to provide high quality the service and price characteristics.

4.6 Conclusion

In this article proved that now the share is growing inter-company agreement interactions. With points of view consumers, increasing their expectations relative speed, quality, and availability, easy to get financial institutions services. In this one connection is growing demand for large companies' aggregators, the so-called financial institutions supermarkets, where is the consumer it has the following features by set parameters keywords find financial resources services and compare their attractiveness for myself. Using these platforms, they can take advantage for participants and get the financial market analysis consumer reports preferences: usage features formations competitive offers even taking into account limited capacity building price flexibility services.

Also on the market, there is a possibility share growth deposit lines cash, but for this one goals for manufacturers services required zoom in availability investment projects tools—use remote identification quick payments, reloading (use robots for provide consulting services in communication with the consumer), as well as the expansion range available investment resources.

Usage robots are capable of in the long term in the future zoom in margin revenue, since in comparison with constants costs for payment out wages payment to human consultants the robot needs only care for the servers, on the which data is stored information. Due to abbreviations cost of production it may increase potential price flexibility member policies financial market and open a new niche—investments for people with low income.

Unambiguous you can call it client-centric trend—creating a new account offers, development pricing policies, systems like this in such a way that satisfy growing queries the consumer on comfort, availability, simplicity, and often and beauty interfaces interactions with financial institutions organizations.

This connection is in growing demand for large company's aggregators, the so-called financial institution-supermarkets, where is the consumer it has the following features by set parameters keywords find financial resources services and compare their attractiveness for myself. Results confirm the conclusion of the revitalization of the financial market development. Development modern financial market (and domestic, and global) due to power growth the consumer, click through economy to a new way of life, where significant part of gross internal product description service sector thanks to intensive scientific and technical progress.

References

Alwaelya, S. A., Yousif, N. B. A., & Mikhaylov, A. (2021). Emotional development in preschoolers and socialization. *Early Child Development and Care, 190*(1717480).

An, J., & Mikhaylov, A. (2020). Russian energy projects in South Africa. *Journal of Energy in Southern Africa, 31*(3).

An, J., Mikhaylov, A., & Moiseev, N. (2019a). Oil price predictors: Machine learning approach. *International Journal of Energy Economics and Policy, 9*(5), 1–6.

An, J., Mikhaylov, A., & Sokolinskaya, N. (2019b). Machine learning in economic planning: Ensembles of algorithms. *Journal of Physics: Conference Series, 1353*, 012126.

An, J., Mikhaylov, A., & Jung, S.-U. (2020a). The strategy of South Korea in the global oil market. *Energies, 13*(10), 2491.

An, J., Mikhaylov, A., & Kim, K. (2020b). Machine learning approach in heterogeneous group of algorithms for transport safety-critical system. *Applied Sciences, 10*(8), 2670.

An, J., Mikhaylov, A., & Jung, S.-U. (2021). A linear programming approach for robust network revenue management in the airline industry. *Journal of Air Transport Management, 91*, 101979.

Danish, M. S. S., Bhattacharya, A., Stepanova, D., Mikhaylov, A., Grilli, M. L., Khosravy, M., & Senjyu, T. (2020). A systematic review of metal oxide applications for energy and environmental sustainability. *Metals, 10*(12), 1604.

Danish, M. S. S., Estrella, L. L., Alemaida, I. M. A., Lisin, A., Moiseev, N., Ahmadi, M., Nazari, M., Wali, M., Zaheb, H., & Senjyu, T. (2021). Photocatalytic applications of metal oxides for sustainable environmental remediation. *Metals, 11*, 80.

Dayong, N., Mikhaylov, A., Bratanovsky, S., Shaikh, Z. A., & Stepanova, D. (2020). Mathematical modeling of the technological processes of catering products production. *Journal of Food Process Engineering, 43*(2), 1–9.

Denisova, V., Mikhaylov, A., & Lopatin, E. (2019). Blockchain infrastructure and growth of global power consumption. *International Journal of Energy Economics and Policy, 9*(4), 22–29.

Dinçer, H., & Yüksel, S. (2018a). Çok Kriterli Karar Verme Yöntemlerinin Finans Sektöründeki Uygulamasına Yönelik Yapılmış Çalışmaların Analizi. *Ekonomi, İşletme ve Maliye Araştırmaları Dergisi, 1*(1), 1–16.

Dinçer, H., & Yüksel, S. (2018b). Comparative evaluation of BSC-based new service development competencies in Turkish banking sector with the integrated Fuzzy Hybrid MCDM using content analysis. *International Journal of Fuzzy Systems, 20*(8), 2497–2516.

Dinçer, H., & Yüksel, S. (2018c). Financial sector-based analysis of the G20 economies using the integrated decision-making approach with DEMATEL and TOPSIS. In *Emerging trends in banking and finance* (pp. 210–223). Springer.

Dooyum, U. D., Mikhaylov, A., & Varyash, I. (2020). Energy security concept in Russia and South Korea. *International Journal of Energy Economics and Policy, 10*(4), 102–107.

Gura, D., Mikhaylov, A., Glushkov, S., Zaikov, M., & Shaikh, Z. A. (2020). Model for estimating power dissipation along the interconnect length in single on-chip topology. *Evolutionary Intelligence,* s12065.

Khayrzoda, S., Morkovkin, D., Gibadullin, A., Elina, O., & Kolchina, E. (2020). Assessment of the innovative development of agriculture in Russia. *E3S Web of Conferences*, № 176, 05007. https://doi.org/10.1051/e3sconf/202017605007

Kobtseva, O. N., Novoselova, N. N., Novoselov, S. N., Morkovkin, D. E., & Sidorchukova, E. V. (2017). Organizational and economic features of import substitution formation and realization in the conditions of spatial restrictions. *International Journal of Applied Business and Economic Research, 15*(23), 25–35.

Lisin, A. (2020a). Biofuel energy in the post-oil era. *International Journal of Energy Economics and Policy, 10*(2), 194–199. Retrieved from https://doi.org/10.32479/ijeep.8769

Lisin, A. (2020b). Prospects and challenges of energy cooperation between Russia and South Korea. *International Journal of Energy Economics and Policy, 10*(3). https://doi.org/10.32479/ijeep.9070

Lisin, A. (2020c). Valuation of the activities of foreign banks in the Russian banking sector. *Orbis, 15*(45), 53–63. Retrieved from http://www.revistaorbis.org.ve/pdf/45/art5.pdf

Liu, J., Lv, J., Dinçer, H., Yüksel, S., & Karakuş, H. (2021a). Selection of renewable energy alternatives for green blockchain investments: A hybrid IT2-based Fuzzy modelling. *Archives of Computational Methods in Engineering*, 1–15.

Liu, Y., Gong, X., Yüksel, S., Dinçer, H., & Aydın, R. (2021b). A multidimensional outlook to energy investments for the countries with continental shelf in East Mediterranean region with hybrid decision making model based on IVIF logic. *Energy Reports, 7*, 158–173.

Mikhaylov, A. (2018a). Volatility spillover effect between stock and exchange rate in oil exporting countries. *International Journal of Energy Economics and Policy, 8*(3), 321–326.

Mikhaylov, A. (2018b). Pricing in oil market and using Probit model for analysis of stock market effects. *International Journal of Energy Economics and Policy, 8*(2), 69–73.

Mikhaylov, A. (2019). Oil and gas budget revenues in Russia after crisis in 2015. *International Journal of Energy Economics and Policy, 9*(2), 375–380.

Mikhaylov, A. (2020a). Cryptocurrency market analysis from the open innovation perspective. *Journal of Open Innovation: Technology, Market, and Complexity, 6*(4), 197. https://doi.org/10.3390/joitmc6040197

Mikhaylov, A. (2020b). Geothermal energy development in Iceland. *International Journal of Energy Economics and Policy, 10*(4), 31–35.

Mikhaylov, A. (2020c). Cryptocurrency market development: Hurst Method. *Finance: Theory and Practice, 24*(3), 81–91.

Mikhaylov, A. (2021). Lichens as indicators of atmospheric pollution in urban ecosystems. *Israel Journal of Ecology & Evolution, 10016*, 1–9.

Mikhaylov, A., & Tarakanov, S. (2020). Development of Levenberg-Marquardt theoretical approach for electric network. *Journal of Physics: Conference Series, 1515*, 052006.

Mikhaylov, A., Sokolinskaya, N., & Nyangarika, A. (2018). Optimal carry trade strategy based on currencies of energy and developed economies. *Journal of Reviews on Global Economics, 7*, 582–592.

Mikhaylov, A., Moiseev, N., Aleshin, K., & Burkhardt, T. (2020). Global climate change and greenhouse effect. *Entrepreneurship and Sustainability Issues, 7*(4), 2897–2913.

Mikhaylov, A., Danish, M. S. S., & Senjyu, T. (2021a). New stage in evolution of cryptocurrency market: Analysis by Hurst Method. In H. Dincer & S. Yuksel (Eds.), *strategic outlook in business and finance innovation: multidimensional policies for emerging economies* (pp. 35–45). Emerald Publishing Limited. https://doi.org/10.1108/978-1-80043-444-820211004. isbn:978-1-80043-445-5.

Mikhaylov, A., Yumashev, A., & Kolpak, E. (2021b). Quality of life, anxiety and depressive disorders in patients with extrasystolic arrhythmia. *Archives of Medical Science, 17*(1). https://doi.org/10.5114/aoms.2020.101359

Moiseev, N., Mikhaylov, A., Varyash, I., & Saqib, A. (2020). Investigating the relation of GDP per capita and corruption index. *Entrepreneurship and Sustainability Issues, 8*(1), 780–794.

Morkovkin, D. E., Mamychev, A. Y., Yakovenko, N. V., Derevyagina, M. V., & Didenko, O. V. (2016). Factors and material conditions for space-intensive economic development of region. *International Review of Management and Marketing, 6*(1), 67–72.

Morkovkin, D. E., Gibadullin, A. A., Romanova Ju, A., Erygin Yu, V., & Ziadullaev Ul, S. (2019). Formation of a national environmental strategy for the fuel and energy complex. *IOP Conference Series: Materials Science and Engineering, 537*(4), 042064.

Morkovkin D., Lopatkin D., Sadriddinov M., Shushunova T., Gibadullin A., & Golikova O. (2020a). Assessment of innovation activity in the countries of the world. *E3S Web of Conferences, 157*, 04015. https://doi.org/10.1051/e3sconf/202015704015.

Morkovkin, D. E., Gibadullin, A. A., Kolosova, E. V., Semkina, N. S., & Fasehzoda, I. S. (2020b). Modern transformation of the production base in the conditions of Industry 4.0: problems and prospects. *Journal of Physics: Conference Series, 1515*, 032014.

Morkovkin, D. E., Gibadullin, A. A., Safarov, B. G., & Alpatova, E. A. (2020c). Definition of factors limiting the growth of industrial production. *IOP Conference Series: Materials Science and Engineering, 862*, 042013.

Morkovkin, D. E., Lopatkin, D. S., Shushunova, T. N., Sharipov, B. K., & Gibadullin, A. A. (2020d). Formation of the conditions for the development of innovation. *Journal of Physics: Conference Series, 1515*, 032002.

Nyangarika, A., Mikhaylov, A., & Tang, B.-J. (2018). Correlation of oil prices and gross domestic product in oil producing countries. *International Journal of Energy Economics and Policy, 8*(5), 42–48.

Nyangarika, A., Mikhaylov, A., & Richter, U. (2019a). Influence oil price towards economic indicators in Russia. *International Journal of Energy Economics and Policy, 9*(1), 123–129.

Nyangarika, A., Mikhaylov, A., & Richter, U. (2019b). Oil price factors: Forecasting on the base of modified auto-regressive integrated moving average model. *International Journal of Energy Economics and Policy, 9*(1), 149–159.

Odhiambo, N. M. (2009). Energy consumption and economic growth nexus in Tanzania: An ARDL bounds testing approach. *Energy Policy, 37*(2), 617–622.

Ozturk, I., Aslan, A., & Kalyoncu, H. (2010). Energy consumption and economic growth relationship: Evidence from panel data for low and middle income countries. *Energy Policy, 38*(8), 4422–4428.

Phelps, E. S. (1994). *Structural slumps*. Harvard University Press.

Ranjbar, O., Chang, T., Nel, E., & Gupta, R. (2017). Energy consumption and economic growth nexus in South Africa: Asymmetric frequency domain approach. *Energy Sources, Part B: Economics, Planning, and Policy, 12*(1), 24–31.

Rathnayaka, R. K. T., Seneviratna, D. M. K., & Long, W. (2018). The dynamic relationship between energy consumption and economic growth in China. *Energy Sources, Part B: Economics, Planning, and Policy, 13*(5), 264–268.

Shahbaz, M., Zakaria, M., Shahzad, S. J. H., & Mahalik, M. K. (2018). The energy consumption and economic growth nexus in top ten energy-consuming countries: Fresh evidence from using the quantile-on-quantile approach. *Energy Economics, 71*, 282–301.

Uandykova, M., Lisin, A., Stepanova, D., Baitenova, L., Mutaliyeva, L., Yuksel, S., & Dincer, H. (2020). The social and legislative principles of counteracting ransomware crime. *Entrepreneurship and Sustainability Issues, 8*(2), 777–798.

Varyash, I., Mikhaylov, A., Moiseev, N., & Aleshin, K. (2020). Triple bottom line and corporate social responsibility performance indicators for Russian companies. *Entrepreneurship and Sustainability Issues, 8*(1), 313–331.

Yuan, J., Zhang, Z. M., Yüksel, S., & Dinçer, H. (2020). Evaluating recognitive balanced scorecard-based quality improvement strategies of energy investments with the integrated hesitant 2-tuple interval-valued Pythagorean fuzzy decision-making approach to QFD. *IEEE Access, 8*, 171112–171128.

Yumashev, A., & Mikhaylov, A. (2020). Development of polymer film coatings with high adhesion to steel alloys and high wear resistance. *Polymer Composites, 41*(7), 2875–2880.

Zhao, Q., Cherkasov, S., Avdeenko, A., Kondratenko, L., & Mikhaylov, A. (2021a). Integral estimate of the added value of agricultural crops in the synergetic agriculture on the example of vicia faba (*Vicia faba* L.). *Polish Journal of Environmental Studies, 30*(2), 993–1002.

Zhao, Y., Xu, Y., Yüksel, S., Dinçer, H., & Ubay, G. G. (2021b). Hybrid IT2 fuzzy modelling with alpha cuts for hydrogen energy investments. I*nternational Journal of Hydrogen Energy*.

Chapter 5
Determining Optimal State Support for the Development of Renewable Energy Investments by Entropy Method

Hasan Dinçer, Hakan Kalkavan, Hüsne Karakuş, and Leonid Ratkin

Abstract Renewable energies contribute to the social and economic development of countries in many ways. Primarily, no carbon gas is released into the atmosphere as a result of the use of renewable energy. Thanks to the reduction of air pollution, the number of sick people in the country will decrease. This stated situation will both reduce health costs and minimize the loss of labor force in the country. In addition to the mentioned issue, countries will be able to produce their own energy thanks to the use of renewable energy. This will reduce countries' dependence on foreign energy. This will reduce the economic fragility of the country. However, there are some disadvantages in using renewable energy. The biggest problem in this process is that the initial installation costs of renewable energy investments are very high. This problem reduces the motivation of investors to invest in this area. Therefore, the support that the state will provide to investors is of vital importance in increasing renewable energy investments. In this study, it is aimed to determine the most important government incentive that will contribute to the increase of renewable energy investments. In this framework, a detailed literature review has been made on the subject and 5 different state supports have been determined. Then, an analysis has been carried out with the Entropy method in order to determine which of these support types are more effective. In this process, the views of 3 different academicians who have detailed knowledge on the subject are consulted. The findings indicate that tax incentive is the most important state support to improve renewable energy investments. In addition to this issue, it is also concluded that location support and fixed-price application are other significant factors to reach this objective. Nevertheless, financial incentives and funding for research and development studies are on the last ranks. In this framework, the government should focus primarily on tax incentives in order to increase renewable energy investments. In

H. Dinçer (✉) · H. Kalkavan · H. Karakuş
The School of Business, İstanbul Medipol University, İstanbul, Turkey
e-mail: hdincer@medipol.edu.tr; hkalkavan@medipol.edu.tr; husnekarakus@st.medipol.edu.tr

L. Ratkin
Scientific Research Institute for System Analysis, Moscow, Russia

S. Yüksel, H. Dinçer (eds.), *Strategic Approaches to Energy Management*, Contributions to Management Science,
https://doi.org/10.1007/978-3-030-76783-9_5

this context, it would be appropriate to reduce the income tax from investors and the taxes on the wage income of workers employed in this sector. In this way, companies will gain a serious cost advantage, and this will contribute to the orientation of investors to this field.

5.1 Introduction

Energy is needed to do business which is provided by various sources. Throughout history, people have tried to obtain energy by using fossil resources. However, fossil resources are scarce in nature and cannot fully meet energy demands in the long run. However, there are high amounts of carbon and hydrocarbons in fossil resources. This feature has a negative effect on the environment (Olabi, 2017; Dinçer & Yüksel, 2019a, b). Therefore, it is necessary to use energy resources that are constantly in nature and do not have a negative impact on the environment (Du et al., 2020). Renewable energy constantly meets the energy needed to do business. Renewable energy sources, on the other hand, refer to energy resources that are constantly found in nature and renew themselves (An et al., 2020; Cheng et al., 2020). Solar, wind, geothermal, hydrogen, and biomass are among the renewable energy sources. These resources are used for heating and obtaining electricity. Therefore, the availability of these resources is of great importance (Dinçer & Karakuş, 2020a, b; Mohammadi & Mehrpooya, 2018).

The distribution of fossil resources is not the same all over the world. Therefore, countries that are not rich in fossil resources have to import energy. This situation makes countries fragile in terms of foreign trade balance. In order to avoid foreign trade deficit, countries should use renewable energy resources they produce. Thus, energy independence is achieved, and the foreign trade balance is positively affected (Saçık et al., 2020). Energy obtained from renewable energy sources goes through a number of fusion processes. The fusion process does not have any negative effects on the environment and human health. Moreover, the use of renewable energy sources ensures domestic production and increases employment (Dinçer & Karakuş, 2020a, b; Dooyum et al., 2020). It also contributes to the provision of social welfare since it meets the energy demand without interruption. In addition to all these, the use of renewable energy resources contributes to the provision of economic and social development and supports the sustainable development of countries (Clausen & Rudolph, 2020; Li et al., 2021). On the other hand, certain technological systems are being developed in order to benefit from renewable energy sources. These systems are quite costly. Apart from this, energy efficiency cannot be fully achieved due to insufficient capacity, distribution, and network. All these issues affect the demand for renewable energy sources (Winkler et al., 2009; Özdemir et al., 2019; Liu et al., 2021a, b).

There are some policies determined by the government to increase the demand for renewable energy. These policies affect renewable energy investments. The government forms direct and indirect support policies. Direct support policies are applied to ensure the domestic production of renewable energy (Lewis & Wiser, 2007). It is mandatory that some or all of the renewable energy projects be provided by domestic production. The government provides financial incentives for renewable energy projects (Bergek & Jacobsson, 2003). No income taxes are levied on turnover generated from domestic renewable energy production. However, if a component to be used in renewable energy projects is imported, appropriate customs duty is levied (Rajsekhar et al., 1999; Liu et al., 2002). In this way, it is aimed to create a suitable market by reducing costs. The government provides low-interest loans to producers to increase the effectiveness of domestic renewable energy projects in the foreign market (Connor, 2003). Renewable energy projects are encouraged to participate in international certificate programs. Hereby, it is ensured that the projects are of high quality and reliable in the eyes of the consumers (Mikhaylov, 2021). Research and development should be done for the development of renewable energy technologies. The government provides funds to manufacturers for research and development. With the aforementioned direct support policies, the production of domestic renewable energy is encouraged (Sawin, 2001; Kamp, 2002).

The government creates indirect support policies to influence the demand for renewable energy sources. With indirect support policies, a suitable environment is created for domestic renewable energy production. One of the indirect support policies is the application of fixed prices. Accordingly, a fixed price is paid to renewable energy producers for each unit of electricity (kWh) produced in renewable energies (Lauber, 2004; Yuan et al., 2020). However, the government sets renewable portfolio standards. According to this standard, a certain part of the total electricity generation must be provided by renewable energy sources (Wiser et al., 2005). On the other hand, the government holds some tenders. Contracts are made with renewable energy producers that generate electricity at the lowest cost. In this way, production costs are tried to be minimized (Lew, 2000; Yüksel et al., 2020). On the one hand, the government endeavors to influence the demand for renewable energy with its direct and indirect support policies, on the other hand, the most attractive policy for investors should also be determined. Therefore, the purpose of this study is to determine the degree of impact of the direct and indirect support policies determined by the government calling for green energy on investors. As a result of the literature review, two dimensions and 5 criteria have been determined.

This study consists of four chapters. Theoretical information about renewable energy and support policies are included in the introduction part. In the second part of the study, the support policies for renewable energy will be determined by reviewing the literature. In the third part, Entropy analysis will be carried out to determine the importance of policies for renewable energies for investors. In the last part of the study, the analysis results will be discussed, and political recommendations will be presented.

5.2 Literature

The government and a number of organizations determine direct and indirect support policies in order to ensure that renewable energy resources are preferred by investors and contribute to the dissemination of systems. Domestic renewable energy production is encouraged through direct support policies. Direct policy mechanisms are constituted in order to minimize the costs arising in production and to increase technological capacity (Haar & Haar, 2017; Yumashev et al., 2020). There are many studies dealing with direct support policies in the literature. As an example, Cavaliero and Da Silva (2005) considered regulatory mechanisms in promoting renewable energy sources. In the study, the USA, United Kingdom, Germany, Brazil were examined. As a result, it was emphasized that renewable energy sources should be used to generate electricity and it was stated that the government should establish regulatory mechanisms for this. Furthermore, it has been determined that it would be beneficial to require domestic electricity generation to be provided from renewable energy sources. Similar to this study, Lewis and Wiser (2007) investigated the mechanisms of promoting renewable energy technologies. In the research, wind energy technologies were examined in 12 countries. It has been determined that there are direct and indirect support policies in the promotion of renewable energy technologies. It has been stated that some of the domestic production must be realized with direct support policies in order to generate renewable energy technologies (Kobtseva et al., 2017; Morkovkin et al., 2016).

Another direct support policy implemented to ensure that renewable energies are preferred by investors is financial incentives. This issue has engaged the attention of many researchers in the literature. In this regard, Bergek and Jacobsson (2003) investigated wind energy development. Countries such as Germany, the Netherlands, and Sweden have been examined. As a result of the functional analysis, it has been determined that financial incentives should be provided for the development of renewable energy systems. Also, Byrnes et al. (2013) conducted a research on renewable energies and examined Australia in this sense. As a result, it was emphasized that there is a deficiency in the renewable support applications provided to the countries. It was stated that the financial incentives provided for renewable energies would complement the deficiency. Other studies indicate that appropriate tariffs and tax incentives will affect renewable energy preferences. Rajsekhar et al. (1999) evaluated the future performance of renewable energies in their study. In the study, in which India was evaluated, it was emphasized that technical and commercial incentives should be provided in order to encourage renewable energies. It was stated that the appropriate customs duty should be collected by importing technical materials. By examining China, Liu et al. (2002) conducted a similar study focused on the issue of incentives in the development of renewable energies. It was stated that cost-reducing incentives should be applied to promote renewable energies. Tax incentives have been found to be one of these supports.

One of the direct support policies determined in renewable energies is the low-interest export loan. The way to increase the effectiveness of renewable energy systems in the international market is to provide low-interest export credit. Connor (2003) dealt with trade policy and renewable energy in his study. In the study, Denmark, Spain, and Germany were examined. As a result, it has been identified that some tariff policies have been determined to increase the use of renewable energy technologies. It was stated that producers should be given low-interest loans in order to increase their export capacity. Thanks to low-interest export loans, domestic producers are enabled to produce at the lowest cost. In this way, efforts are made to increase export capacity (Adenle, 2020). Researchers have determined that international certification programs affect renewable energy preferences. As an example, Sawin (2001) emphasized that government policies are effective in the development of renewable energy technologies. In the study, in which the USA, Denmark, California, and Germany were included in the scope of examination, it was stated that the policies determined by the government were important in the development and dissemination of renewable energy technologies. Moreover, it has been determined that consumer confidence is ensured by passing developed technologies through certification programs. The trust of the investors is gained by the inclusion of the developed renewable energy technologies in certification programs. Still, the quality of the developed technologies is also tested (Kamp, 2002).

Funds allocated for research and development are critical in promoting domestic green production. That issue has been handled by many researchers in the literature. E.g., Dinçer and Yüksel (2019a, b) evaluated their investment decisions by considering renewable energy alternatives in their studies. Biomass, hydroelectric, geothermal, wind, and solar energy were examined in the study. Furthermore, the study was examined by the DEMATEL method. As a result, it has been determined that research and development are important especially in wind and solar energy investment decisions. Funds for research and development in renewable energies are significant in investment decisions (Wang et al., 2020; Zhao et al., 2021).

In renewable energies, after domestic production is encouraged, a suitable market environment for renewable energy production should be established. The government creates indirect support mechanisms for this. It is noticed in the literature that one of the indirect support mechanisms is the fixed price application. E.g., Rowlands (2005) investigated the effects of a fixed price tariff for solar photovoltaic systems. In the study, in which Canada was examined, it was stated that policies such as renewable portfolio standards were applied in the promotion of renewable energies. However, another way of promoting renewable energies is determined to be a fixed price tariff. By fixed-price application, a fixed price is paid for each unit of electricity produced. Costs are brought under control with the fixed-price paid application (Sijm, 2002). Similar to the result of this study, Lauber (2004) evaluated the guarantee tariffs for renewable energy in his study. In this regard, countries such as the USA, England, and Germany were examined. As a result, it was stated that guarantee tariffs and renewable portfolio standards should be created together in order to accelerate the transition to renewable energy. It has been found that both policies complement each other.

One of the indirect renewable energy policies is renewable portfolio standards. In the literature, this issue has been the focus of attention for researchers. E.g., Wiser et al. (2005) evaluated renewable portfolio standards. In this research, it has been determined that renewable portfolio standards should be applied to control and evaluate the renewable energy supply. Another study with similar results was conducted by Markard and Truffer (2006). In the study examining five European countries, it was emphasized that policymakers set some rules in order to increase electricity generation from renewable energies. One of these rules is that some part of electricity generation is carried out with renewable energy sources. For this, it was underlined that renewable portfolio standards are important. Government tenders ensure that renewable energy technologies are affordable. Highly energy-efficient and low-cost systems are used through tenders. This matter affects investments in renewable energies (Lew, 2000). In this context, one of the indirect support mechanisms is government tenders. Dobrotkova et al. (2018) explained the importance of tenders for solar energy in the study, in which developing countries were examined. As a result, it has been determined that governments are going through auctions in order to minimize the costs for solar panels. It was identified that renewable energy prices were also taken under control with the auction.

According to the conclusions obtained in the literature review, it has been determined that there are many support policies in the promotion of renewable energies which have direct and indirect effects. Further, it has been specified that direct support policies encourage domestic renewable energy production. It is noted that direct support policies are mandatory domestic production, financial incentives, appropriate customs and tax incentives, low-interest export credit, participation in international certification programs, and funding for research and development. However, it has been ascertained that indirect support policies are fixed-price applications, renewable portfolio standards, and government auctions. It is stated in the literature that an appropriate market environment for renewable energy generation is endeavored to be created with indirect support policies. In the studies, it is emphasized that direct and indirect support mechanisms reduce renewable energy costs and increase renewable energy preferences. There are only a few studies in the literature that identify which support policies are how important in the eyes of investors. Therefore, in this study, it is recommended to determine the degree of impact of the direct and indirect support policies decided by the government calling green energy on investors. Further, this study will be supported by the Entropy method which is thought to contribute to the literature.

5.3 Analysis Results

In this study, it is aimed to determine the optimal state support required to increase renewable energy investments. In this framework, 5 different criteria were determined as a result of detailed literature review. Details of these factors are given in Table 5.1.

Table 5.1 The details of criteria

Criteria	References
Financial incentives (C1)	Bergek and Jacobsson (2003); Byrnes et al. (2013); Connor (2003); Adenle (2020); Zander et al. (2019); Husein and Chung (2018)
Tax incentives (C2)	Rajsekhar et al. (1999); Liu et al. (2002); Cavaliero and Da Silva (2005); Lewis and Wiser (2007); Yang et al. (2019); Miller and Carriveau (2018)
Funding for Research and Development studies (C3)	Dinçer and Yüksel (2019a, b); Torani et al. (2016); Wang et al. (2020); Sawin (2001); Kamp (2002); Zafar et al. (2019); Carbajo and Cabeza (2018)
Fixed-Price application (C4)	Rowlands (2005); Sijm (2002); Lauber (2004); Wiser et al. (2005); Mitchell et al. (2006); Nojavan et al. (2017); McPherson and Tahseen (2018)
Location support (C5)	Lew (2000); Dobrotkova et al. (2018); Lewis and Wiser (2007); Markard and Truffer (2006); Dominguez et al. (2018); Furmankiewicz et al. (2020)

Table 5.2 The details of evaluation scales

Influence level	Influence number
No (N)	1
Low (L)	2
Medium (M)	3
High (H)	4
Very high (VH)	5

As can be seen from Table 5.1, 5 different support types that can be applied by the state have been determined in order to increase renewable energy investments. In this context, the state can attract investors' attention by offering low interest loans. On the other hand, the tax reduction will also contribute to the reduction of the costs of companies that will invest in renewable energy. Furthermore, government support for research and development activities for renewable energy will help the development of these energy types in the country. In addition, a fixed price advantage can be made with companies before investment. In this way, the risk of companies not generating income will be minimized. Finally, the ground support to be provided by the state will also contribute to reducing the costs of investors. These criteria are evaluated by 3 different experts. These people consist of academicians with at least 14 years of experience in the subject. Experts have taken into account the 5-point evaluation scale while evaluating the limbs, and the details are given in Table 5.2.

Moreover, the details of expert evaluations with respect to the criteria are demonstrated on Table 5.3.

In the analysis process, Entropy methodology is taken into consideration. For this purpose, firstly, pairwise comparison matrix is created by considering the average values of expert opinions. This matrix is stated on Table 5.4.

Just then, this matrix is normalized by dividing all values to the sum of the columns. Table 5.5 gives information about the normalized matrix.

Table 5.3 The evaluations of decision makers

Criteria	C1	C2	C3	C4	C5
Decision Maker 1					
C1	0	4	3	5	4
C2	2	0	2	3	3
C3	2	2	0	3	3
C4	1	1	2	0	2
C5	2	2	2	3	0
Decision Maker 2					
C1	0	2	3	3	2
C2	2	0	3	2	2
C3	2	2	0	2	2
C4	2	2	2	0	2
C5	2	2	3	2	0
Decision Maker 3					
C1	0	3	2	3	3
C2	2	0	3	3	3
C3	3	3	0	4	4
C4	2	1	2	0	2
C5	2	2	2	3	0

Table 5.4 Pairwise comparison matrix

Criteria	C1	C2	C3	C4	C5
C1	0.00	3.00	2.67	3.67	3.00
C2	2.00	0.00	2.67	2.67	2.67
C3	2.33	2.33	0.00	3.00	3.00
C4	1.67	1.33	2.00	0.00	2.00
C5	2.00	2.00	2.33	2.67	0.00

Table 5.5 Normalized matrix

Criteria	C1	C2	C3	C4	C5
C1	0.00	0.35	0.28	0.31	0.28
C2	0.25	0.00	0.28	0.22	0.25
C3	0.29	0.27	0.00	0.25	0.28
C4	0.21	0.15	0.21	0.00	0.19
C5	0.25	0.23	0.24	0.22	0.00

After that, the entropy value is calculated. By using this value, the weights of the criteria are identified as in Table 5.6.

Table 5.6 indicates that tax incentive is the most important state support to improve renewable energy investments. In addition to this issue, it is also concluded that location support and fixed-price application are other significant factors to reach this objective. Nonetheless, financial incentives and funding for research and development studies are on the last ranks.

Table 5.6 Weights of the criteria

Criteria	Weights
Financial incentives (C1)	0.1933
Tax incentives (C2)	0.2206
Funding for Research and Development studies (C3)	0.1931
Fixed-Price application (C4)	0.1950
Location support (C5)	0.1978

5.4 Discussion and Conclusion

Renewable energies provide social and economic benefits to countries. Thanks to the use of renewable energy, the carbon emission problem is reduced. This situation also helps to prevent the air pollution problem. In this way, people will get sick less and this will increase their quality of life. On the other hand, this mentioned issue also contributes to the economic development of the country. In a country where the number of sick people is decreasing, there will be an increase in the workforce. This will help increase industrial production in the country. In addition, with the reduction of diseases, there will be a serious decrease in health expenditures in the country. This situation will positively affect the country's budget balance.

However, there are some disadvantages in the use of renewable energy. The most important problem in this process is that the initial cost of investments is very high. This situation negatively affects the profitability of renewable energy investments. Therefore, many investors in the country cannot focus on this area. It is clear that the state has a very serious role in this process. Thanks to some incentives to be provided by the state to renewable energy investors, it will be possible to minimize the problems in these investments. For example, the low interest loan facility to be given to investors by the state will reduce the cost of investments. In addition, tax cuts will also provide investors with similar advantages.

In this study, it is aimed to determine the most optimal state support to be provided in order to increase renewable energy investments. In this context, a comprehensive literature review has been carried out. As a result of this analysis, 5 different types of support that can be provided by the state to renewable energy investors were determined. After that, it was determined which types of support should be given more priority by using the Entropy method. It is identified that tax incentive is the most important state support to improve renewable energy investments. In addition to this issue, it is also concluded that location support and fixed-price application are other significant factors to reach this objective. Nonetheless, financial incentives and funding for research and development studies are on the last ranks.

Considering these results, the government should focus primarily on tax incentives in order to increase renewable energy investments. In this context, a serious reduction can be made in the income tax of investors. This will significantly contribute to increasing profitability as it will reduce the cost of investments. In addition to the mentioned issue, it would be appropriate to reduce the taxes collected from workers' income. This will help reduce labor costs. On the other hand, research and development activities on renewable energy may also be exempt from tax. This situation will play an important role in the development of the sector.

References

Adenle, A. A. (2020). Assessment of solar energy technologies in Africa-opportunities and challenges in meeting the 2030 agenda and sustainable development goals. *Energy Policy, 137*, 111180.

An, J., Mikhaylov, A., & Richter, U. H. (2020). Trade war effects: Evidence from sectors of energy and resources in Africa. *Heliyon, 6*(12), e05693.

Bergek, A., & Jacobsson, S. (2003). The emergence of a growth industry: A comparative analysis of the German, Dutch and Swedish wind turbine industries. In *Change, transformation and development* (pp. 197–227). Physica.

Byrnes, L., Brown, C., Foster, J., & Wagner, L. D. (2013). Australian renewable energy policy: Barriers and challenges. *Renewable Energy, 60*, 711–721.

Carbajo, R., & Cabeza, L. F. (2018). Renewable energy research and technologies through responsible research and innovation looking glass: Reflexions, theoretical approaches and contemporary discourses. *Applied Energy, 211*, 792–808.

Cavaliero, C. K. N., & Da Silva, E. P. (2005). Electricity generation: Regulatory mechanisms to incentive renewable alternative energy sources in Brazil. *Energy Policy, 33*(13), 1745–1752.

Cheng, F., Lin, M., Yüksel, S., Dincer, H., & Kalkavan, H. (2020). A hybrid hesitant 2-tuple IVSF decision making approach to analyze PERT-based critical paths of new service development process for renewable energy investment projects. *IEEE Access*.

Clausen, L. T., & Rudolph, D. (2020). Renewable energy for sustainable rural development: Synergies and mismatches. *Energy Policy, 138*, 111289.

Connor, P. M. (2003). *National innovation, industrial policy and renewable energy technology*.

Dinçer, H., & Karakuş, H. (2020a). Yenilenebilir Enerjinin Sürdürülebilir Ekonomik Kalkınma Üzerindeki Etkisi: BRICS ve MINT Ülkeleri Üzerine Karşılaştırmalı Bir Analiz. *Ekonomik ve Sosyal Araştırmalar Dergisi, 1*(1), 100–123.

Dinçer, H., & Karakuş, H. (2020b). Yenilenebilir Enerji Yatırımları ile İstihdam Arasındaki İlişkinin Belirlenmesi: G7 Ülkeleri Üzerine Ekonometrik Bir Analiz. *İstatistik ve Uygulamalı Bilimler Dergisi, 1*(1), 40–49.

Dinçer, H., & Yüksel, S. (2019a). Balanced scorecard-based analysis of investment decisions for the renewable energy alternatives: A comparative analysis based on the hybrid fuzzy decision-making approach. *Energy, 175*, 1259–1270.

Dinçer, H., & Yüksel, S. (2019b). Multidimensional evaluation of global investments on the renewable energy with the integrated fuzzy decision-making model under the hesitancy. *International Journal of Energy Research, 43*(5), 1775–1784.

Dobrotkova, Z., Surana, K., & Audinet, P. (2018). The price of solar energy: Comparing competitive auctions for utility-scale solar PV in developing countries. *Energy Policy, 118*, 133–148.

Dominguez, O. D. M., Kasmaei, M. P., Lavorato, M., & Mantovani, J. R. S. (2018). Optimal siting and sizing of renewable energy sources, storage devices, and reactive support devices to obtain a sustainable electrical distribution systems. *Energy Systems, 9*(3), 529–550.

Dooyum, U. D., Mikhaylov, A., & Varyash, I. (2020). Energy security concept in Russia and South Korea. *International Journal of Energy Economics and Policy, 10*(4), 102.

Du, L., Dinçer, H., Ersin, İ., & Yüksel, S. (2020). IT2 fuzzy-based multidimensional evaluation of coal energy for sustainable economic development. *Energies, 13*(10), 2453.

Furmankiewicz, M., Hewitt, R. J., Janc, K., & Kazak, J. K. (2020, December). Europeanisation of energy policy and area-based partnerships: Regional diversity of interest in renewable energy sources in local development strategies in Poland. In *IOP Conference Series: Earth and Environmental Science* (Vol. 609, No. 1, p. 012033). IOP Publishing.

Haar, L. N., & Haar, L. (2017). An option analysis of the European Union renewable energy support mechanisms. *Economics of Energy & Environmental Policy, 6*(1).

Husein, M., & Chung, I. Y. (2018). Optimal design and financial feasibility of a university campus microgrid considering renewable energy incentives. *Applied Energy, 225*, 273–289.

Kamp, L. M. (2002). *Learning in wind turbine development: A comparison between the Netherlands and Denmark.* Universiteit Utrecht, Faculteit Ruimtelijke Wetenschappen.

Kobtseva, O. N., Novoselova, N. N., Novoselov, S. N., Morkovkin, D. E., & Sidorchukova, E. V. (2017). Organizational and economic features of import substitution formation and realization in the conditions of spatial restrictions. *International Journal of Applied Business and Economic Research, 15*(23), 25–35.

Lauber, V. (2004). REFIT and RPS: Options for a harmonised community framework. *Energy Policy, 32*(12), 1405–1414.

Lew, D. J. (2000). Alternatives to coal and candles: Wind power in China. *Energy Policy, 28*(4), 271–286.

Lewis, J. I., & Wiser, R. H. (2007). Fostering a renewable energy technology industry: An international comparison of wind industry policy support mechanisms. *Energy Policy, 35*(3), 1844–1857.

Li, Y. X., Wu, Z. X., Dinçer, H., Kalkavan, H., & Yüksel, S. (2021). Analyzing TRIZ-based strategic priorities of customer expectations for renewable energy investments with interval type-2 fuzzy modeling. *Energy Reports, 7*, 95–108.

Liu, W. Q., Gan, L., & Zhang, X. L. (2002). Cost-competitive incentives for wind energy development in China: Institutional dynamics and policy changes. *Energy Policy, 30*(9), 753–765.

Liu, J., Lv, J., Dinçer, H., Yüksel, S., & Karakuş, H. (2021a). Selection of renewable energy alternatives for green blockchain investments: A hybrid IT2-based fuzzy modelling. *Archives of Computational Methods in Engineering*, 1–15.

Liu, Y., Gong, X., Yüksel, S., Dinçer, H., & Aydın, R. (2021b). A multidimensional outlook to energy investments for the countries with continental shelf in East Mediterranean region with hybrid decision making model based on IVIF logic. *Energy Reports, 7*, 158–173.

Markard, J., & Truffer, B. (2006). The promotional impacts of green power products on renewable energy sources: Direct and indirect eco-effects. *Energy Policy, 34*(3), 306–321.

McPherson, M., & Tahseen, S. (2018). Deploying storage assets to facilitate variable renewable energy integration: The impacts of grid flexibility, renewable penetration, and market structure. *Energy, 145*, 856–870.

Mikhaylov, A. (2021). Development of Friedrich von Hayek's theory of private money and economic implications for digital currencies. *Terra Economicus, 19*(1).

Miller, L., & Carriveau, R. (2018). A review of energy storage financing—Learning from and partnering with the renewable energy industry. *Journal of Energy storage, 19*, 311–319.

Mitchell, C., Bauknecht, D., & Connor, P. M. (2006). Effectiveness through risk reduction: A comparison of the renewable obligation in England and Wales and the feed-in system in Germany. *Energy Policy, 34*(3), 297–305.

Mohammadi, A., & Mehrpooya, M. (2018). A comprehensive review on coupling different types of electrolyzer to renewable energy sources. *Energy, 158*, 632–655.

Morkovkin, D. E., Mamychev, A. Y., Yakovenko, N. V., Derevyagina, M. V., & Didenko, O. V. (2016). Factors and material conditions for space-intensive economic development of region. *International Review of Management and Marketing, 6*(1), 67–72.

Nojavan, S., Zare, K., & Mohammadi-Ivatloo, B. (2017). Application of fuel cell and electrolyzer as hydrogen energy storage system in energy management of electricity energy retailer in the presence of the renewable energy sources and plug-in electric vehicles. *Energy Conversion and Management, 136*, 404–417.

Olabi, A. G. (2017). *Renewable energy and energy storage systems.*

Özdemir, Ö., Hobbs, B. F., van Hout, M., & Koutstaal, P. (2019). *Capacity vs energy subsidies for renewables: Benefits and costs for the 2030 EU power market (No. EPRG1911).*

Rajsekhar, B., Van Hulle, F., & Jansen, J. C. (1999). Indian wind energy programme: Performance and future directions. *Energy Policy, 27*(11), 669–678.

Rowlands, I. H. (2005). Envisaging feed-in tariffs for solar photovoltaic electricity: European lessons for Canada. *Renewable and Sustainable Energy Reviews, 9*(1), 51–68.

Saçık, S. Y., Yokuş, N., Alagöz, M., & Yokuş, T. (2020). Optimum renewable energy investment planning in terms of current deficit: Turkey model. *Energies, 13*(6), 1509.

Sawin, J. L. (2001). *The role of government in the development and diffusion of renewable energy technologies: Wind power in the United States, California, Denmark and Germany, 1970-2000.*

Sijm, J. P. M. (2002). *The performance of feed-in tariffs to promote renewable electricity in European countries* (Vol. 7). Energy Research Centre of the Netherlands ECN.

Torani, K., Rausser, G., & Zilberman, D. (2016). Innovation subsidies versus consumer subsidies: A real options analysis of solar energy. *Energy Policy, 92*, 255–269.

Wang, Q., Li, S., & Pisarenko, Z. (2020). Heterogeneous effects of energy efficiency, oil price, environmental pressure, R&D investment, and policy on renewable energy-evidence from the G20 countries. *Energy, 118322.*

Winkler, H., Hughes, A., & Haw, M. (2009). Technology learning for renewable energy: Implications for South Africa's long-term mitigation scenarios. *Energy Policy, 37*(11), 4987–4996.

Wiser, R., Porter, K., & Grace, R. (2005). Evaluating experience with renewables portfolio standards in the United States. *Mitigation and Adaptation Strategies for Global Change, 10* (2), 237–263.

Yang, X., He, L., Xia, Y., & Chen, Y. (2019). Effect of government subsidies on renewable energy investments: The threshold effect. *Energy Policy, 132*, 156–166.

Yuan, J., Zhang, Z. M., Yüksel, S., & Dinçer, H. (2020). Evaluating recognitive balanced scorecard-based quality improvement strategies of energy investments with the integrated hesitant 2-tuple interval-valued Pythagorean fuzzy decision-making approach to QFD. *IEEE Access, 8*, 171112–171128.

Yüksel, S., Dinçer, H., & Uluer, G. S. (2020). The role of technological development on renewable energy usage: An econometric analysis for G7 countries. In *Handbook of research on sustainable supply chain management for the global economy* (pp. 136–153). IGI Global.

Yumashev, A., Ślusarczyk, B., Kondrashev, S., & Mikhaylov, A. (2020). Global indicators of sustainable development: Evaluation of the influence of the human development index on consumption and quality of energy. *Energies, 13*(11), 2768.

Zafar, M. W., Shahbaz, M., Hou, F., & Sinha, A. (2019). From nonrenewable to renewable energy and its impact on economic growth: The role of research & development expenditures in Asia-Pacific Economic Cooperation countries. *Journal of Cleaner Production, 212*, 1166–1178.

Zander, K. K., Simpson, G., Mathew, S., Nepal, R., & Garnett, S. T. (2019). Preferences for and potential impacts of financial incentives to install residential rooftop solar photovoltaic systems in Australia. *Journal of Cleaner Production, 230*, 328–338.

Zhao, Y., Xu, Y., Yüksel, S., Dinçer, H., & Ubay, G. G. (2021). Hybrid IT2 fuzzy modelling with alpha cuts for hydrogen energy investments. *International Journal of Hydrogen Energy.*

Chapter 6
Energy Flow Analysis from Russia to South Korea

Mafura Uandykova and Tomonobu Sengyu

Abstract The objective of this study is devoted to energy flow analysis on the back of cooperation of South Korea and Russian Federation in the fuel and energy sector. An analysis of this topic and related prospects was done through the study of documents and projects that have direct relation to such cooperation as well as data regarding export and import of energy resources, concepts and theories of development in this field. The cooperation in the energy sector between South Korea and the Russian Federation is one of the most promising areas of development between two countries and today has a serious potential to expand and deepen, due to both economic and political factors. The document defines the main problems of the fuel and energy complex of the Russian Federation, goals and objectives of energy sector development, priorities of state energy policy, global trends in energy consumption and changes in energy markets, and international relations regarding energy sector.

6.1 Introduction

The geographical proximity coupled with favorable political climate, which, as we see, is forming today on Korean Peninsula, allow us to seriously talk about the possibilities of trilateral format of cooperation and promotion of joint projects in the oil and gas sector between Russia and Korea such as the Trans-Korean Railway and the Trans-Korean gas pipeline, as well as projects for joint exploration of energy resources in the Far East and East Siberia.

M. Uandykova (✉)
Narxoz University, Almaty, Republic of Kazakhstan

T. Sengyu
Department of Electrical and Electronics Engineering, University of the Ryukyus, Okinawa, Japan
e-mail: b985542@tec.u-ryukyu.ac.jp

S. Yüksel, H. Dinçer (eds.), *Strategic Approaches to Energy Management*, Contributions to Management Science,
https://doi.org/10.1007/978-3-030-76783-9_6

It is with the advent of the current President of South Korea, Moon Jae In under the auspices of the "new northern policy" and the "new energy policy" and thanks to his efforts to establish an inter-Korean dialogue today, more than ever before, the Russian Federation and South Korea are opening up real prospects for increasing the share of energy exports from Russia into South Korea and investments from South Korea in projects in the Far East regarding the oil and gas infrastructure.

Foreign policy of the Russian Federation on reorientation to the East also favors the strengthening of cooperation with South Korea (Odhiambo, 2009; Ozturk et al., 2010; Phelps, 1994; Ranjbar et al., 2017; Rathnayaka et al., 2018; Shahbaz et al., 2018).

According to the British oil and gas company BP, in 2016, the Republic Korea ranks 9th in the world in energy consumption, accounting for 2.2% of global consumption. Moreover, in the Republic of South Korea there are only insignificant coal reserves, which, according to BP estimates in 2016, amounted to 326 million tons. For comparison: the reserves of fossil coal of the Russian Federation in the same year amounted to 162 billion tons.

Obviously, the volume of mining does not allow to satisfy the needs of South Korean industry as well as the housing sector in energy resources. Thus, South Korea is forced to rely almost entirely on imports. It is the main energy resources such as crude oil, liquefied natural gas (LNG), coal, anthracites and others. Moreover, the Republic of South Korea is on the list of five leading countries by imports of LNG, crude oil and refined products. Actually in 2016, the dependence of South Korea on energy imports reached 94.7%.

According to the same report, of all energy resources, most of which, as demonstrated above, is imported from abroad, 61.4% used in the industrial sector, 18.9% accounted for transport, 17%—in housing and commercial sector, which means consumption of energy resources in private households and other premises, and the remaining 2.8% were applied in the public sector (Moiseev, 2017c; Moiseev & Akhmadeev, 2017).

Crude oil accounts for up to 40.1% of imports, up to 99% of which used in the refining industry. In this case, the main Middle East countries are exporters—they account for up to 85.9% of the total import of crude oil. Of these, Saudi Arabia exports 30.1%, Kuwait—14.8%, Iraq—12.8%. Then, Asian states follow the export of oil to the Republic of South Korea (6.7%) and Africa (2.8%).

Despite the rather scarce oil reserves in South Korea, thanks to imports, the state manages to remain one of the world leaders in oil refining: there are ten largest refineries on its territory—SK Energy Co., Ltd. Ulsan Refinery, GS Caltex Yeosu Refinery and S-Oil Ulsan Refinery, occupying 2nd, 4th and 10th places in the world rank respectively. Moreover, these refineries not only differ in large volumes of oil refining, but also high quality of the product due to the use of advanced technologies of raw materials purification.

Refined products and imported petroleum products mainly go to meet the domestic needs of the country, which in percentage terms is 61.6% for 2015, and 4% goes for international bunkering, that is, refueling vessels to provide movement and other needs of the ship transportation. At this, making the necessary calculations according to the available data on the statistical website KOSIS agencies, of the total number of petroleum products imported and directly processed in Korea, 34.3% is exported.

Moreover, the Republic of Korea exports 1343 million barrels per day, being the 8th largest exporter of petroleum products in the world for 2016. From the resulting petroleum products to mainly gasoline and jet fuel are exported, while imported petroleum products are naphtha, or so-called naphtha, used as a raw material for the petrochemical industry in the production of olefins, as well as fuel (Moiseev, 2017a, b; Moiseev & Sorokin, 2018).

Bitumen is second in terms of import volume (25.7% of total energy imports)—solid or tar-like products, representing a mixture of hydrocarbons and their derivatives. Bitumen exporters are countries such as Australia, Russia and Indonesia.

Accordingly, 38.2% of bitumen goes to final consumption and is used mainly in the industrial sector (98.3%)—metallurgical industry—and in housing and commercial sector (1.7%)—for residential heating.

The third place in the import structure is occupied by LNG (15.4%), the main exporters which are Qatar, Australia and Indonesia. The vast majority of LNG goes to electricity generation and its actual regasification—the process converting LNG from liquid back to gaseous state. In the Republic of South Korea is located four regasification plants. Regarding the remaining energy resources, figures the following: nuclear energy imports account for 11.6% (uranium imports from Russia, Canada and Australia) and anthracite (highest grade fossil humus coal metamorphism)—2.1% (China, Australia and Russia).

6.2 Literature Review

The complementary nature of the economies of two countries, due to the dependence of one on the export of energy resources, and the other on the imports, allows us to say that the development of energy cooperation is today a priority for these states under consideration and serves the purpose of providing both energy in particular and national security in general (Nyangarika et al., 2019a, b; Yuan et al., 2020; Zhao et al., 2021a, b).

The object of this study is energy cooperation of Russian Federation (RF) and the Republic of Korea (RK).

The authors of this work achieved their goals through consistent solving of the following tasks:

1. to identify and analyze the methodological basis of energy cooperation of the Republic of South Korea and the Russian Federation, including the concepts, theories and strategies of the parties involved (Denisova et al., 2019);
2. to study the factors contributing to the development of energy cooperation of the Republic of South Korea and the Russian Federation (Mikhaylov, 2018a, b);
3. to find out what is the institutional base of energy cooperation of the Republic of South Korea and the Russian Federation and give it an analysis;

4. to consider the main joint projects of the Republic of South Korea and the Russian Federation in the field of energy cooperation, including active, suspended and those in development (Meynkhard, 2019a, 2020);
5. to identify and analyze conditions and problems that impede strengthening and expanding energy cooperation between the Republic of South Korea and the Russian Federation (Mikhaylov, 2019; Mikhaylov et al., 2020);
6. to analyze the current political situation in the Republic of South Korea and the Russian Federation and identify

how does it contribute or hinder the deepening of energy cooperation (Meynkhard, 2019b).

This final qualification work covers the time period from 1991 to 2018, due primarily to the fact that officially diplomatic relations between the Russian Federation and the Republic of South Korea were established in 1991, and throughout over the next few years, major contracts were signed, following by signing the relevant memoranda that served as the institutional core for development of energy cooperation between two states. It was the 90s to be the beginning of fuel and energy cooperation by increasing energy exports and the development of some joint projects in this area (Dayong et al., 2020).

6.3 Methods

The document defines the main problems of the fuel and energy complex of the Russian Federation, goals and objectives of energy sector development, priorities of state energy policy, global trends in energy consumption and changes in energy markets, and also international relations regarding energy sector (Mikhaylov et al., 2018; Nyangarika et al., 2018).

According to this strategy, a significant increase in export volumes is expected of oil and gas in the Asia-Pacific region (APR) by the end of this period: more than a third of all gas exports will be accounted as exports into this region by the year 2035. Also, despite environmental restrictions, coal supplies from Russia due to its high quality will expand into APR. Nevertheless, this is unlikely to affect the Republic of South Korea according to the "new energy policy" (Uandykova et al., 2020; Danish et al., 2020, 2021; Lisin, 2020a, b, c; An et al., 2021).

President Moon Jae In, as presented in the next section, has planned to reduce the use of coal and the transition to less environmentally harmful sources of energy (An et al., 2020a, b).

One of the main tasks of Russian Federation is to improve the production structure of fuel and energy complex, which includes the creation of new energy industrial complexes, which would stimulate the development of regions such as

Siberia, the Far East and the Arctic along with diversification of the geographical structure of exports with a bias in the Asia-Pacific region (An et al., 2019a, b).

The development of energy infrastructure in these regions is also one of priorities of state energy policy. Thus, the main directions of development of the energy sector will be to export diversification, increasing the share of Asia-Pacific importing countries, improving the creation of new energy infrastructure projects in Siberia, at the Far East and the Arctic, as well as attracting investment in the energy sector. Considering emphasis on Asia-Pacific; favorable conditions are created in the near future for the development of energy cooperation with the Republic of South Korea, since both sides, as it will be demonstrated below, have complementary goals and interests in energy sphere (Mikhaylov, 2020; Mikhaylov & Tarakanov, 2020).

6.4 Results

The new president of South Korea put forward his policy towards settlement of the cooperation with the Russian Federation and the energy sector as a whole. This course presented in a number of documents, dubbed “Korean policy Peninsula”, “New Northern Policy” and “New Energy Policy”.

The first document is of interest to our work in connection with the fact that, the main objective in the foreign policy of the administration is to establish peace and prosperity on the Korean peninsula, of which the task of creating the so-called “new economic community” provided close economic cooperation between the Republic of South Korea and the North Korea and ensuring communication with the nearest states in NEA—the creation of “three economic zones”: it is assumed that one of which will pass through the east coast of the Korean Peninsula and will link Peninsula states with the Russian Federation.

Within the framework of such an economic zone, it is planned implement triangular cooperation projects as the most promising in regarding the withdrawal of the North Korea from economic isolation. However, even in the case of suspension of the rapprochement of the two Korean states, which we observe during 2017–2018, Moon Jae In expressed his intentions to develop bilateral cooperation with the Russian Federation (Branch, 1993).

“New Northern Policy” is a speech by Moon Jae In in the Third East Economic Forum (WEF), held on September 6–7, 2017, which contains main objectives and activities of the South Korean administration regarding development of relations with the “northern countries”, and, first of all, with Russia. The president emphasized that the Far Eastern region for centuries has been an extremely a promising region, in cooperation with which, East Asian states will be able to achieve global prosperity and that such cooperation seems possible under the current Russian government (Balestra & Nerlove, 1966; Davis, 2008, 2011).

He also emphasized on the common historical ancestry of coexistence and experience of Russian and Korean peoples and that they should aim towards the basis for further strengthen of collaboration (Cameron, 1985). Moon Jae In made an important statement that during his presidential term, he would like to draw closer to Russia and strengthen existing communication.

It is in the field of energy cooperation that new opportunities for the Russian Federation to expand the supply of liquefied natural gas (LNG) in South Korea, that is connected with the "new energy policy" of Moon Jae In presented in May 2017, which implies a transition from nuclear energy and energy received in coal burning process, to increase the share of LNG and renewable energy sources.

The Korean Gas Corporation (KOGAS) is considering the possibility of increasing LNG imports from Russia and building Trans-Korean gas pipeline, as well as increased investment in exploration of coastal gas deposits in Sakhalin and the Arctic (Brown, 2001; Gura et al., 2020; Dooyum et al., 2020).

The goals and objectives of the Russian and South Korean political courses can be called complementary, since, for example, for the Russian Federation one of the priority tasks described in "Energy Strategy-2035" (ES-2035), is the development of the Far East and Siberia as in energy as well as infrastructure plans, while Moon Jae In expresses serious intentions to actively participate in the development of the above regions.

Also, the Russian Federation seeks to diversify energy importing countries and increase the share of imports to the Asia-Pacific countries, and South Korea. Thus, the foreign policy of the current South Korean administrations and policies of the Russian Federation to "reorient Russia to the East" create favorable conditions for deepening cooperation (Moiseev et al., 2020).

It is based on goals and objectives, represented in the "new northern politics", "new politics in Korean peninsula", " the new energy policy" from the South Korean side and on "Energy Strategy-2035" from the Russian side. The Russian Federation, having significant energy potential, and South Korea, seeking to diversify imports of energy resources, increasing the share of LNG imports and strengthening ties with the "northern states", have enormous potential at this stage to expand the cooperation primarily in the energy sector.

This is precisely the essence of the "new northern policy". President believes Putin's Far Eastern policy and "the new northern politics" have much in common and that for the development of the Far East, the Republic of South Korea is an ideal partner who is deeply interested in the development of the region and, accordingly, will take an active part in this. Moreover, it is joint development of the Far Eastern region by the efforts of the Russian Federation and South Korea, according to Moon Jae In, will become the main lever in solving the North Korean nuclear issue by North Korea involvement in tripartite cooperation projects (Morkovkin et al., 2019, 2020a, b, c, d).

Moon Jae In paid special attention to cooperation primarily in implementation of joint projects, which for one reason or another were suspended or all remained on paper. Here, most likely, we are talking about Hassan Rajin projects, the construction of the Trans-Korean gas pipeline and participation of South Korean oil and gas companies in the exploration of energy sources in the Far East and Siberia. The President expressed confidence that the more there will be successful projects, the stronger the trust between the two states, which in turn will lead to the creation of even more joint long-term projects.

In order to coordinate the achievement of the goals and objectives of the "new northern policy" was created a "Committee on Northern Economic Cooperation", personally controlled by the President of the Republic of South Korea (Mikhaylov et al., 2021a, b).

The first meeting of the Committee was held on December 7, 2017 with the aim to develop a detailed program of cooperation between South Korea and the Russian Federation. It is assumed that this cooperation program should be based on "9 bridges"—a concept, proposed by Moon Jae In as part of his "new northern policy". These "bridges" represent such areas of cooperation as shipbuilding, marine development ports and harbors, development of the Northern Sea Route, railway transport, gas, electricity, labor market, agriculture and fisheries.

It is worth noting that theoretical developments and concepts on the topic of energy cooperation of countries today is not well described and the authors have given the most relevant of the classical theories and approaches. One of the later is the concept of energy security and it actively developed by domestic and foreign researchers (Mikhaylov et al., 2021a; Mikhaylov, 2021).

In the course of our work, we will operate with a neoliberal approach, proclaiming the interdependence of countries as a determining factor for their cooperation, as well as "ES-2035"—from the Russian Federation and the "new northern policy" and "New energy policy"—by the current administration of the Republic of South Korea – in as the methodological basis on which it is built and will develop energy cooperation in the short and medium term.

6.5 Conclusion

As we can see, almost the entire economy of the country and the welfare of the Korean people depend on well-established and stable energy supplies (Varyash et al., 2020; Zhao et al. 2021a, b; An & Mikhaylov, 2020).

Thus, from the above data, we can conclude that imported crude oil is more suited to the needs of the refining industry, from which it follows that the demand for this type of energy in areas not concerning industry, is insignificant and its increase is unlikely to be observed in the future, given the demographic situation in the region—a reduction in the number of population, a large proportion of the elderly, as well as government measures, aimed at lowering the energy intensity of the economy as a whole. Besides, natural gas, coal and nuclear make up the most demand for energy. The share of oil and other liquid energy resources will continue to decrease and according to estimates for 2014 to 2018, their share in the total energy consumption will decrease from 40 to 34% (Alwaelya et al., 2020; Yumashev & Mikhaylov, 2020).

Coal as a primary source of energy is used to generate electricity (61.8% of all imported and mined coal in the Republic of South Korea).

Consequently, energy cooperation, its development and diversification of imports are a key area of foreign economic and foreign policy activities of the government, relevant ministries and private companies. In this regard, the Russian side should look for ways to strengthen such cooperation, since; obviously, it will be mutually beneficial and long-term cooperation, taking into account geographic and geological factors (Mikhaylov et al., 2021b; Liu et al., 2021a, b; Cheng et al., 2020).

References

Alwaelya, S. A., Yousif, N. B. A., & Mikhaylov, A. (2020). Emotional development in preschoolers and socialization. *Early Child Development and Care, 190*(1717480).

An, J., & Mikhaylov, A. (2020). Russian energy projects in South Africa. *Journal of Energy in Southern Africa, 31*(3).

An, J., Mikhaylov, A., & Moiseev, N. (2019a). Oil price predictors: Machine learning approach. *International Journal of Energy Economics and Policy, 9*(5), 1–6.

An, J., Mikhaylov, A., & Sokolinskaya, N. (2019b). Machine learning in economic planning: Ensembles of algorithms. *Journal of Physics: Conference Series, 1353*, 012126.

An, J., Mikhaylov, A., & Jung, S.-U. (2020a). The strategy of South Korea in the global oil market. *Energies, 13*(10), 2491.

An, J., Mikhaylov, A., & Kim, K. (2020b). Machine learning approach in heterogeneous group of algorithms for transport safety-critical system. *Applied Sciences, 10*(8), 2670.

An, J., Mikhaylov, A., & Jung, S.-U. (2021). A linear programming approach for robust network revenue management in the airline industry. *Journal of Air Transport Management, 91*, 101979.

Balestra, P., & Nerlove, M. (1966). Pooling gross section and time series data in the estimation of a dynamic model: The demand for natural gas. *Econometrica, 34*(3), 585–612.

Branch, E. (1993). Short run income elasticity of demand for residential electricity using consumer expenditure. *Energy Journal, 14*(4), 111–121.

Brown, M. (2001). Market failures and barriers as a basis for clean energy policies. *Energy Policy, 29*(14), 1197–1207.

Cameron, T. A. (1985). A nested logit model of energy conservation activity by owners of existing single family. *Review of Economics and Statistics, 67*(2), 205–211.

Cheng, F., Lin, M., Yüksel, S., Dİncer, H., & Kalkavan, H. (2020). A hybrid hesitant 2-tuple IVSF decision making approach to analyze PERT-based critical paths of new service development process for renewable energy investment projects. *IEEE Access*.

Danish, M. S. S., Bhattacharya, A., Stepanova, D., Mikhaylov, A., Grilli, M. L., Khosravy, M., & Senjyu, T. (2020). A systematic review of metal oxide applications for energy and environmental sustainability. *Metals, 10*(12), 1604.

Danish, M. S. S., Estrella, L. L., Alemaida, I. M. A., Lisin, A., Moiseev, N., Ahmadi, M., Nazari, M., Wali, M., Zaheb, H., & Senjyu, T. (2021). Photocatalytic applications of metal oxides for sustainable environmental remediation. *Metals, 11*, 80.

Davis, L. (2008). Durable goods and residential demand for energy and water: Evidence from a field trial. *RAND Journal of Economics, 39*(2), 530–546.

Davis, L. (2011), Evaluating the slow adoption of energy efficient investments: Are renters less likely to have energy efficient appliances? In *The design and implementation of US climate policy* (pp. 301–316). University of Chicago Press.

Dayong, N, Mikhaylov, A, Bratanovsky, S, Shaikh, Z. A., & Stepanova, D. (2020). Mathematical modeling of the technological processes of catering products production. Journal of Food Process Engineering, 43(2), 1–9.

Denisova, V., Mikhaylov, A., & Lopatin, E. (2019). Blockchain infrastructure and growth of global power consumption. *International Journal of Energy Economics and Policy, 9*(4), 22–29.

Dooyum, U. D., Mikhaylov, A., & Varyash, I. (2020). Energy security concept in Russia and South Korea. *International Journal of Energy Economics and Policy, 10*(4), 102–107.

Gura, D., Mikhaylov, A., Glushkov, S., Zaikov, M., & Shaikh, Z. A. (2020). Model for estimating power dissipation along the interconnect length in single on-chip topology. *Evolutionary Intelligence*, s12065.

Lisin, A. (2020a). Biofuel energy in the post-oil era. *International Journal of Energy Economics and Policy, 10*(2), 194–199. https://doi.org/10.32479/ijeep.8769

Lisin, A. (2020b). Prospects and challenges of energy cooperation between Russia and South Korea. *International Journal of Energy Economics and Policy, 10*(3). https://doi.org/10.32479/ijeep.9070

Lisin, A. (2020c). Valuation of the activities of foreign energy in the Russian emerging sector. *Orbis, 15*(45), 53–63. Retrieved from http://www.revistaorbis.org.ve/pdf/45/art5.pdf

Liu, J., Lv, J., Dinçer, H., Yüksel, S., & Karakuş, H. (2021a). Selection of renewable energy alternatives for green blockchain investments: A hybrid IT2-based fuzzy modelling. *Archives of Computational Methods in Engineering*, 1–15.

Liu, Y., Gong, X., Yüksel, S., Dinçer, H., & Aydın, R. (2021b). A multidimensional outlook to energy investments for the countries with continental shelf in East Mediterranean region with hybrid decision making model based on IVIF logic. *Energy Reports, 7*, 158–173.

Meynkhard, A. (2019a). Energy efficient development model for regions of the Russian federation: Evidence of crypto mining. *International Journal of Energy Economics and Policy, 9*(4), 16–21.

Meynkhard, A. (2019b). Fair market value of bitcoin: Halving effect. Investment management and financial innovations. *Management and Financial Innovations, 16*(4), 72–85. https://doi.org/10.21511/imfi.16(4).2019.07

Meynkhard, A. (2020). Priorities of Russian energy policy in Russian-Chinese relations. *International Journal of Energy Economics and Policy, 10*(1), 65–71.

Mikhaylov, A. (2018a). Volatility spillover effect between stock and exchange rate in oil exporting countries. *International Journal of Energy Economics and Policy, 8*(3), 321–326.

Mikhaylov, A. (2018b). Pricing in oil market and using probit model for analysis of stock market effects. *International Journal of Energy Economics and Policy, 8*(2), 69–73.

Mikhaylov, A. (2019). Oil and gas budget revenues in Russia after crisis in 2015. *International Journal of Energy Economics and Policy, 9*(2), 375–380.

Mikhaylov, A. (2020). Cryptocurrency market analysis from the open innovation perspective. *Journal of Open Innovation: Technology, Market, and Complexity, 6*(4), 197. https://doi.org/10.3390/joitmc6040197

Mikhaylov, A. (2021). Lichens as indicators of atmospheric pollution in urban ecosystems. *Israel Journal of Ecology & Evolution, 10016*, 1–9.

Mikhaylov, A., & Tarakanov, S. (2020). Development of Levenberg-Marquardt theoretical approach for electric network. *Journal of Physics: Conference Series, 1515*, 052006.

Mikhaylov, A., Sokolinskaya, N., & Nyangarika, A. (2018). Optimal carry trade strategy based on currencies of energy and developed economies. *Journal of Reviews on Global Economics, 7*, 582–592.

Mikhaylov, A., Moiseev, N., Aleshin, K., & Burkhardt, T. (2020). Global climate change and greenhouse effect. *Entrepreneurship and Sustainability Issues, 7*(4), 2897–2913.

Mikhaylov, A., Danish, M. S. S., & Senjyu, T. (2021a). New stage in evolution of cryptocurrency market: Analysis by Hurst method. In H. Dincer & S. Yuksel (Eds.), *Strategic outlook in business and finance innovation: Multidimensional policies for emerging economies* (pp. 35–45). Emerald Publishing Limited. https://doi.org/10.1108/978-1-80043-444-820211004. isbn:978-1-80043-445-5.

Mikhaylov, A., Yumashev, A., & Kolpak, E. (2021b). Quality of life, anxiety and depressive disorders in patients with extrasystolic arrhythmia. *Archives of Medical Science, 17*(1). https://doi.org/10.5114/aoms.2020.101359

Moiseev, N. (2017a). Forecasting time series of economic processes by model averaging across data frames of various lengths. *Journal of Statistical Computation and Simulation, 87*(17), 3111–3131.

Moiseev, N. (2017b). p-value adjustment to control type I errors in linear regression models. *Journal of Statistical Computation and Simulation, 87*(9), 1701–1711.

Moiseev, N. (2017c). Linear model averaging by minimizing mean-squared forecast error unbiased estimator. *Model Assisted Statistics and Applications, 11*(4), 325–338.

Moiseev, N., & Akhmadeev, B. (2017). Agent-based simulation of wealth, capital and asset distribution on stock markets. *Journal of Interdisciplinary Economics, 29*(2), 176–196.

Moiseev, N., & Sorokin, A. (2018). Interval forecast for model averaging methods. *Model Assisted Statistics and Applications, 18*(2), 125–138.

Moiseev, N., Mikhaylov, A., Varyash, I., & Saqib, A. (2020). Investigating the relation of GDP per capita and corruption index. *Entrepreneurship and Sustainability Issues, 8*(1), 780–794.

Morkovkin, D. E., Gibadullin, A. A., Romanova Ju, A., Erygin Yu, V., & Ziadullaev Ul, S. (2019). Formation of a national environmental strategy for the fuel and energy complex. *IOP Conference Series: Materials Science and Engineering, 537*(4), 042064.

Morkovkin, D., Lopatkin, D., Sadriddinov, M., Shushunova, T., Gibadullin, A., & Golikova, O. (2020a). Assessment of innovation activity in the countries of the world. *E3S Web of Conferences, 157*, 04015. https://doi.org/10.1051/e3sconf/202015704015

Morkovkin, D. E., Gibadullin, A. A., Kolosova, E. V., Semkina, N. S., & Fasehzoda, I. S. (2020b). Modern transformation of the production base in the conditions of industry 4.0: Problems and prospects. *Journal of Physics: Conference Series, 1515*, 032014.

Morkovkin, D. E., Gibadullin, A. A., Safarov, B. G., & Alpatova, E. A. (2020c). Definition of factors limiting the growth of industrial production. *IOP Conference Series: Materials Science and Engineering, 862*, 042013.

Morkovkin, D. E., Lopatkin, D. S., Shushunova, T. N., Sharipov, B. K., & Gibadullin, A. A. (2020d). Formation of the conditions for the development of innovation. *Journal of Physics: Conference Series, 1515*, 032002.

Nyangarika, A., Mikhaylov, A., & Tang, B.-J. (2018). Correlation of oil prices and gross domestic product in oil producing countries. *International Journal of Energy Economics and Policy, 8*(5), 42–48.

Nyangarika, A., Mikhaylov, A., & Richter, U. (2019a). Influence oil price towards economic indicators in Russia. *International Journal of Energy Economics and Policy, 9*(1), 123–129.

Nyangarika, A., Mikhaylov, A., & Richter, U. (2019b). Oil price factors: Forecasting on the base of modified auto-regressive integrated moving average model. *International Journal of Energy Economics and Policy, 9*(1), 149–159.

Odhiambo, N. M. (2009). Energy consumption and economic growth nexus in Tanzania: An ARDL bounds testing approach. *Energy Policy, 37*(2), 617–622.

Ozturk, I., Aslan, A., & Kalyoncu, H. (2010). Energy consumption and economic growth relationship: Evidence from panel data for low and middle income countries. *Energy Policy, 38*(8), 4422–4428.

Phelps, E. S. (1994). *Structural slumps*. Harvard University Press.

Ranjbar, O., Chang, T., Nel, E., & Gupta, R. (2017). Energy consumption and economic growth nexus in South Africa: Asymmetric frequency domain approach. *Energy Sources, Part B: Economics, Planning, and Policy, 12*(1), 24–31.

Rathnayaka, R. K. T., Seneviratna, D. M. K., & Long, W. (2018). The dynamic relationship between energy consumption and economic growth in China. *Energy Sources, Part B: Economics, Planning, and Policy, 13*(5), 264–268.

Shahbaz, M., Zakaria, M., Shahzad, S. J. H., & Mahalik, M. K. (2018). The energy consumption and economic growth nexus in top ten energy-consuming countries: Fresh evidence from using the quantile-on-quantile approach. *Energy Economics, 71*, 282–301.

Uandykova, M., Lisin, A., Stepanova, D., Baitenova, L., Mutaliyeva, L., Yuksel, S., & Dincer, H. (2020). The social and legislative principles of counteracting ransomware crime. *Entrepreneurship and Sustainability Issues, 8*(2), 777–798.

Varyash, I., Mikhaylov, A., Moiseev, N., & Aleshin, K. (2020). Triple bottom line and corporate social responsibility performance indicators for Russian companies. *Entrepreneurship and Sustainability Issues, 8*(1), 313–331.

Yuan, J., Zhang, Z. M., Yüksel, S., & Dinçer, H. (2020). Evaluating recognitive balanced scorecard-based quality improvement strategies of energy investments with the integrated hesitant 2-tuple interval-valued Pythagorean fuzzy decision-making approach to QFD. *IEEE Access, 8*, 171112–171128.

Yumashev, A., & Mikhaylov, A. (2020). Development of polymer film coatings with high adhesion to steel alloys and high wear resistance. *Polymer Composites, 41*(7), 2875–2880.

Zhao, Q., Cherkasov, S., Avdeenko, A., Kondratenko, L., & Mikhaylov, A. (2021a). Integral estimate of the added value of agricultural crops in the synergetic agriculture on the example of vicia faba (*Vicia faba* L.). *Polish Journal of Environmental Studies, 30*(2), 993–1002.

Zhao, Y., Xu, Y., Yüksel, S., Dinçer, H., & Ubay, G. G. (2021b). Hybrid IT2 fuzzy modelling with alpha cuts for hydrogen energy investments. *International Journal of Hydrogen Energy.*

Chapter 7
Factors Causing Delay in the Installation of Nuclear Power Plants

Serhat Yüksel, Alexey Mikhaylov, and Gözde Gülseven Ubay

Abstract Nuclear energy is a type of energy that has many advantages. Generating uninterrupted energy, not releasing carbon gas to the atmosphere and obtaining high amounts of electrical energy are examples to be given to this issue. These factors have increased the demands of countries for nuclear power plants. However, there is also a group against nuclear energy due to certain factors. Issues such as the explosion risk of the nuclear power plant and how to destroy radioactive wastes are accepted as the main reasons for this idea. On the other hand, in order for nuclear power plants to operate effectively, the country must have the necessary technological infrastructure and qualified personnel. These factors cause countries not to give priority to nuclear power plants. This study focuses on the factors that delay the establishment of nuclear power plants. In this context, first of all, a detailed literature review has been carried out on the subject. As a result of this study, 5 basic factors that will delay the establishment of nuclear power plants were determined. After that, an analysis has been conducted with the fuzzy DEMATEL method in order to determine which of the relevant factors are more important. The findings indicate that the energy reserve amount of the countries has been determined as the most important factor in opening nuclear power plants. Additionally, the expertise level and public acceptance are other significant issue in this regard. On the other hand, policy changes and technological infrastructure have lower importance for this situation. This result shows that countries resort to nuclear energy in order to reduce their dependence on foreign energy.

S. Yüksel (✉) · G. G. Ubay
The School of Business, İstanbul Medipol University, İstanbul, Turkey
e-mail: serhatyuksel@medipol.edu.tr; gozde.ubay@std.medipol.edu.tr

A. Mikhaylov
Research Center of Monetary Relations, Financial University Under the Government of the Russian Federation, Moscow, Russia

S. Yüksel, H. Dinçer (eds.), *Strategic Approaches to Energy Management*, Contributions to Management Science,
https://doi.org/10.1007/978-3-030-76783-9_7

7.1 Introduction

With the Industrial Revolution, in which the industry was developing rapidly, and human power was replaced by machine power, the demand for energy has increased rapidly on a global scale (Ağbulut et al., 2019). With the necessity of every aspect of life, the energy factor is considered among the most basic inputs in the production process for the realization of economic and social development today. However, many countries cannot meet this increasing energy demand. The unbalanced distribution of energy resources and the gradual decrease in reserves with this increasing demand have pushed countries into different pursuits (An et al., 2020; Zhao et al., 2021). During the industrial revolution, the energy need, which was met with fossil fuels, increased the use of nuclear energy first and then the use of renewable energy resources with the effect of technological developments (Mikhaylov, 2020; Liu et al., 2021a,b). Although investments in renewable energy have started to increase in recent years, nuclear energy has been one of the most frequently preferred energy generation methods of countries for many years due to the efficient energy it creates and its environmental friendliness (Poinssot et al., 2016).

Nuclear reaction occurs when any atomic nucleus is altered by colliding with another physical entity such as alpha particles, gamma rays, neutrons, protons or any atom (Nowotny et al., 2016). Fission and fusion, two of these nuclear reactions, are of particular interest as they release large amounts of energy. Today, these two are only used for electricity generation from the fission reaction. The widespread use of nuclear power plants started with the oil crisis in the early 1970s (Cheng et al., 2020; Yumashev et al., 2020). Countries that do not have oil and other hydrocarbon resources have turned to nuclear power plants in order to reduce their dependence on these resources and to ensure their energy supply security (Renn & Marshall, 2016). While nuclear power plants were put into operation rapidly all over the world, although there was a slowdown with the Three Mile Island in the USA in 1979 and the Chernobyl accidents in Soviet Russia in 1986, nuclear power plants continue to be established all over the world.

Nuclear power plants have been the subject of discussion in lots of the countries for many years. The reason for this is that although this energy has many advantages, its disadvantages are also serious. Nuclear power is cheap and sustainable (Kok & Benli, 2017). They do not emit greenhouse gases like other fossil fuels (Du et al., 2020), so it is accepted as more environment friendly (Vossen, 2020). They are not affected by changes in weather conditions, they can generate electrical energy 24/7. Although these power plants require high technology, uranium, which mainly used in electricity production, can be found in many countries. For this reason, most countries can reduce their energy dependency by producing their own energy (Dooyum et al., 2020; Yüksel et al., 2020). However, despite these advantages, the use of nuclear energy cannot reach the desired level exactly. One of the most important reasons for this is that the wastes generated during nuclear energy production are quite dangerous. Storage or disposal of these wastes poses a danger to both the local community and nature. Besides, the consequences of accidents that may occur in these power plants are highly devastating. The immediate and long-term consequences of nuclear power plant explosions have adversely affected the lives of many people (Ho & Kristiansen, 2019; Zafar et al., 2019).

Although these features of nuclear energy affect people positively or negatively, there are some factors that delay or prevent the establishment of nuclear power plants in countries. Politics may come at the beginning of these elements. Energy is an irreplaceable tool used in all areas of life (Xie et al., 2020; Markard et al., 2020). For this reason, all countries need energy directly. However, it is not always easy to meet this needed energy. For this reason, in many countries, the energy issue is included in the policy. Throughout history, many governments have used energy policies as a choice tool (Jenkins et al., 2016; Nojavan et al., 2017). However, as the governments changed, projects were paused or canceled, especially since nuclear power plants are long-term investments and it takes years to achieve results. Besides, it is one of the other factors delaying the public approval of the establishment of nuclear power plants. Especially after nuclear accidents, anti-nuclear energy actions are carried out in many places (Dinçer et al., 2020; Yüksel & Çağlayan, 2020). Since energy is an issue that concerns the whole country, it is extremely important to include the public in energy policies (Baron & Herzog, 2020). Lastly, the technology level of the country is extremely important in the establishment of nuclear power plants. Generating energy in nuclear power plants requires a high level of knowledge and technological infrastructure (Li et al., 2017; Sarkodie & Adams, 2018). For this reason, countries that do not have sufficient technology will either have to import this technology or will try to develop in this area. In both cases, it will delay the implementation of projects.

For these reasons, in this study, it is aimed to determine the factors that cause a delay in the installation of nuclear power plants. In this context, firstly, by scanning a large literature, five different factors that cause a delay were found. Afterwards, by using these factors an analysis is made by applying fuzzy DEMATEL method. Finally, some suggestions are presented to contribute to the acceleration of the nuclear power plant installation. There are some novelties in this study. First, risks are analyzed in general in the nuclear energy literature. In addition, nuclear energy policies in countries have been frequently discussed. However, there are very few studies analyzing the reasons for the delay in the establishment of nuclear power plants. Another contribution of the study to literature is related to the methodology used. Although the Fuzzy DEMATEL method has been considered in many different subjects, it was used for the first time in this study for nuclear energy. Hence, importance order of these can be explained in more detail and an impact-relation map can be created (Dinçer & Yüksel, 2020). Moreover, by combining these methods within a fuzzy logic, more accurate results can be taken (Zhu et al., 2020).

This study has four different sections. The first section in the study contains general information on the nuclear energy and its features. In the second part of the study, similar studies are explained by conducting a huge analyze on the literature. In the third section of the study, an analysis is made by using fuzzy DEMATEL method. And finally, in the last section, some suggestions in order to fasten the installations of nuclear power plants are given.

7.2 Literature

Many studies have been conducted in the literature on nuclear energy and nuclear power plants. The vast majority of these studies focused on which issues should be considered to accelerate these investment projects. At this point, public acceptance in nuclear energy investments has been examined by many researchers. It is extremely important for countries to obtain public acceptance in order to implement nuclear energy projects. For this reason, efforts should be made to increase the support of the public to the project when starting such investments. As mentioned by Wang et al., one of the primary steps to ensure public acceptance is to inform the public about the issue. In many studies, it is seen that public knowledge is a key for public acceptance. As Odonkor and Adams (2020) mentioned in their study, to get a successful implementation, social acceptance of nuclear energy is critical. However, although public knowledge positively affects public acceptance to a point, the negative impact of nuclear accidents in the past on the public continues in many places. As Roh and Kim (2017) by using one-way ANOVA, public perception of risk on nuclear energy has progressed badly, and in order to change public's perspective necessary steps are needed to taken. Environmental beliefs are also an important factor that affects the public to accept these types of projects. As Wang et al. (2019) examined in their study that takes place in China, it is seen that environmental beliefs and nuclear power acceptance are positively correlated. However, when there is a place attachment, this correlation turns to a negative direction.

Another factor that can cause a delay in the installation of nuclear power plants is political changes. Energy is used in all areas of life and it is the duty of states to provide this energy to their people (Li et al., 2021; Yuan et al., 2020). Since nuclear energy projects are long-term projects, they often face the danger of being delayed or disappearing amid policy changes (Prăvălie & Bandoc, 2018; Wang et al., 2019). The policies to be implemented by most governments on energy or the rules put into effect in this field are different. Therefore, there are delays in establishing nuclear power plants in countries that do not have a long-term energy policy (Azam et al., 2021). The governments that are the leading countries and the policies followed by these governments are of great importance for energy investments. For example, As Richter and Wurster (2016) examined, it is seen that as countries become democratic, nuclear power is replaced by greener energies, with more public influence in the administration. In this case, it can lead to the suspension of existing projects.

Energy resources reserves of countries may delay the establishment of nuclear power plants by these countries (Ağbulut, 2019). Countries that have a better alternative to nuclear energy will abandon the high cost of nuclear energy and turn to their own resources. For example, a country that has a lot of coal reserves can enter nuclear energy projects later, because it is both cheap and easier to obtain. This process can continue as long as the energy demand can be met. Although nuclear energy is an efficient form of energy generation, it is a very troublesome business with both security measures, waste management and facility maintenance. For this reason, when countries find a type of energy that they can access and operate more easily, they can shelve the issue of nuclear energy. In addition, the elements used in nuclear energy production are not equally distributed in every country. For this reason, for countries where these elements are not available or scarce, it may become costly to obtain the element. As a result, it will be possible for countries to turn to other energy sources and delay their nuclear energy installation (Ozturk, 2017).

In addition to these, some technical requirements may cause delay in the establishment of nuclear power plants. Technological infrastructure and expertise in the subject area are among these requirements. Nuclear energy is obtained with high technology materials and high technology is needed in the use of this energy, under the auspices of waste (Ho et al., 2019). For this reason, countries that have not reached a certain level in technology may switch to nuclear energy later than normal. Another factor that may cause delay in the establishment of nuclear power plants is the level of expertise of the country on the subject (Abdullah & Najib, 2016). Nuclear power plants require high technology as well as experts who will ensure the continuity of the facilities that will use this technology. At this point, there are two options that countries can follow. The first is to import the technology and experts required for nuclear energy from other countries. However, this process will both create a financial burden for the country and will not be able to fully ensure the energy independence of the country (Prăvălie & Bandoc, 2018). Technologically insufficient countries or countries that do not have sufficient expertise have the opportunity to bring themselves to a sufficient level in this regard by giving importance to education and development. However, since these situations cannot occur in a short time, they may cause some delays.

7.3 Analysis Results

In this study, it is aimed to determine the most important factors delaying the establishment of nuclear reactors. In this framework, a detailed literature review is made, and 5 different factors are determined. These factors are details on Table 7.1.

Table 7.1 states that there are 5 different criteria that may delay the installation of the nuclear power plants. First of all, the energy policies of countries can affect this process. In this context, if a country's policy is to focus on renewable energy sources, nuclear energy investments may not be prioritized. In addition, the inadequate procedures in the country can hinder this process. Secondly, the amount of reserves available in countries can also affect nuclear energy policies. In this context,

Table 7.1 The details of criteria

Criteria	Criteria no
Policy changes	C1
Countries' their own reserves	C2
Expertise level	C3
Technological infrastructure	C4
Public acceptance	C5

Table 7.2 Experts' evaluation details

Criteria	C1	C2	C3	C4	C5
Decision maker (DM) 1					
C1	–	L	L	L	N
C2	H	–	H	H	H
C3	M	L	–	L	M
C4	M	L	M	–	M
C5	M	L	M	L	–
Decision maker (DM) 2					
C1	–	N	L	L	M
C2	VH	–	H	VH	VH
C3	M	N	–	M	L
C4	L	N	L	–	L
C5	M	N	L	L	–
Decision maker (DM) 3					
C1	–	N	M	L	L
C2	VH	–	VH	VH	VH
C3	L	L	–	M	M
C4	L	N	L	–	M
C5	M	L	L	L	–

Table 7.3 The details of triangular fuzzy numbers

Influence level	Triangular Fuzzy Numbers		
No (N)	0	0	0.25
Low (L)	0	0.25	0.5
Medium (M)	0.25	0.5	0.75
High (H)	0.5	0.75	1
Very high (VH)	0.75	1	1

if a country has sufficient energy reserves, it can throw nuclear power plants into the background. On the other hand, qualified personnel to work in nuclear power plants are needed. If qualified personnel are insufficient in a country, this can delay the use of nuclear energy. The adequacy of the country's technological infrastructure is similarly important. Finally, if the citizens living in the country have a bad perspective on nuclear energy, this will delay the installation of nuclear power plants.

An analysis has been conducted with the fuzzy DEMATEL approach in order to identify those that are more important than these factors. In this context, 3 different experts are asked to evaluate the criteria. These experts consist of academics who have researched nuclear energy and have at least 13 years of experience. The people evaluated these criteria by considering 5 different scales which are no (N), low (L), medium (M), high (H) and very high (VH) (Zhang et al., 2020; Dinçer & Yüksel, 2020; Delen et al., 2020; Yüksel & Dinçer, 2020; Xie et al., 2020; Korsakienė et al., 2020). The evaluations of the experts are given in Table 7.2.

After that, these evaluations are converted into triangular fuzzy numbers. The details of these fuzzy numbers are demonstrated on Table 7.3.

The details of the triangular fuzzy numbers for different decision makers are stated on Table 7.4.

By considering these numbers, direct relation matrix is generated. The details of these matrix are shown in Table 7.5.

In the next step, this matrix is normalized. In this process, the values are divided to the maximum of the total value in the rows. The details of this new matrix are illustrated on Table 7.6.

Just then, the total relation matrix is generated and depicted in Table 7.7.

After that, the defuzzification process has been conducted and the weights of the criteria are identified. The details are given on Table 7.8.

Table 7.8 indicates that the criterion of the countries' own reserves is the most important factor. In addition to this issue, the expertise level and public acceptance are other significant issues. However, policy changes and technological infrastructure have lower weights in comparison with others.

7.4 Discussion and Conclusion

In this study, factors delaying the establishment of nuclear power plants are emphasized. For this purpose, first of all, a detailed literature review has been made on the subject. As a result of this examination, 5 different criteria are determined that could affect this issue. Then, an analysis has been carried out with the fuzzy DEMATEL method in order to determine which of these factors are more important. It is concluded that the energy reserve amount of the countries has been determined as the most important factor in opening nuclear power plants. Furthermore, the expertise level and public acceptance are other significant issue in this regard. On the other hand, policy changes and technological infrastructure have lower importance for this situation.

As it can be seen from the analysis results obtained, nuclear energy studies slow down when there is an energy reserve in the country. In other words, in countries that do not have their own energy reserves, nuclear power plant establishment processes are progressing faster. This situation indicates that nuclear power plants are preferred mainly for economic reasons. In this context, countries that do not have their own energy reserves depend on other countries for energy. This situation causes the country to experience both economic and social problems. The country that imports the energy from abroad makes the payment in foreign currency. This situation causes the country's exchange rate risk to increase. Moreover, high foreign exchange payments made for energy negatively affect the current account balance of the country. In addition to economic factors, the country, which has to buy energy from abroad, becomes politically dependent on these countries.

Considering these issues, it is important both socially and economically for countries to be independent in energy. In this context, countries develop various energy strategies to produce their own energy. The main purpose here is to ensure the security of the country's energy supply. Setting up nuclear power plants is also a

Table 7.4 Triangular fuzzy numbers for different decision makers

Criteria	C1			C2			C3			C4			C5		
Triangular Fuzzy numbers of decision maker 1 for direct relation matrix															
C1	0.00	0.00	0.00	0.00	0.25	0.50	0.00	0.25	0.50	0.00	0.25	0.50	0.00	0.00	0.25
C2	0.50	0.75	1.00	0.00	0.00	0.00	0.50	0.75	1.00	0.50	0.75	1.00	0.50	0.75	1.00
C3	0.25	0.50	0.75	0.00	0.25	0.50	0.00	0.00	0.00	0.00	0.25	0.50	0.25	0.50	0.75
C4	0.25	0.50	0.75	0.00	0.25	0.50	0.25	0.50	0.75	0.00	0.00	0.00	0.25	0.50	0.75
C5	0.25	0.50	0.75	0.00	0.25	0.50	0.25	0.50	0.75	0.00	0.25	0.50	0.00	0.00	0.00
Triangular fuzzy numbers of decision maker 2 for direct relation matrix															
C1	0.00	0.00	0.00	0.00	0.00	0.25	0.00	0.25	0.50	0.00	0.25	0.50	0.25	0.50	0.75
C2	0.75	1.00	1.00	0.00	0.00	0.00	0.50	0.75	1.00	0.75	1.00	1.00	0.75	1.00	1.00
C3	0.25	0.50	0.75	0.00	0.00	0.25	0.00	0.00	0.00	0.25	0.50	0.75	0.00	0.25	0.50
C4	0.00	0.25	0.50	0.00	0.00	0.25	0.00	0.25	0.50	0.00	0.00	0.00	0.00	0.25	0.50
C5	0.25	0.50	0.75	0.00	0.00	0.25	0.00	0.25	0.50	0.00	0.25	0.50	0.00	0.00	0.00
Triangular fuzzy numbers of decision maker 3 for direct relation matrix															
C1	0.00	0.00	0.00	0.00	0.00	0.25	0.25	0.50	0.75	0.00	0.25	0.50	0.00	0.25	0.50
C2	0.75	1.00	1.00	0.00	0.00	0.00	0.75	1.00	1.00	0.75	1.00	1.00	0.75	1.00	1.00
C3	0.00	0.25	0.50	0.00	0.25	0.50	0.00	0.00	0.00	0.25	0.50	0.75	0.25	0.50	0.75
C4	0.00	0.25	0.50	0.00	0.00	0.25	0.00	0.25	0.50	0.00	0.00	0.00	0.25	0.50	0.75
C5	0.25	0.50	0.75	0.00	0.25	0.50	0.00	0.25	0.50	0.00	0.25	0.50	0.00	0.00	0.00

Table 7.5 Direct relation matrix

Criteria	C1			C2			C3			C4			C5		
C1	0.00	0.00	0.00	0.00	0.08	0.33	0.08	0.33	0.58	0.00	0.25	0.50	0.08	0.25	0.50
C2	0.67	0.92	1.00	0.00	0.00	0.00	0.58	0.83	1.00	0.67	0.92	1.00	0.67	0.92	1.00
C3	0.17	0.42	0.67	0.00	0.17	0.42	0.00	0.00	0.00	0.17	0.42	0.67	0.17	0.42	0.67
C4	0.08	0.33	0.58	0.00	0.08	0.33	0.08	0.33	0.58	0.00	0.00	0.00	0.17	0.42	0.67
C5	0.25	0.50	0.75	0.00	0.17	0.42	0.08	0.33	0.58	0.00	0.25	0.50	0.00	0.00	0.00

Table 7.6 Normalized matrix

Criteria	C1			C2			C3			C4			C5		
C1	0.00	0.00	0.00	0.00	0.02	0.08	0.02	0.08	0.15	0.00	0.06	0.13	0.02	0.06	0.13
C2	0.17	0.23	0.25	0.00	0.00	0.00	0.15	0.21	0.25	0.17	0.23	0.25	0.17	0.23	0.25
C3	0.04	0.10	0.17	0.00	0.04	0.10	0.00	0.00	0.00	0.04	0.10	0.17	0.04	0.10	0.17
C4	0.02	0.08	0.15	0.00	0.02	0.08	0.02	0.08	0.15	0.00	0.00	0.00	0.04	0.10	0.17
C5	0.06	0.13	0.19	0.00	0.04	0.10	0.02	0.08	0.15	0.00	0.06	0.13	0.00	0.00	0.00

Table 7.7 Total relation matrix

Criteria	C1			C2			C3			C4			C5		
C1	0.00	0.04	0.19	0.00	0.03	0.19	0.02	0.11	0.30	0.00	0.09	0.28	0.02	0.09	0.29
C2	0.19	0.34	0.57	0.00	0.04	0.21	0.16	0.30	0.54	0.17	0.31	0.53	0.18	0.32	0.55
C3	0.05	0.15	0.38	0.00	0.06	0.23	0.00	0.05	0.22	0.04	0.14	0.35	0.04	0.15	0.37
C4	0.02	0.12	0.34	0.00	0.03	0.20	0.02	0.12	0.32	0.00	0.04	0.19	0.04	0.14	0.34
C5	0.06	0.16	0.38	0.00	0.05	0.22	0.02	0.12	0.33	0.00	0.10	0.31	0.00	0.05	0.21

Table 7.8 The weights of the criteria

Criteria	Weights
Policy changes	0.1887
Countries' their own reserves	0.2329
Expertise level	0.1983
Technological infrastructure	0.1827
Public acceptance	0.1973

way for countries to generate their own energy. Thanks to nuclear energy, countries that can produce their own energy will not have to import energy from abroad. On the other hand, the use of nuclear energy also has some risks, such as the possibility of explosion and the generation of radioactive waste. Therefore, it is observed that countries that have their own energy reserves generally do not prefer this type of energy.

References

Abdullah, L., & Najib, L. (2016). Sustainable energy planning decision using the intuitionistic fuzzy analytic hierarchy process: Choosing energy technology in Malaysia. *International Journal of Sustainable Energy, 35*(4), 360–377.

Ağbulut, Ü. (2019). Turkey's electricity generation problem and nuclear energy policy. *Energy Sources, Part A: Recovery, Utilization, and Environmental Effects, 41*(18), 2281–2298.

Ağbulut, Ü., Ceylan, İ., Gürel, A. E., & Ergün, A. (2019). The history of greenhouse gas emissions and relation with the nuclear energy policy for Turkey. *International Journal of Ambient Energy, 2019*, 1–9.

An, J., Mikhaylov, A., & Richter, U. H. (2020). Trade war effects: Evidence from sectors of energy and resources in Africa. *Heliyon, 6*(12), e05693.

Azam, A., Rafiq, M., Shafique, M., Zhang, H., & Yuan, J. (2021). Analyzing the effect of natural gas, nuclear energy and renewable energy on GDP and carbon emissions: A multi-variate panel data analysis. *Energy, 219*, 119592.

Baron, J., & Herzog, S. (2020). Public opinion on nuclear energy and nuclear weapons: The attitudinal nexus in the United States. *Energy Research & Social Science, 68*, 101567.

Cheng, F., Lin, M., Yüksel, S., Dİncer, H., & Kalkavan, H. (2020). A hybrid hesitant 2-tuple IVSF decision making approach to analyze PERT-based critical paths of new service development process for renewable energy investment projects. *IEEE Access*.

Delen, D., Dorokhov, O., Dorokhova, L., Dinçer, H., & Yüksel, S. (2020). Balanced scorecard-based analysis of customer expectations for cosmetology services: A hybrid decision modeling approach. *Journal of Management Analytics, 7*(4), 532–563.

Dinçer, H., & Yüksel, S. (2020). Defining the strategic impact-relation map for the innovative investments based on IT2 fuzzy DEMATEL: Evidence from the European tourism sector. In *Handbook of research on creating sustainable value in the global economy* (pp. 103–124). IGI Global.

Dinçer, H., Yüksel, S., Çağlayan, Ç., & Uluer, G. S. (2020). The contribution of nuclear energy investment on sustainable financial and economic development. *Journal of Financial Economics and Banking, 1*(1), 39–51.

Dooyum, U. D., Mikhaylov, A., & Varyash, I. (2020). Energy security concept in Russia and South Korea. *International Journal of Energy Economics and Policy, 10*(4), 102.

Du, L., Dinçer, H., Ersin, İ., & Yüksel, S. (2020). IT2 fuzzy-based multidimensional evaluation of coal energy for sustainable economic development. *Energies, 13*(10), 2453.
Ho, S. S., & Kristiansen, S. (2019). *Environmental debates over nuclear energy: Media, communication, and the public.*
Ho, S. S., Leong, A. D., Looi, J., & Chuah, A. S. (2019). Online, offline, or word-of-mouth? Complementary media usage patterns and credibility perceptions of nuclear energy information in Southeast Asia. *Energy Research & Social Science, 48*, 46–56.
Jenkins, K., Heffron, R. J., & McCauley, D. (2016). The political economy of energy justice: A nuclear energy perspective. In *The Palgrave Handbook of the international political economy of energy* (pp. 661–682). Palgrave Macmillan.
Kok, B., & Benli, H. (2017). Energy diversity and nuclear energy for sustainable development in Turkey. *Renewable Energy, 111*, 870–877.
Korsakienė, R., Raišienė, A. G., Dinçer, H., Yüksel, S., & Aleksejevec, V. (2020). Strategic mapping of eco-innovations and human factors: Business projects' success revisited. In *Strategic Outlook for innovative work behaviours* (pp. 1–19). Springer.
Li, M. J., Zhu, H. H., Guo, J. Q., Wang, K., & Tao, W. Q. (2017). The development technology and applications of supercritical CO_2 power cycle in nuclear energy, solar energy and other energy industries. *Applied Thermal Engineering, 126*, 255–275.
Li, Y. X., Wu, Z. X., Dinçer, H., Kalkavan, H., & Yüksel, S. (2021). Analyzing TRIZ-based strategic priorities of customer expectations for renewable energy investments with interval type-2 fuzzy modeling. *Energy Reports, 7*, 95–108.
Liu, J., Lv, J., Dinçer, H., Yüksel, S., & Karakuş, H. (2021a). Selection of renewable energy alternatives for green blockchain investments: A hybrid IT2-based fuzzy modelling. *Archives of Computational Methods in Engineering, 2021*, 1–15.
Liu, Y., Gong, X., Yüksel, S., Dinçer, H., & Aydın, R. (2021b). A multidimensional outlook to energy investments for the countries with continental shelf in East Mediterranean region with hybrid decision making model based on IVIF logic. *Energy Reports, 7*, 158–173.
Markard, J., Bento, N., Kittner, N., & Nunez-Jimenez, A. (2020). Destined for decline? Examining nuclear energy from a technological innovation systems perspective. *Energy Research & Social Science, 67*, 101512.
Mikhaylov, A. (2020). Geothermal energy development in Iceland. *International Journal of Energy Economics and Policy, 10*(4), 31.
Nojavan, S., Zare, K., & Mohammadi-Ivatloo, B. (2017). Application of fuel cell and electrolyzer as hydrogen energy storage system in energy management of electricity energy retailer in the presence of the renewable energy sources and plug-in electric vehicles. *Energy Conversion and Management, 136*, 404–417.
Nowotny, J., Hoshino, T., Dodson, J., Atanacio, A. J., Ionescu, M., Peterson, V., Prince, K. E., Yamawaki, M., Bak, T., Sigmund, W., Veziroglu, T. N., & Alim, M. A. (2016). Towards sustainable energy. Generation of hydrogen fuel using nuclear energy. *International Journal of Hydrogen Energy, 41*(30), 12812–12825.
Odonkor, S. T., & Adams, S. (2020). An assessment of public knowledge, perception and acceptance of nuclear energy in Ghana. *Journal of Cleaner Production, 122279.*
Ozturk, I. (2017). Measuring the impact of alternative and nuclear energy consumption, carbon dioxide emissions and oil rents on specific growth factors in the panel of Latin American countries. *Progress in Nuclear Energy, 100*, 71–81.
Poinssot, C., Bourg, S., & Boullis, B. (2016). Improving the nuclear energy sustainability by decreasing its environmental footprint. Guidelines from life cycle assessment simulations. *Progress in Nuclear Energy, 92*, 234–241.
Prăvălie, R., & Bandoc, G. (2018). Nuclear energy: Between global electricity demand, worldwide decarbonisation imperativeness, and planetary environmental implications. *Journal of Environmental Management, 209*, 81–92.
Renn, O., & Marshall, J. P. (2016). Coal, nuclear and renewable energy policies in Germany: From the 1950s to the "Energiewende". *Energy Policy, 99*, 224–232.

Richter, T., & Wurster, S. (2016). Policy diffusion among democracies and autocracies: A comparison of trade reforms and nuclear energy policy. *Global Policy, 7*(4), 541–547.
Roh, S., & Kim, D. (2017). Effect of Fukushima accident on public acceptance of nuclear energy (Fukushima accident and nuclear public acceptance). *Energy Sources, Part B: Economics, Planning, and Policy, 12*(6), 565–569.
Sarkodie, S. A., & Adams, S. (2018). Renewable energy, nuclear energy, and environmental pollution: Accounting for political institutional quality in South Africa. *Science of the Total Environment, 643*, 1590–1601.
Vossen, M. (2020). Nuclear energy in the context of climate change: A frame analysis of the Dutch print media. *Journalism Studies*, 1–20.
Wang, S., Wang, J., Lin, S., & Li, J. (2019). Public perceptions and acceptance of nuclear energy in China: The role of public knowledge, perceived benefit, perceived risk and public engagement. *Energy Policy, 126*, 352–360.
Xie, Y., Peng, Y., Yüksel, S., Dınçer, H., Uluer, G. S., Çağlayan, Ç., & Li, Y. (2020). Consensus-based public acceptance and mapping of nuclear energy investments using spherical and pythagorean fuzzy group decision making approaches. *IEEE Access, 8*, 206248–206263.
Yuan, J., Zhang, Z. M., Yüksel, S., & Dinçer, H. (2020). Evaluating recognitive balanced scorecard-based quality improvement strategies of energy investments with the integrated hesitant 2-tuple interval-valued Pythagorean fuzzy decision-making approach to QFD. *IEEE Access, 8*, 171112–171128.
Yüksel, S., & Çağlayan, Ç. (2020). The factors affecting the investors' decisions: A study on nuclear energy investments. *Economics Literature, 2*(2), 177–185.
Yüksel, S., & Dinçer, H. (2020). SERVQUAL-based performance analysis of agricultural financing in E-banking industry: An evaluation by IT2 fuzzy decision-making model. In *Tools and techniques for implementing international E-Trading tactics for competitive advantage* (pp. 21–41). IGI Global.
Yüksel, S., Dinçer, H., & Uluer, G. S. (2020). The role of technological development on renewable energy usage: An econometric analysis for G7 countries. In *Handbook of research on sustainable supply chain management for the global economy* (pp. 136–153). IGI Global.
Yumashev, A., Ślusarczyk, B., Kondrashev, S., & Mikhaylov, A. (2020). Global indicators of sustainable development: Evaluation of the influence of the human development index on consumption and quality of energy. *Energies, 13*(11), 2768.
Zafar, M. W., Shahbaz, M., Hou, F., & Sinha, A. (2019). From nonrenewable to renewable energy and its impact on economic growth: The role of research & development expenditures in Asia-Pacific Economic Cooperation countries. *Journal of Cleaner Production, 212*, 1166–1178.
Zhang, G., Zhou, S., Xia, X., Yüksel, S., Baş, H., & Dincer, H. (2020). Strategic mapping of youth unemployment with interval-valued intuitionistic hesitant fuzzy DEMATEL based on 2-tuple linguistic values. *IEEE Access, 8*, 25706–25721.
Zhao, Y., Xu, Y., Yüksel, S., Dinçer, H., & Ubay, G. G. (2021). Hybrid IT2 fuzzy modelling with alpha cuts for hydrogen energy investments. *International Journal of Hydrogen Energy.*
Zhu, L., Hu, L., Yüksel, S., Dinçer, H., Karakuş, H., & Ubay, G. G. (2020). Analysis of strategic directions in sustainable hydrogen investment decisions. *Sustainability, 12*(11), 4581.

Chapter 8
A New Approach of Energy Financing: The Yields of Green Bonds in Emerging Economies

Musa Gün and Melih Kutlu

Abstract This study focuses on a relatively new financial asset, namely the green bond which is particularly used to finance environmentally friendly and climate stimulating projects. From a point of business view, the financial performance of assets naturally affects the investment decisions of actors including policymakers, governments and investors. Focusing on this angle, this study aims at analyzing and comparing the yields of green bonds and equivalent conventional peers in emerging economies including Brazil, China, Czech Republic, Korea, Malaysia and Taiwan. In this context, firstly, green financing and green bonds introduced. Afterwards, based on the literature review, we employ a matching method to test the yield differences among the bonds. The analyses have resulted in mixed findings. The descriptive statistics show that on average green bonds offer a higher yield than conventional bonds in China, Czech Republic and Malaysia, whereas they provide lower yields in Brazil, Korea and Taiwan. However, considering the findings as a whole, the differences between the yields are not statistically significant.

8.1 Introduction

Energy management, which is expressed as a set of activities covering processes from ensuring the efficient use of energy resources, energy investments and planning these operations, has been the subject of several studies. The aim of energy management is not only to achieve cheaper energy resources but also to minimize the

M. Gün (✉)
The Faculty of Economics and Administrative Sciences, Recep Tayyip Erdoğan University, Merkez/Rize, Turkey
e-mail: musa.gun@erdogan.edu.tr

M. Kutlu
The Faculty of Economics, Administrative and Social Sciences, Samsun University, Canik/Samsun, Turkey
e-mail: melih.kutlu@samsun.edu.tr

S. Yüksel, H. Dinçer (eds.), *Strategic Approaches to Energy Management*, Contributions to Management Science,
https://doi.org/10.1007/978-3-030-76783-9_8

environmental effects of these resources. The energy management issue has been more marked in developing countries compared to developed countries since these countries have a faster ability of growth than the developed ones plus have more potential for energy investments and improvements (Wang, 2020). Also, from a business perspective, energy management is of significance especially for economies that highly depend on foreign energy supplies as it has direct economic effects such as high share and cost of energy in the total production of enterprises (Bayar & Gavriletea, 2019).

Considering the scarcity of energy resources and the damage caused by fossil fuels to the environment, it is seen that sustainability-themed investments and the financing tools for those investments are growing rapidly in the world (Ielasi et al., 2018). According to the Institute of International Finance (IIF) (2021), the sustainable debt issuance, which was around $350 billion in 2019, exceeded $650 billion by the end of 2020. Green bonds have the largest share among this issuance, which makes up approximately 50% of the total issuance.

On the other hand, neither the public nor the private sector, especially in underdeveloped or developing countries, cannot afford green projects that require a huge amount of investment. Meanwhile, the high risk and low return possibility of these projects are the main factors that affect the interest of public or private investors entering such long-term financing projects (Yoshino & Taghizadeh-Hesary, 2018). Many studies (see, e.g., Banga, 2019; Weber & Saravade, 2019) have been carried out on both the economic and environmental contributions of green bonds within the scope of green projects, which especially aim to meet and to achieve the commitments of the 2030 Agenda for Sustainable Development and the United Nations Framework Convention on Climate Change's 2015 Paris Accord.

However, only a small portion of studies have been dealt with the return of these financial assets, which is the main factor affecting the investment decisions of participants including policymakers, governments as well as investors. Focusing on this angle, this study aims at discussing the basic research question: Are green bonds different from conventional bonds in terms of yield? To be able to fill the gap in the literature and to guide the potential investors about alternative investment opportunities, this study aims to analyse the financial performance of the green bonds and compare the return profile of green bonds with conventional peers over the selected sample in emerging economies. For these purposes, we employ a matching method to test whether the yields of green bonds differ from the yields of conventional bonds. In our empirical research covering the period from January 2018 to February 2021, we show that on average green bonds offer a higher yield than conventional bonds in China, Czech Republic and Malaysia, whereas they provide lower yields in Brazil, Korea and Taiwan. However, considering as a whole, the difference between the yields is not statistically significant.

The rest of this study is organized as follows: Section Two presents a general framework of green financing; Section Three provides the literature about the topic; Section Four gives the methodology and data as well as the study findings, and Section Five involves discussion of the findings and the conclusion.

8.2 Conceptual Framework of Green Finance and Green Bonds

Today, people are more demanding of companies and governments to reduce environmental, health and safety risks. Thus, companies and governments that integrated into corporate social responsibility and sustainability are more trusted by investors. In the framework of sustainability, green finance is a prominent topic of financial markets since it gives a link between global environmental priorities and the financial system. Green finance supports green entrepreneurship activities. It also supports the development of green financial products and green investors. Moreover, it contributes to the evaluation of environmental risks and effective management of the emission trade market. Definitions regarding green finance are:

German Corporation for International Cooperation GmbH (Schaefer, 2011): "Green Finance is a strategic approach to incorporate the financial sector in the transformation process towards low-carbon and resource-efficient economies, and in adaptation to climate change."

PBC and UNEP (2015): "Green finance policy refers to a series of policy and institutional arrangements to attract private capital investments into green industries such as environmental protection, energy conservation and clean energy through financial services including lending, private equity funds, bonds, shares and insurance."

Although green finance and sustainable finance look alike, the concepts differ from each other. We can distinguish between these two concepts in terms of their contribution to sustainable development goals. Sustainable development goals comprise environmental, social, economic and other sub-goals. Sustainable finance is a broader concept that contributes to all dimensions of sustainable development (Berrou et al., 2019). On the other hand, green finance usually contributes to the environmental dimension of these goals. The environmental dimension covers climate change mitigation, climate change adaptation and other environmental objectives.

Considering the definitions of green finance and its contribution to sustainable development goals, the issues of climate change and related carbon emission reduction are important in terms of green finance. According to Breitenfellner et al. (2020), since carbon reduction is a global-scale concept, it needs global scale investments. Differences in numbers, calculations, scenarios and assumptions make it difficult to calculate the size of the investment. However, no matter how difficult it is to calculate the size of the investment, investments directed to emerging and developing countries.

8.2.1 Green Financial Markets and Green Bonds

Environmental funds weather derivatives or bonds, which are traded in green financial markets, are used to protect companies from unexpected natural disasters with carbon emission reduction. The intermediary task of raising funds and allocating funds in green financial markets is for environmental protection's capital movement. Green finance markets enable the efficient use of various production factors. It also supports economic development with financial leverage (Wang & Zhi, 2016). The green finance system has goals such as adding climate change risk to financial decisions for low carbon emissions, preventing the risk caused by climate change from disrupting financial stability, and contributing to the development of organizations such as green investment banks and products such as green bonds (Meltzer, 2016).

Green bonds, one of the green finance instruments, briefly defined as the bonds that create resources for the financing of projects which provide positive benefits to the environment and the climate (Rakić & Mitić, 2012). The key feature that distinguishes these bonds from other bonds is that their proceeds are used in green projects (Horsch & Richter, 2017). The concept of the green project covers the topics such as energy efficiency, renewable energy, pollution control and prevention, management of natural resources, conservation of biodiversity, and clean transportation (Banga, 2019). Although green bonds can be used for all green projects, it is generally focused on renewable energy and energy efficiency. According to Bartels et al. (2016), more than half of the funds provided from these bonds are used in the areas of energy efficiency and renewable energy projects.

Green bonds are conceptually similar to conventional bonds. However, green bonds returns are used only for investment in green projects (Baulkaran, 2019). Additionally, green bonds are less liquid than conventional bonds. Green bonds investors are predominantly institutional actors. While profitability and cash flows are important for conventional bond investors, green bond investors also consider environmental factors (Hyun et al., 2019). In other words, green bonds provide financial returns to investors as well as ensuring environmental benefits. With these two outstanding advantages, green bonds are increasingly attracting the attention of investors. To classify the green finance investors in terms of return expectations, it can be expressed that philanthropic investors just expect a social return while socially responsible investors expect economic return with social impact. On the other hand, commercial investors only consider the economic return of financial instruments (Noh, 2019).

Green bonds contribute to government policies and the ability of companies to manage climate-related risks. For this reason, it attracts the attention of not only investors interested in climate, environment and sustainability issues, but also all other investors. Within the framework of the 2015 Paris Climate Agreement, it is aimed to make economies resistant to climate changes. In line with these purposes, green bonds attract the attention of financial institutions such as mutual funds and insurance companies (Reboredo, 2018). The green bonds also support clean energy financing and can refinance the projects in the long term after they realize projects. The first green bond was issued in 2007 as a climate awareness bond by the European Investment Bank (2021).

However, regulations on green bonds are also needed. Central banks, in particular, need to establish regulations and set criteria for financing projects and company classification (Dikaui & Volz, 2018). The international guidelines brought out by the International Capital Market Association (ICMA) for green bonds are called the "Green Bond Principles" (GBP). These were first drafted in 2014 and have since gone through periodic eight updates. The Climate Bond Initiative classifies the bond market associated with green bonds as a "climate-aligned bond universe". According to GBP, "green bonds are any type of bond instrument where the proceeds will be exclusively applied to finance or refinance new or existing eligible green projects". Similarly, World Bank (2015) defines green bonds as "a debt security that is issued to raise capital specifically to support climate-related or environmental projects". The projects related to renewable energy, energy efficiency, pollution prevention and control, environmental sustainability, terrestrial/aquatic biodiversity conservation, transportation (electric/hybrid/public/rail), water and wastewater management, climate change, eco-friendly production/technology and green buildings can be financed by green bonds.

8.3 Literature Review

The major objectives of the green bond market are to empower and organize the debt markets to be able to support environmental sustainability projects. Green bonds, which aim to provide capital and facilitate investments for environmental projects, attract considerable attention from a broad range of audience including investors, decision-makers and academicians.

Especially after the announcement of the "Green Bond Principles" by the International Capital Market Association in 2014 and the Paris Climate Agreement of 2015, many studies have been started to be carried out on the subject. In this framework, early studies generally highlight the importance and role of green bonds to realize sustainable economic growth (Clapp, 2014; DuPont et al., 2015; Wang & Zhi, 2016). Similarly, Shishlov et al. (2016), Caldecott (2017), and Glomsrød and Wei (2018) emphasise the benefits and impacts of green investment assets. Furthermore, they outline the key factors and barriers that prevent emerging economies from using this novel financial asset.

On the one hand, it is a fact that investors who invest in green bonds care about the environment and climate change. In the other hand, from a financial perspective, they also consider the risk-return levels of the assets (Preclaw & Bakshi, 2015). Therefore, all actors like investors and issuing institutions as well as nations, want to know about the risk and return profiles of the green bonds while they are making investment decisions. Regarding our research focus on the financial performance of green bonds, Östlund (2015) is among the first scholars who analyses and compares the returns of green bonds with conventional bonds. The findings of the study demonstrate that the spread between green bonds and conventional bonds are statistically significant just for the year 2015. However, the test results for the overall dataset do not produce enough evidence that there is a meaningful difference between the yields of bonds. The study consequences are supported by Preclaw and Bakshi (2015)'s research insisting that green bonds are not markedly different from conventional bonds.

In the empirical study conducted by Petrova (2016), panel data and time-series analyses have been implemented to determine the financial performance of green bonds towards conventional bonds. The data of the analyses include two green bond indices and three benchmarks of conventional bond indices in the United States debt market during the period between 2008 and 2016. The statistical results of the research show that green bonds slightly overperform conventional bonds, but the difference between the yields is not significant.

Corresponding researches including those of Nofsinger and Varma (2014), Becchetti et al. (2015) and Reboredo et al. (2017) also analyse whether the returns of various green instruments such as bonds, mutual funds or equity indices are different from conventional counterparts. Broadly reporting, the studies show mixed results in these comparisons and the findings confirm that there is no consensus. In some cases, the authors point to the fact that green instruments are more attractive than others and could protect investors against price fluctuations. In other cases, for instance, Reboredo et al. (2017) report that green funds underperformed notwithstanding the findings of Petrova (2016). As the financial returns have decisive effects on investment decisions, the lower performance of green instruments could be explained by the investors' willingness to pay for environmental concerns. Supporting this conclusion, Karpf and Mandel (2018) explain the lower performance of green bonds by environmental considerations.

Labelling a bond as green certificates that the issuer firm or the institution has sustainable projects and thus this critical green bond announcement increases the issuer's visibility so that it positively affects the stock returns of the issuer. Examining the stock price reaction to green bond disclosure, Tang and Zhang (2020) could not find sufficient proof that green bonds are offered at a lower yield than conventional similar bonds. However, the study results claim that the issuance of a green bond contributes to the existent stockholders' wealth due to the increasing awareness of the issuer with the announcement of green bonds. Similarly, Flammer (2020) investigates the relationship between green bond issuance and stock market reactions. The main conclusions from these analyses present that investors' response is positive to the announcements of green instrument issuances.

Despite the lower returns, demonstrated by some empirical investigations and bank reports such as HSBC (2016), Shurey (2017) and Zerbib (2017), the green bonds are more desirable than conventional counterparts due to their positive signalling effects like good credit rating about the issuing institutions' commitments towards the environment (Flammer, 2020). Therefore, investors respond positively to the issuance of green bonds.

A related strand of the literature investigates the relationship between green and other financial instruments within the framework of their financial performances and volatility behaviours as well as tries to determine the environmental sensitivity of green bonds. In this context, the primary result of Nanayakkara and Colombage (2019) shows that green bonds are issued and traded at a premium compared with conventional bonds. Their results confirm the findings from other researchers such as and Wulandari et al. (2018). More recently, Kanamura (2020) analyses the relationship between green bonds, energy commodities and conventional bonds in the U.S. market. The statistical outputs of the study demonstrate that green bonds outperform conventional peers. This finding is consistent with the results of Jin et al. (2020) in that green bonds perform well and could be an applicable hedging mechanism in the energy commodity market.

The extant literature review and the aforementioned discussions indicate that a plethora of researches has especially focused on developed countries. Furthermore, the studies on whether green bonds achieve superior performance than conventional ones have resulted in mixed empirical findings. In this chapter, therefore, we attempt to fill the gap in particular for emerging economies by employing a comparative in-depth analysis. Thus, the results of the study contribute to the investors, policymakers and scholars by elucidating the importance of green bonds in funding environmentally friendly projects and also in managing the risk-return profiles in their investment decisions.

8.4 Data and Methodology

This part introduces the data set and the empirical results. We investigate the green bonds performances and compare the yields of green bonds with equivalent conventional bonds in emerging economies including Brazil, Czech Republic, China, Korea, Malaysia and Taiwan by using a matching method. To restate that the main question and the hypothesis are whether green bonds perform superior to their conventional counterparts.

We ground the study model on those of Kreander et al. (2005), Renneboog et al. (2008), Bauer et al. (2005), Zerbib (2017), Reboredo et al. (2017), Hyun et al. (2019) and Nanayakkara and Colombage (2019) in which various versions of the matching method are used to examine and to compare the yields as well as pricing of green and conventional bonds. We perform a matching method to test whether the yields of green bonds differ from the yields of equivalent conventional bonds. This method is also identified as a model-free technique. It is especially preferred to investigate and compare the characteristics of identical financial assets. We attempt to match the sample green bonds with similar conventional bonds.

8.4.1 Sample Set and Introductory Information

Green bonds generally traded over-the-counter market. Therefore, the data for the eligible green bonds are limited. Compiled data related to green bonds is available only after 2015, together with the common market sense in consequence of the announcement of "Green Bond Principles" by the International Capital Market Association.

Among lots of bonds, we choose only currently active bonds, which are being traded or issued in the period from January 1, 2018, to February 28, 2021, with similar conditions such as maturity, bond structure, coupon payment and currency type for each country. The data used in the study obtained from the database of Thomson Reuters Eikon & Datastream.

To determine the sample set, we first select the entire sample of active bonds (symbolised by the letter *N* in Table 8.1). Then, we filter the data set for each country by the bond type, either green bond or conventional bond. At first, we try to determine the green bonds, then match these bonds with equivalent conventional peers. Among the entire sample, we focus on 22 and 48 remaining green bonds and their conventional matching pairs, respectively. Table 8.1 shows the sample set construction and introductory information about green and conventional bonds.

The initial sample composes of 73 and 307 green and conventional bonds, respectively. Within the scope of the empirical analysis, we select those whose data can be accessed. The sample-set for each country in Table 8.1 also presents the issued amounts, currency types, bond types, coupon types, coupon rates, issuer type and maturity.

The amount issued column shows the minimum and maximum values of multiple samples. For example, the issued amounts of 100,000 and 600,000 are the minimum and maximum amounts of the four green bonds selected in Brazil. The general inference about the issued amount is that conventional bonds are issued at a higher amount than green bonds. The currency column indicates that it issues all set of sample bonds in local currencies of the countries. In our sample, bond types are either index-linked (IXL) or straight (STR). The index-linked bond type means that the cash flow mostly adjusted according to a price index, whereas the straight type intends fixed cash flows. Besides these, bond types could include a bond with a warrant, convertible, zero-coupon, and floating rate. And, the coupon types comprises several methods of coupon payments such as fixed, floating, inflation-linked, complex coupon, resettable coupon, step-up or step-down coupon. Our sample consists of the inflation-linked coupon (INLX) and fixed coupon (FIX) payments. Likewise, the coupon rate column presents the minimum and maximum rates of the selected samples. Lastly, issuer types include various sort of bonds such as sovereign, sub-sovereign, agency, supranational, financial institution, and corporate.

Table 8.1 Sample set construction and introductory information

Country	Bond	N/(n)	Amount issued	Currency	Bond type	Coupon type	Coupon rate	Issuer type	Maturity (year)
Brazil	Green	4 (4)	100,000–681,000	C	IXL	INLX	4.28%–6.78%	CORP	10–15
	Conventional	64 (1)	900,000	C	IXL	INLX	3.54%	CORP	10
Czech Republic	Green	1 (1)	530,000	CK	STR	FIX	3.00%	FIN	3
	Conventional	13 (2)	2,550,000–5000,000	CK	STR	FIX	1.25%–2.00%	FIN	5
China	Green	21 (11)	200,000–1,500,000	CH	STR	FIX	3.86%–7.8%	CORP FIN	1–7
	Conventional	78 (9)	500,000–20,000,000	CH	STR	FIX	1.99%–7.3%	CORP FIN	2–5
Korea	Green	42/(4)	60,000,000–90,000,000	KW	STR	FIX	1.87%–2.38%	CORP	3–5
	Conventional	98 (14)	10,000,000–250,000,000	KW	STR	FIX	1.00%–2.47%	CORP	2–7
Malaysia	Green	4 (1)	8000	M$	STR	FIX	8.00%	FIN	1
	Conventional	16 (4)	5000–100,000	M$	STR	FIX	3.05%–7.00%	FIN	1
Taiwan	Green	1 (1)	3,300,000	TW	STR	FIX	1.1%	CORP	5
	Conventional	38 (18)	260,000–8000,000	TW	STR	FIX	0.88%–1.55%	CORP	5

Currency: *C* Brazil Cruzeiro; *CK* Czech Republic Koruna; *CH* China Yuen Renminbi; *KW* South Korea Won; *M$* Malaysia Ringgit; *TW* New Taiwan $; Bond Type: *IXL* Index Linked; *STR* Straight; Coupon Type: *INLX* Inflation Linked Coupon; *FIX* Fixed-Coupon; Issuer Type: *CORP* Corporate; *FIN* Financial Institution (Banks and Insurers)

8.4.2 Empirical Results

In the analysis process, we empirically examine whether the yields of green bonds are statistically different from the yields of equivalent conventional bonds. After matching the green and conventional bonds through their distinctive properties including currency, bond type, coupon type, issuer type, and maturity for each country, we apply post hoc tests to explore differences between the yield means.

In contrast to bid-ask yields, current yield, liquidity proxies or spreads preferred in previous studies (see, e.g., Hachenberg & Schiereck, 2018; Hyun et al., 2019), we make use of redemption yields to compare the bonds. From a point of investors' view, redemption yield, which is also known as the yield to maturity, implicitly reflects all characteristics of a bond including cash flow changes, current market price, par value as well as term to maturity. Moreover, redemption yield is a quite useful method for investment decision since it expressed as an annual rate and enable to compare bonds that have discrete features. It also avoids the heterogeneity problem among bonds. The average redemption yield for each group of bonds and the results of test statistics on mean difference are displayed in Table 8.2 below.

Table 8.2 Descriptive statistics and test of average yield difference

							Test of mean difference (P-values)	
Country	N	Bond	Min. (%)	Max. (%)	Mean (%)	Std. dev.	Within each group	Between groups
Brazil	85	Green	2.00	4.19	2.68	0.472	0.000***–1.000	0.000***
		Conventional	4.14	5.51	4.79	0.303	–	
Czech Republic	78	Green	4.50	6.60	5.43	0.459	–	0.000***
		Conventional	3.99	5.45	4.70	0.381	0.000***	
China	362	Green	2.97	7.87	5.34	1.164	0.001***–0.863	0.000***–0.995
		Conventional	1.93	11.34	4.83	2.153	0.000***–1.000	
Korea	457	Green	1.09	1.95	1.57	0.163	0.000***–0.987	0.000***–1.000
		Conventional	0.71	10.39	1.93	1.065	0.000***–0.977	
Malaysia	83	Green	2.73	2.87	2.79	0.037	–	0.000***–0.077*
		Conventional	1.95	2.83	2.56	0.203	0.000***	
Taiwan	977	Green	0.43	0.99	0.72	0.146	–	0.000***–0.996
		Conventional	0.42	1.18	0.79	0.164	0.000***–0.994	

Note: If the number of sample bonds is one, the value within each group is not calculated. Also, if there are fewer than three groups the significance of Anova test results is used instead of post hoc tests

***, **, and * indicate significance at the 1%, 5%, and 10% levels, respectively

Table 8.2 gives the empirical results country by country. The numbers of observations (N) are in daily frequency. Firstly, green and conventional bonds are considered as two different groups and mean differences are tested within each group. For example, the result of the observation with 85 daily data in Brazil shows that the green bonds, which are stated in Table 8.1, have 2%, 4.19% and 2.68% of minimum, maximum, and average yields respectively. When we look at the mean values, it is seen that the yields of green bonds are higher than conventional bonds in China, Czech Republic and Malaysia. The mean difference test results within each group for green bonds in Brazil (0.000***–1.000) show that the yields of some green bonds in the sample are statistically different from others (0.000***) while some of them are almost the same (1.000). Similarly, the yield of conventional bonds shows statistically mixed results among themselves. For instance, the analyses performed on 18 conventional bonds with 977 daily data for Taiwan indicate that the yields of some bonds in the sample are significantly different (0.000***), from each other, while the yields of some bonds are very close to each other (0.994).

After within-group tests, we test the yield differences between green and conventional bonds. When we compare the yields of green bonds with equivalent conventional pairs, the findings are mixed as seen from the p-values (*the mean difference between group statistics*). In Brazil and Czech Republic, the difference between the yields of green and conventional bonds are statistically significant, whereas in other countries, this difference is significant for some pairs and it is insignificant for others. Considering these findings, it can explicitly be stated that the average yields of green bonds in Czech Republic are higher than conventional bonds, while these yields are lower in Brazil. In brief, there is no clear evidence in other countries whether green bonds show higher or lower performance than equivalent conventional ones or not.

8.5 Conclusion

Investors naturally look for high performing assets to invest in. Green bonds have the advantage of offering financial performance as much as conventional bonds but also have positive signalling effects, including the reward of reputation and good credit rating (Hyun et al., 2019; Flammer, 2020). Green bonds may also be preferred because of environmental considerations despite their lower performance (Karpf & Mandel, 2018). Investors prefer to buy green bonds and hold them until maturity rather than trade. This study aims to analyse and compare the yields of green and conventional bonds in emerging economies including Brazil, Czech Republic, China, Korea, Malaysia and Taiwan. To this end, the study first introduces the concept of green finance and green bonds, then it presents the literature review which is discussing the financial performance of green bonds and equivalent conventional bonds in various markets. We examine green and equivalent conventional bonds over a sample period from January 1, 2018, to February 28, 2021, by using a matching method.

The yields of green bonds as a financial performance indicator have a decisive role in investment decisions as well as in the success of green projects. In particular, we attempt to respond to two essential problems: (1) do green bonds offer investors different levels of yields? (2) do green bonds propose higher or lower financial performance than equivalent conventional bonds? The solutions to these problems direct us to analyse the yields of green and conventional bonds. The empirical analysis demonstrates that the results are mixed in line with the findings of Nofsinger and Varma (2014), Becchetti et al. (2015) and Reboredo et al. (2017) studying especially social responsible investments and assets. Besides, the test results of the mean difference that we obtain for the sample of Brazil and the Czech Republic indicate that the yields of green bonds and equivalent conventional bonds are statistically different from each other. In contrast to some studies and bank reports such as HSBC (2016), Preclaw and Bakshi (2015), Shurey (2017) and Zerbib (2017), green bonds show higher yields compared to conventional peers for the sample in China and Malaysia. However, descriptive statistics show that green bonds performed slightly worse than conventional bonds in Korea and Taiwan. Broadly summarizing, as a whole, we could not assert that there is a significant difference between the yields.

In a conclusion, the nascent market of green finance and the lack of unique standards in this market limit the eligibility of data as well as cause global investors to stay away from the market. In this context, it is hoped that besides the literature contribution, this study will help attract investors' interest in green bonds.

References

Banga, J. (2019). The green bond market: A potential source of climate finance for developing countries. *Journal of Sustainable Finance & Investment, 9*(1), 17–32.

Bartels, W., Kurznack, L., Briaut, L., & Krimphoff, J. (2016). *Mainstreaming the green bond market: Pathways toward common standards*, KPMG advisory N.V., Netherlands. Accessed January 28, 2021, from https://assets.kpmg.com/content/dam/kpmg/lu/pdf/mainstreaming-the-green-bond-market-kpmg-wwf-2016.pdf

Bauer, R., Koedijk, K., & Otten, R. (2005). International evidence on ethical mutual fund performance and investment style. *Journal of Banking & Finance, 29*(7), 1751–1767.

Baulkaran, V. (2019). Stock market reaction to green bond issuance. *Journal of Asset Management, 20*(5), 331–340.

Bayar, Y., & Gavriletea, M. D. (2019). Energy efficiency, renewable energy, economic growth: Evidence from emerging market economies. *Quality & Quantity, 53*(4), 2221–2234.

Becchetti, L., Ciciretti, R., Dalò, A., & Herzel, S. (2015). Socially responsible and conventional investment funds: Performance comparison and the global financial crisis. *Applied Economics, 47*(25), 2541–2562.

Berrou, R., Dessertine, P., & Migliorelli, M. (2019). An overview of green finance. In M. Migliorelli & P. Dessertine (Eds.), *The rise of green finance in Europe* (pp. 3–29). Palgrave Macmillan.

Breitenfellner, A., Hasenhüttl, S., Lehmann, G., & Tschulik, A. (2020). Green finance–opportunities for the Austrian financial sector. *Stability and Security, 40*, 45–61.

Caldecott, B. (2017). Introduction to special issue: Stranded assets and the environment. *Journal of Sustainable Finance & Investment, 7*(1), 1–13.

Clapp, C. (2014). Climate finance: Capitalising on green investment trends. *The Way Forward in International Climate Policy*, 44–48.

Dikaui, S., & Volz, U. (2018). Central banking, climate change and green finance. *ADBI Working Paper No. 867*. Tokyo: Asian Development Bank Institute. Accessed from https://www.adb.org/publications/central-banking-climate-change-and-greenfinance

DuPont, C., Levitt, J., & Bilmes, L. (2015). Green bonds and land conservation: The evolution of a new financing tool. *HKS Working Paper*, No. 072.

European Investment Bank. (2021). *Climate awareness bonds*. Accessed February 7, 2021, from https://www.eib.org/en/investor_relations/cab/index.htm

Flammer, C. (2020). Corporate green bonds. *Journal of Financial Economics*, Forthcoming. Accessed from https://ssrn.com/abstract=3125518

Glomsrød, S., & Wei, T. (2018). Business as unusual: The implications of fossil divestment and green bonds for financial flows, economic growth and energy market. *Energy for Sustainable Development, 44*, 1–10.

Hachenberg, B., & Schiereck, D. (2018). Are green bonds priced differently from conventional bonds? *Journal of Asset Management, 19*(6), 371–383.

Horsch, A., & Richter, S. (2017). Climate change driving financial innovation: The case of green bonds. *The Journal of Structured Finance, 23*(1), 79–90.

HSBC. (2016). *Green Bond report*. Accessed February 1, 2021, from https://www.hsbc.com/-/files/hsbc/investors/fixedincomesecurities/green-bond-reports/pdfs/161006-green-bonds-report.pdf

Hyun, S., Park, D., & Tian, S. (2019). Differences between green bonds versus conventional bonds: An empirical exploration. In J. Sachs, W. T. Woo, N. Yoshino, & F. Taghizadeh-Hesary (Eds.), *Handbook of green finance: Energy security and sustainable development* (pp. 127–154). Springer Nature.

Ielasi, F., Rossolini, M., & Limberti, S. (2018). Sustainability-themed mutual funds: An empirical examination of risk and performance. *Journal of Risk Finance, 19*(3), 247–261.

Institute of International Finance. (2021). *Sustainable debt monitor: Up, up and away*. Institute of International Finance Reports. Accessed from https://www.iif.com/Publications

Jin, J., Han, L., Wu, L., & Zeng, H. (2020). The hedging effect of green bonds on carbon market risk. *International Review of Financial Analysis, 71*, 101509.

Kanamura, T. (2020). Are green bonds environmentally friendly and good performing assets? *Energy Economics, 88*, 104767.

Karpf, A., & Mandel, A. (2018). The changing value of the 'green' label on the US municipal bond market. *Nature Climate Change, 8*(2), 161–165.

Kreander, N., Gray, R. H., Power, D. M., & Sinclair, C. D. (2005). Evaluating the performance of ethical and non-ethical funds: A matched pair analysis. *Journal of Business Finance & Accounting, 32*(7-8), 1465–1493.

Meltzer, J. P. (2016). Financing low carbon, climate resilient infrastructure: The role of climate finance and green financial systems. *Global Economy and Development, Working Paper, 96*. Accessed from https://ssrn.com/abstract=2841918

Nanayakkara, M., & Colombage, S. (2019). Do investors in green bond market pay a premium? Global evidence. *Applied Economics, 51*(40), 4425–4437.

Nofsinger, J., & Varma, A. (2014). Socially responsible funds and market crises. *Journal of Banking & Finance, 48*, 180–193.

Noh, H. J. (2019). Financial strategy to accelerate green growth. In J. Sachs, W. T. Woo, N. Yoshino, & F. Taghizadeh-Hesary (Eds.), *Handbook of green finance: Energy security and sustainable development* (pp. 37–61). Springer Nature.

Östlund, E. (2015). *Are investors rational profit maximizers or do they exhibit a green preference. Evidence from the green bond market* (Master's Thesis in Economics, No. 21875), Stockholm School of Economics, Stockholm.

People's Bank of China (PBC) and United Nations Environment Programme (UNEP). (2015). *Establishing China's green financial system: Theoretical framework of green finance report*. People's Bank of China Research Bureau and UNEP.

Petrova, A. (2016). *Green bonds: Lower returns or higher responsibility?* Radboud University Master Thesis (S4461746), Nijmegen School of Management, Nijmegen.

Preclaw, R., & Bakshi, A. (2015*). The cost of being green. Report*, Barclays Credit Research, New York.

Rakić, S., & Mitić, P. (2012). *Green banking–green financial products with special emphasis on retail banking products*. Educons University, Sremska Kamenica.

Reboredo, J. C. (2018). Green bond and financial markets: Co-movement, diversification and price spillover effects. *Energy Economics, 74*, 38–50.

Reboredo, J. C., Quintela, M., & Otero, L. A. (2017). Do investors pay a premium for going green? Evidence from alternative energy mutual funds. *Renewable and Sustainable Energy Reviews, 73*, 512–520.

Renneboog, L., Ter Horst, J., & Zhang, C. (2008). The price of ethics and stakeholder governance: The performance of socially responsible mutual funds. *Journal of Corporate Finance, 14*(3), 302–322.

Schaefer, J. (2011). *Green finance an innovative approach to fostering sustainable economic development and adaptation to climate change report.* The Deutsche Gesellschaft für Internationale Zusammenarbeit GmbH.

Shishlov, I., Morel, R., & Cochran, I. (2016). Beyond transparency: Unlocking the full potential of green bonds. *Institute for Climate Economics*, 1–28.

Shurey, D. (2017). Investors are willing to pay a "green" premium. *Bloomberg Bloomberg New Energy Finance Report*, 1–8.

Tang, D. Y., & Zhang, Y. (2020). Do shareholders benefit from green bonds? *Journal of Corporate Finance, 61*, 101427.

Wang, H. K. H. (2020). *Renewable energy Management in emerging economies: Strategies for growth*. Routledge.

Wang, Y., & Zhi, Q. (2016). The role of green finance in environmental protection: Two aspects of market mechanism and policies. *Energy Procedia, 104*, 311–316.

Weber, O., & Saravade, V. (2019). *Green Bonds: Current development and their future*. Centre for International Governance Innovation Papers No. 210, Waterloo, Canada.

World Bank. (2015). What are Green Bonds? *Working Paper, No. 99662*. Washington, DC. Accessed from http://documents.worldbank.org/curated/en/400251468187810398/What-are-green-bonds

Wulandari, F., Schäfer, D., Stephan, A., & Sun, C. (2018). Liquidity risk and yield spreads of green bonds. *DIW Discussion Papers, No.* 1728, Deutsches Institut für Wirtschaftsforschung.

Yoshino, N., & Taghizadeh-Hesary, F. (2018). Alternatives to private finance: Role of fiscal policy reforms and energy taxation in development of renewable energy projects. In V. Anbumozhi, K. Kalirajan, & F. Kimura (Eds.), *Financing for low-carbon energy transition: Unlocking the potential of private capital*. Springer.

Zerbib, O. D. (2017). The green bond premium. *Working Paper*. Accessed from https://ssrn.com/abstract=2890316 or https://doi.org/10.2139/ssrn.2890316

Chapter 9
Modelling the Volatility Spillovers Among Energy Stock Returns in Developed, Developing and Fragile Economies Using EGARCH Analysis

Merter Akıncı and Gönül Yüce Akıncı

Abstract The main motivation of the paper is to examine the volatility and spillover dynamics of the stock returns of energy companies operating in the oil and gas sectors in stock markets of American, European, Asian and fragile countries using EGARCH analysis for the period from 10.15.2015 to 01.29.2021. The results reveal that the current term conditional variance of the energy stock returns are affected by the past term economic shocks, and the effects of the volatility persistence of the previous terms on the current term conditional variance continue in the long run. Besides, it is found that the negative shocks affect more strongly the return volatilities of energy stocks in comparison with positive shocks, suggesting the existence of leverage impact. In addition, the results also point out the presence of the spillover impacts of the return volatilities from the energy stocks of developed and developing countries to those of fragile countries.

9.1 Introduction

The energy and energy sector is an indispensable resource in the economic and social development and social welfare of countries. It is a fact that energy has an important share in production activities. As well as energy production, energy demand has also been steadily increasing due to the trade opportunities, population growth, industrialization, urbanization and globalization. The rapid increase in the world population and the rise of the middle class around the world lead to an increase in the demand for energy. Increasing energy prices with the increasing energy

M. Akıncı
Ünye Faculty of Economics and Administrative Sciences, Department of Economics, Ordu University, Altınordu/Ordu, Turkey

G. Y. Akıncı (✉)
Ünye Faculty of Economics and Administrative Sciences, Department of Business Administration, Ordu University, Altınordu/Ordu, Turkey

S. Yüksel, H. Dinçer (eds.), *Strategic Approaches to Energy Management*, Contributions to Management Science,
https://doi.org/10.1007/978-3-030-76783-9_9

demand of the world and the tendency of energy resources to be depleted throughout the world make the sector even more important. The macroeconomic effects to be caused by fluctuations in the energy sector and unstable market structures manifest themselves not only in real markets but also in the financial markets.

Oil prices are generally determined externally with the establishment of the Organization of Petroleum Exporting Countries (OPEC) and take its place in the agenda as a factor that starts to make its destructive effects more and more (Burbidge & Harrison, 1984; Akıncı et al., 2013). Besides, even if increases in oil prices are an important source of income for oil-exporting countries, it can put downward pressures on total supply for countries that use energy input as a production factor in their production processes (Darby, 1982; Akıncı et al., 2013). Due to rising prices and fluctuating returns, the financial systems are negatively affected on the one hand, and the macroeconomic stability of the economies of countries, which are in great need of energy, is shaken on the other hand. The unstable market structures triggered by the financial crises make the theoretical relations between macroeconomic indicators to be damaged, and the confidence in the Keynesian system leads to a loss in the late 1970s. The Keynesian understanding of economics has completely failed to explain the stagflation phenomenon caused by the oil crises. This failure in question has been brought about a structural change in all economic systems and the hegemony of the neo-classical school began in the 1980s. Neo-classical economics, which has strengthened its place in the world economic systems with the third wave phenomenon of globalization and the motto of "there is no other alternative", has deregulated both financial and real markets and made countries dependent and connected to each other. As a result, the spillover and contagion effects of volatilities and instabilities in financial and real markets have accelerated.

It is important in terms of how the global developments in the financial markets are perceived by the companies that have investments in energy production. The diversified, balanced and efficient generation portfolio in terms of energy resources are one of the competitive advantages in the energy sector. In this sense, energy input causes technology policies to be directed to areas that will increase energy efficiency, and efforts to find new energy resources are evaluated on the basis of main requirements in order to turn the wheels of economies in a sustainable dimension. These requirements have gained momentum especially with the effect of oil shocks in the 1970s, and countries have focused on increasing energy production and efficiency with the help of new technological developments and tried to reduce input costs. This process is designed to both increase production and efficiency by means of new technologies and to pave the way for accelerating energy savings, leading to a decrease in consumption. However, the main point that should not be overlooked here is to reduce the total energy consumption without any deterioration in living standards.

With the increase in energy sector in the world, the subject of energy investments has started to gain importance. In this context, in order for the energy investment to be evaluated in a sound manner, the investment amount, share in the market, duration, profitability, cost technology and financing should also be taken into account (Çağıl, 2012). Therefore, the main motivation of the paper is to examine the volatility dynamics of the stock returns of energy companies operating in the oil and gas sectors in the developed, developing and fragile country groups. In addition, the volatility spillovers of the energy stock returns from three American, ten European and seven Asian stock markets to stock markets of the nine fragile countries are investigated using EGARCH analysis for the period from 10.15.2015 to 01.29.2021. The sections of the study can be listed as follows: Following the introduction part, the studies in the literature will be summarized in the second section, and in the third section, the methodological information, data set and econometric model will be introduced. Following the fourth section, in which econometric application findings will be presented, the study will be concluded with a fifth and final section, where an overall assessment and policy recommendations will be presented.

9.2 Literature Review

The energy sector and the analysis of volatilities in this sector, which affect both the profit motives of individual investors and macroeconomic factors as a whole, has led some applied researches. One of the most common views that emerged following the oil shocks of the 1970s is that volatilities in energy prices and returns and their spillover effects are much more dominant than those of other commodities (Regnier, 2007). As emphasized by Sadorsky (2001) and Oberndorfer (2009), the unique characteristics of the energy sector and the fact that the inputs it uses and the output it produces are both homogeneous and subject to change in international markets cause price/return volatilities and spillover effects in this sector to come to the fore. In addition, the fact that the capital intensity of the sector is relatively high and the capital is a mobile factor can increase the volatility dominance of the sector.

In the light of these explanations, determining the price and return volatility of various companies operating in the energy sector and calculating the spillover and contagion effects of these volatilities has become a very interesting topic for energy policy and portfolio diversification process. In addition, determining price and return volatility and volatility spillover in the energy sector enables asset pricing processes to be carried out with great precision. The fact that individual investors reorganize their portfolios depending on the fluctuations in the energy sector and, in a macroeconomic context, policy makers redesign the economic system in a way to eliminate volatilities, requires detailed analysis of the sector. Therefore, Table 9.1 presents summary information about some studies in the literature in the context of the subject mentioned.

Table 9.1 Literature review summary

Author	Country	Time span	Method	Main findings
Fleming and Ostdiek (1999)	The USA	1982–1997	Rolling Estimation and GMM	The results show that an abnormal increase occurs in price volatility of crude oil for 3 weeks following the foundation of NYMEX. It is also noted that price volatilities increase in long-term.
Hammoudeh et al. (2004)	The USA	1995:06–2001:10	Johansen Cointegration and VEC Analysis	It is asserted that there is a volatility effect of oil futures market on the oil sector stocks and the impact of a volatility-dampening on the other stocks has been observed.
Regnier (2007)	The USA	1945:01–2005:08	Time Series and Mann-Whitney Rank-Sum Tests	The findings of the analysis point out that the volatility of energy prices are higher than those of products sold by domestic producers.
Chiou and Lee (2009)	The USA	1992:01–2006:11	Threshold Cointegration Analysis and ARJI Model	The results showing the negative effect of oil prices on stock returns indicate that oil price volatilities has an asymmetry effect for oil contracts of spot and future markets.
Wang and Wu (2012)	The USA	1992:07–2011:04	Univariate and Multivariate GARCH Models	The analyses indicate both the existence of the volatilities of energy prices and asymmetry effects. Besides, it is also shown that the linkage between oil and products of other market is determined by volatility spillover.
Arouri et al. (2012)	Europe	1998:01–2009:12	VAR-GARCH	The results of the analysis reflect that there is a statistically significant volatility spillover impact between energy sector stock returns and oil prices. Besides, it is noted that the conditional volatility of the returns of

(continued)

Table 9.1 (continued)

Author	Country	Time span	Method	Main findings
				energy sector is determined by its previous news and volatilities.
Salisu and Fasanya (2013)	The USA	2000:01–2012:03	GARCH Analyses with Structural Breaks	The results show the presence of volatility persistence and asymmetry effects of oil prices. The authors also point out that investors react according to news because of existence the leverage effects.
Awartani and Maghyereh (2013)	Gulf Cooperation Council Countries	2004–2012	KPSS Variance Decomposition and Simple and Dynamic Conditional Correlation	The results of the analysis indicate that volatility transmission of oil prices and returns are bi-directional. Besides, it is noted that the return and price volatility effects from oil markets to other stock markets are higher than *vice versa*.
Zhang and Wang (2014)	China and World Oil Markets	2001:12–2013:12	Rolling Estimation	A bi-directional and asymmetrical return and volatility spillover effects between China and other oil markets have been found. Besides, it is mentioned that the energy sector effects of China on world energy markets have been increased.
Guesmi and Fattoum (2014)	Five Oil-Importing and Four Oil Exporting Countries	2000:03–2010:03	Multivariate GJR-DCC-GARCH Analysis	The findings of the analysis indicate that market co-movements among countries increase the coefficients of conditional correlation positively and global business cycle volatilities give rise to oil shocks. Besides, it is pointed out that there is a positive correlation between oil prices and stock markets.
Lin et al. (2014)	Ghana, West Africa, Nigeria	2000:01–2010:12	VAR-GARCH, AGARCH and DCC-GARCH	The presence of volatility spillover effects has been found in analysis and

(continued)

Table 9.1 (continued)

Author	Country	Time span	Method	Main findings
				interdependence nexus between the returns of stock markets and of oil has been detected. Lastly, it is mentioned the presence of asymmetric effects.
Jouini and Harrathi (2014)	Gulf Cooperation Council Countries	2005:06–2011:03	BEKK-GARCH	The results of the analysis revealing the shock and volatility relationships in oil and stock markets reveal that volatility spillover impacts occur in GCC. Besides, the exhibition of asymmetry linkage is mentioned in conditional variance.
Bouri (2015)	Lebanon	1998:01–2014:05	VAR-GARCH	A weak but positive one-way directional causal linkage volatility and return pass-through effect from oil prices to stock market has been found in the analysis.
Lin and Li (2015)	The USA, Europe and Japan	1992:01–2012:12	VEC-MGARCH	The findings support the existence of price spillover effect from oil to gas, but not *vice versa*. Besides, it is also noted that oil price volatility has a spillover effect to gas and *vice versa* in US and Europe. In Japan, volatility has been found to be independent in oil and gas markets.
Ergen and Rizvanoghlu (2016)	The USA	2001:01–2013:11	Augmented GARCH Model	The results of the analysis indicate that natural gas futures volatility is higher on Mondays and during winters. It is also pointed out that highness of the volatility is said to be associate with storage level divergence and seasonal temperatures. Besides, weather shocks on volatility is found.

(continued)

Table 9.1 (continued)

Author	Country	Time span	Method	Main findings
Gen et al. (2016)	North American and European Countries	1998:01–2016:05	Ensemble Decomposition and VAR	It is noted that the shock effect of oil prices gets weaker in North America while it gets stronger in Europe. In addition, short- and mid-term volatilities caused by oil price shocks has been observed in North America and Europe. Although a positive change of oil price shocks in trend of gas market has been found in North America, a negative change of oil price shocks in trend of gas market has been detected in Europe.
Tsuji (2018)	The USA, Canada, Australia and Russia	2000:01–2017:08	MEGARCH	A one-way return pass-through between oil equities and oil futures and a two-way volatility spillovers between oil equities have been observed from the results of the analysis. Besides, downside risk has been found important due to their association with leverage impact.

9.3 Dataset, Model and Econometric Methodology

The main aim of this work is to examine the volatility dynamics of the stock returns of energy companies operating in the oil and gas sectors in the developed, developing and fragile country groups. In addition, the volatility spillovers of the energy stock returns from three American, ten European and seven Asian stock markets to stock markets of the nine fragile countries are investigated using EGARCH analysis for the period of 10.15.2015–01.29.2021. Average return values of all energy companies operating in the oil and gas sectors in the relevant stock markets are calculated over the daily returns and these series are used in the analysis. In order to eliminate the excesses in the calculated average series, Extreme Value Analysis are applied and the series is provided to have a general mean value. To estimate the daily returns of the energy stocks, the daily closing prices of the energy stocks are used.

The data set are available at https://tr.investing.com/stock-screener/?sp=country::63|sector::9|industry::14|equityType::a. The daily return series of the energy stocks can be calculated as follows:

$$R_t = \ln(P_t) - \ln(P_{t-1}) \tag{9.1}$$

where R_t shows the return of the energy stocks on day t, $\ln(P_t)$ and $\ln(P_{t-1})$ present the natural logarithm of the closing price of the energy stocks on day t and t^{-1}, respectively.

In the GARCH models volatility is assumed to be symmetrical in reply to positive and negative shocks and that's why GARCH analysis are disqualified for detecting the asymmetric nexus in variance system. However, real world experiences show that volatilities that emerge in response to positive and negative economic shocks may be asymmetrical (Özden, 2008, p. 344; Songül, 2010, p. 18; Yüce Akıncı & Akıncı, 2020, p. 307). Accordingly, EGARCH analysis developed by Nelson (1991) can be used to estimate the asymmetric impacts of the shocks and the leverage effects can be modeled econometrically in the context of energy stocks. In order to investigate the asymmetric or leverage impacts, EGARCH models are generally applied. In addition, EGARCH analysis are widely used if asymmetric impacts of good and bad news are tried to estimate. Consequently, it may be noted that EGARCH models in which the asymmetry impacts in the volatility structure are considered is an econometric analysis where the conditional variance is established based on both magnitudes and signs of lagged error terms.

EGARCH analysis developed by Nelson (1991) may be written as follow:

$$\ln\left(\sigma_{Y,t}^2\right) = \omega + \sum_{k=1}^{p} \beta_1 \ln\left(\sigma_{t-1}^2\right) + \sum_{i=1}^{q} \alpha_1 \left|\frac{\varepsilon_{t-1}}{\sigma_{t-1}}\right| + \sum_{s=1}^{r} \gamma_1 \frac{\varepsilon_{t-1}}{\sigma_{t-1}} + \psi_1 \ln\left(U_{X,t}\right) \tag{9.2}$$

Because EGARCH model considers the logarithm of variances and positive or negative shocks are exponential, the conditional variance will always take the positive values. Besides, the equation numbered (9.2) also expresses of no restrictions on the parameters of ω, α, β and γ. In addition, σ_t^2 presents the following period estimated variance and it is known as the conditional variance. α implies the impacts of the past period shocks on the current period conditional variance. β which indicates the volatility resistance explains the persistence of past period shocks on the current period conditional variance. Leverage impact showing the effect of good or bad news on future volatility is represented by γ parameter. If the coefficient of γ parameter is zero ($\gamma_s = 0$), existence of a symmetrical nexus among variables is noted. As it can be understood, an asymmetric nexus is said if the coefficient of γ is different from zero ($\gamma_s \neq 0$). If γ takes positive values, the impact of shocks on conditional variance is anticipated to be $\alpha + \gamma$ and if γ takes negative values, in other words the leverage effect exists, the impact of shocks on conditional variance is anticipated to be $-\alpha + \gamma$ (Enders, 2015; Korap, 2010; Yüce Akıncı & Akıncı, 2020). The parameter of $U_{X,\,t}$ presents the squares of error terms obtained from EGARCH analysis for the energy stock of X. The volatility spillover impact is estimated by considering the statistical significance of the coefficient of ψ parameter. Volatility spillover impact from one energy stock (for instance X) to another (for instance Y) is said if the coefficient of ψ is statistically significant. Lastly, ε presents the white-noise error term.

9.4 The Results of the Econometric Analysis

To examine the volatility movements of the energy stock returns by applying EGARCH model, stationary level of the returns has been estimated. To this end, Table 9.2 points out the findings of ADF unit root test of the energy stock return series. The results show that the energy stock returns are stationary at level and but their significance levels differ.

Following ADF unit root test, ARIMA model and the conditional mean equation can be estimated. To this end, the optimum energy stock models are estimated, and the findings of the conditional mean and variance equations are pointed out in Table 9.3. The results reflecting no autocorrelation show the presence of the ARCH effect in the residuals.

ARCH and GARCH models are insufficient in estimating asymmetry impacts in variance. Therefore, it is required to use EGARCH analysis developed by Nelson (1991) to estimate the asymmetry impacts of the shocks on volatility. Table 9.4 show the analysis findings of ARMA and EGARCH models for the energy stock return. Besides, Table 9.5 points out the estimation results of ARMA and EGARCH models and it also reveals the spillover impacts from the American, European and Asian energy stocks to the energy stocks of stock markets of the nine fragile countries.

As seen in Tables 9.4 and 9.5, since the coefficients of α showing the shock impacts are generally positive, it can be argued that the return volatilities of the energy stocks are affected by the economic conditions and shocks. It is possible to say that the shock effects especially originating from the American stock markets are much more dominant. Therefore, it may be noted that shocks experienced in the economic system have a volatility impact on the returns of energy stocks. Besides, the coefficients of β reflecting the persistence of volatility process take both negative and positive values. Negative and positive values of the persistence of volatility process point out that the volatilities experienced in the previous terms bring about negative and positive impact on the current term conditional variance, respectively. As it can be seen in Tables 9.4 and 9.5, since the coefficients of β parameters are mostly positive and therefore their net effect are positive, it can be argued that the effects of the volatility persistence of the previous terms on the current term conditional variance continue in the long run. Considering the shock and volatility persistence impacts as a whole, the economic shocks are seen to increase the volatility persistence. The fact that the shock effects of energy stocks are much

Table 9.2 The findings of ADF unit root test

The results of ADF Unit Root test at level			
Variable	Intercept	Trend & Intercept	None
American Stock Markets			
*R*US (S&P 500)	−14.622(0)*** [0.000]	−14.115(0)*** [0.000]	−14.262(0)*** [0.000]
*R*Mexico (S&P/BMV IPC)	−15.221(0)*** [0.000]	−15.228(0)*** [0.000]	−15.843(0)*** [0.000]
*R*Canada (S&P/TSX)	−14.227(0)*** [0.000]	−14.260(0)*** [0.000]	−14.220(0) *** [0.000]
European Stock Markets			
*R*Germany (DAX)	−13.701(0)*** [0.000]	−13.554(0)*** [0.000]	−13.554(0)*** [0.000]
*R*UK (FTSE 100)	−13.551(0)*** [0.000]	−13.582(0)*** [0.000]	−13.523(0)*** [0.000]
*R*France (CAC40)	−14.049(0)*** [0.000]	−14.017(0)*** [0.000]	−14.013(0)*** [0.000]
*R*Spain (IBEX 35)	−14.151(0)*** [0.000]	−14.880(0)*** [0.000]	−14.882(0)*** [0.000]
*R*Italy (FTSE MIB)	−12.770(0)*** [0.000]	−12.703(0)*** [0.000]	−13.252(0)*** [0.000]
*R*Switzerland (SMI)	−14.117(0)*** [0.000]	−14.015(0)*** [0.000]	−14.773(0)*** [0.000]
*R*Portugal (PSI 20)	−12.881(0)*** [0.000]	−12.775(0)*** [0.000]	−12.805(0)*** [0.000]
*R*Belgium (BEL 20)	−15.554(0)*** [0.000]	−15.498(0)*** [0.000]	−15.474(0)*** [0.000]
*R*Austria (ATX Prime)	−12.734(0)*** [0.000]	−12.711(0)*** [0.000]	−12.662(0)*** [0.000]
*R*Sweden (OMX S30)	−15.224(0)*** [0.000]	−15.247(0)*** [0.000]	−15.191(0)*** [0.000]
Asian Stock Markets			
*R*Japan (Nikkei 225)	−12.112(0)*** [0.000]	−12.171(0)*** [0.000]	−12.115(0)*** [0.000]
*R*Australia (S&P/ASX 200)	−15.226(0)*** [0.000]	−15.278(0)*** [0.000]	−15.188(0)*** [0.000]
*R*China (China A50)	−12.484(0)*** [0.000]	−12.455(0)*** [0.000]	−12.414(0)*** [0.000]
*R*Hong Kong (Hang Seng)	−12.474(0)*** [0.000]	−12.487(0)*** [0.000]	−12.602(0)*** [0.000]
*R*South Korea (KOSPI)	−13.364(0)*** [0.000]	−12.386(0)*** [0.000]	−12.333(0)*** [0.000]
*R*Taiwan (Taiwan Weighted)	−14.116(0)*** [0.000]	−14.877(0)*** [0.000]	−14.882(0)*** [0.000]
*R*Sri Lanka (CSE All-Share)	−10.117(0)*** [0.000]	−10.463(0)*** [0.000]	−10.336(0)*** [0.000]

(continued)

Table 9.2 (continued)

The results of ADF Unit Root test at level						
Variable	Intercept		Trend & Intercept		None	
Stock Markets of the Fragile Countries						
*R*Brazil (Bovespa)	−13.226(0)*** [0.000]		−13.252(0)*** [0.000]		−13.254(0)*** [0.000]	
*R*India (BSE Sensex)	−13.353(0)*** [0.000]		−13.378(0)*** [0.000]		−13.402(0)*** [0.000]	
*R*Indonesia (IDX Composite)	−15.776(0)*** [0.000]		−15.885(0)*** [0.000]		−15.776(0)*** [0.000]	
*R*South Africa (FTSE/JSE)	−13.226(0)*** [0.000]		−13.220(0)*** [0.000]		−13.303(0)*** [0.000]	
*R*Argentina (S&P/BYMA)	−11.118(0)*** [0.000]		−11.221(0)*** [0.000]		−11.116(0)*** [0.000]	
*R*Pakistan (Karachi Meezan 30)	−11.262(0)*** [0.000]		−11.284(0)*** [0.000]		−11.669(0)*** [0.000]	
*R*Egypt (EGX 30)	−11.779(0)*** [0.000]		−11.759(0)*** [0.000]		−11.757(0)*** [0.000]	
*R*Qatar (Doha All-Share)	−12.402(0)*** [0.000]		−12.411(0)*** [0.000]		−12.482(0)*** [0.000]	
*R*Turkey (BIST 100)	−14.227(0)*** [0.000]		−14.268(0)*** [0.000]		−14.363(0)*** [0.000]	
Critical values	1%	−3.456	1%	−3.985	1%	−2.580
	5%	−2.877	5%	−3.429	5%	−1.952
	10%	−2.575	10%	−3.142	10%	−1.628

Note: *R* shows the return of the energy stocks. The values in parenthesis point out the optimum lag lengths using the AIC over a maximum of 18 lag lengths. The values in square brackets reflect the probabilities of the coefficient

*** Shows the stationary of the variable at the significance level of 1%

more dominant in American stock markets causes the return volatility persistence to occur at much higher levels in these stock markets. Moreover, the negative coefficients of γ implying the leverage impact show that the shocks experienced in economic systems have an asymmetric impact on the return volatilities of the energy stocks. Because the coefficients of γ are negative, it may be noted that the negative shocks affect more strongly the return volatilities of energy stocks in comparison with positive shocks, suggesting the existence of leverage impact. In general, it can be argued that bad news affects the return volatilities of energy stocks in all major stock markets.

Besides, the spillover impacts of the return volatilities of energy stocks from the American, European and Asian stock markets to the fragile countries' stock markets can be observed in the results of Table 9.5. The findings reflect that return volatilities experienced in the American, European and Asian energy stocks create a spillover impact on the energy stocks of the fragile countries. In this context, it can be said that the energy stock returns of the fragile countries are generally affected by the return volatilities experienced in the American, European and Asian stock markets and the

Table 9.3 The diagnosis findings of the conditional mean equation

Stock market	Optimum ARMA and EGARCH models	The diagnosis tests		AIC value of ARMA
American Stock Markets				
*R*US	ARMA(1,1)/EGARCH (2,2)	BG(5): 3.553 (0.488)	ARCH(5): 42.116*** (0.000)	−4.115
		BG(10): 12.248 (0.261)	ARCH(10): 56.442*** (0.000)	
*R*Mexico	ARMA(1,1)/EGARCH (1,2)	BG(5): 3.110 (0.426)	ARCH(5): 27.573*** (0.002)	−3.529
		BG(10): 10.085 (0.774)	ARCH(10): 33.144*** (0.000)	
*R*Canada	ARMA(1,1)/EGARCH (1,2)	BG(5): 2.101 (0.795)	ARCH(5): 32.220*** (0.000)	−4.111
		BG(10): 5.175 (0.553)	ARCH(10): 41.663*** (0.000)	
European Stock Markets				
*R*Germany	ARMA(1,1)/EGARCH (2,1)	BG(5): 0.542 (0.774)	ARCH(5): 34.771*** (0.000)	−3.991
		BG(10): 7.337 (0.491)	ARCH(10): 40.053*** (0.000)	
*R*UK	ARMA(1,1)/EGARCH (1,2)	BG(5): 2.242 (0.713)	ARCH(5): 26.825*** (0.000)	−3.078
		BG(10): 13.414 (0.394)	ARCH(10): 34.443*** (0.000)	
*R*France	ARMA(1,1)/EGARCH (1,1)	BG(5): 2.339 (0.875)	ARCH(5): 37.656*** (0.000)	−4.127
		BG(10): 10.117 (0.672)	ARCH(10): 50.117*** (0.000)	
*R*Spain	ARMA(1,1)/EGARCH (2,1)	BG(5): 1.792 (0.823)	ARCH(5): 21.337*** (0.000)	−3.233
		BG(10): 10.623 (0.449)	ARCH(10): 34.129*** (0.000)	
*R*Italy	ARMA(1,1)/EGARCH (1,1)	BG(5): 3.343 (0.714)	ARCH(5): 37.883*** (0.000)	−4.110
		BG(10): 10.858 (0.465)	ARCH(10): 51.297*** (0.000)	
*R*Switzerland	ARMA(1,1)/EGARCH (2,1)	BG(5): 3.556 (0.652)	ARCH(5): 21.776** (0.067)	−3.282
		BG(10): 10.499 (0.411)	ARCH(10): 30.337* (0.006)	

(continued)

Table 9.3 (continued)

Stock market	Optimum ARMA and EGARCH models	The diagnosis tests		AIC value of ARMA
*R*Portugal	ARMA(1,1)/EGARCH (1,3)	BG(5): 3.017 (0.647)	ARCH(5): 30.413*** (0.000)	−4.775
		BG(10): 11.112 (0.395)	ARCH(10): 44.719*** (0.000)	
*R*Belgium	ARMA(1,1)/EGARCH (1,2)	BG(5): 1.337 (0.878)	ARCH(5): 22.330*** (0.005)	−3.735
		BG(10): 8.257 (0.577)	ARCH(10): 34.772*** (0.000)	
*R*Austria	ARMA(1,1)/EGARCH (1,1)	BG(5): 2.774 (0.811)	ARCH(5): 33.552*** (0.000)	−5.884
		BG(10): 12.270 (0.383)	ARCH(10): 52.774*** (0.000)	
*R*Sweden	ARMA(1,2)/EGARCH (1,2)	BG(5): 3.253 (0.683)	ARCH(5): 42.873*** (0.000)	−5.699
		BG(10): 9.749 (0.523)	ARCH(10): 56.681*** (0.000)	
Asian Stock Markets				
*R*Japan	ARMA(2,1)/EGARCH (1,2)	BG(5): 3.775 (0.882)	ARCH(5): 18.887** (0.027)	−3.114
		BG(10): 8.585 (0.785)	ARCH(10): 21.539** (0.018)	
*R*Australia	ARMA(2,2)/EGARCH (1,3)	BG(5): 8.824 (0.776)	ARCH(5): 20.774** (0.017)	−3.667
		BG(10): 15.671 (0.180)	ARCH(10): 32.882*** (0.001)	
*R*China	ARMA(3,1)/EGARCH (2,2)	BG(5): 4.550 (0.871)	ARCH(5): 33.779*** (0.000)	−4.899
		BG(10): 9.273 (0.526)	ARCH(10): 42.113*** (0.000)	
*R*Hong Kong	ARMA(1,2)/EGARCH (1,2)	BG(5): 6.990 (0.743)	ARCH(5): 31.556*** (0.000)	−4.990
		BG(10): 14.117 (0.350)	ARCH(10): 44.669*** (0.000)	
*R*South Korea	ARMA(1,1)/EGARCH (2,1)	BG(5): 0.990 (0.952)	ARCH(5): 33.222*** (0.000)	−4.562
		BG(10): 7.550 (0.880)	ARCH(10): 47.443*** (0.000)	
*R*Taiwan	ARMA(1,3)/EGARCH (1,3)	BG(5): 5.775 (0.798)	ARCH(5): 20.322** (0.019)	−3.544
		BG(10): 9.882 (0.683)	ARCH(10): 31.776*** (0.000)	

(continued)

Table 9.3 (continued)

Stock market	Optimum ARMA and EGARCH models	The diagnosis tests		AIC value of ARMA
*R*Sri Lanka	ARMA(1,3)/EGARCH (3,1)	BG(5): 8.116 (0.787)	ARCH(5): 33.274*** (000)	−4.556
		BG(10): 13.672 (0.424)	ARCH(10): 47.253*** (0.000)	
Stock Markets of the Fragile Countries				
*R*Brazil	ARMA(1,2)/EGARCH (1,2)	BG(5): 2.759 (0.852)	ARCH(5): 25.887*** (0.000)	−3.880
		BG(10): 11.556 (0.511)	ARCH(10): 37.710*** (0.000)	
*R*India	ARMA(2,1)/EGARCH (1,3)	BG(5): 10.884 (0.614)	ARCH(5): 31.657*** (0.000)	−4.993
		BG(10): 16.098 (0.471)	ARCH(10): 46.503*** (0.000)	
*R*Indonesia	ARMA(2,3)/EGARCH (1,1)	BG(5): 8.258 (0.727)	ARCH(5): 22.334** (0.011)	−3.662
		BG(10): 15.994 (0.521)	ARCH(10): 34.111*** (0.000)	
*R*South Africa	ARMA(1,1)/EGARCH (1,3)	BG(5): 8.247 (0.721)	ARCH(5): 25.330*** (0.001)	−3.877
		BG(10): 14.664 (0.686)	ARCH(10): 36.713*** (0.000)	
*R*Argentina	ARMA(2,2)/EGARCH (2,1)	BG(5): 1.871 (0.912)	ARCH(5): 18.612* (0.078)	−3.117
		BG(10): 9.334 (0.686)	ARCH(10): 29.240*** (0.000)	
*R*Pakistan	ARMA(1,2)/EGARCH (1,2)	BG(5): 3.606 (0.744)	ARCH(5): 31.881*** (0.000)	−4.886
		BG(10): 10.555 (0.577)	ARCH(10): 44.764*** (0.000)	
*R*Egypt	ARCH(3,1)/EGARCH (1,3)	BG(5): 2.799 (0.883)	ARCH(5): 14.990* (0.075)	−3.417
		BG(10): 8.114 (0.736)	ARCH(10): 26.582*** (0.007)	
*R*Qatar	ARMA(2,2)/EGARCH (1,1)	BG(5): 4.383 (0.699)	ARCH(5): 33.112*** (0.000)	−5.127
		BG(10): 13.171 (0.488)	ARCH(10): 48.233*** (0.000)	

(continued)

Table 9.3 (continued)

Stock market	Optimum ARMA and EGARCH models	The diagnosis tests		AIC value of ARMA
*R*Turkey	ARMA(1,1)/EGARCH (1,2)	BG(5): 2.187 (0.896)	ARCH(5): 16.225* (0.064)	−3.011
		BG(10): 10.733 (0.512)	ARCH(10): 22.336** (0.018)	

Note: *R* shows the return of the energy stocks. BG is the Breusch-Godfrey autocorrelation test and ARCH is the autoregressive conditional heteroscedasticity test
***, ** and * imply the significance at 1%, 5% and 10%, respectively

return instabilities of energy stocks are transferred among them. Since the coefficients of ψ are positive, the existence of spillover effects can be noted. In addition, the dominance of the spillover impacts originating from the American stock markets can be noted. It is impossible to deny the role of neoliberal economic policies, which have made world markets interdependent and interdependent due to financial integration and financial deregulation policies since 1980s, in the spread of shock effects and the return volatilities persistence that emerges due to economic shocks. Lastly, the findings of the analysis show no ARCH impacts in the models.

9.5 Conclusion and Policy Implications

The main aim of this work is to examine the volatility and spillover dynamics of the stock returns of energy companies operating in the oil and gas sectors in the developed, developing and fragile country groups using EGARCH analysis for the period from 10.15.2015 to 01.29.2021. The results reveal that the current term conditional variance of the energy stock returns are impacted by the past term economic shocks, and the effects of the volatility persistence of the previous terms on the current term conditional variance continue in the long run. Besides, the existence of leverage impact is found. In addition, the results also point out the presence of the spillover impacts of the return volatilities from the energy stocks of developed and developing countries to the energy stocks of fragile countries.

The energy sector, which is the backbone of economic systems, is one of the factors that create economic and financial instability due to its unique organizational form. Organized mostly as oligopoly and cartel, this sector is in a dominant position in determining both national and international energy prices. Especially in the oil and natural gas sector, oligopolistic companies cause high and unpredictable volatilities in energy prices and returns. Price and return volatilities in energy spread all over the world through financial markets and unstable market conditions affect economic systems. Therefore, energy volatilities have widespread effects on real markets as well as financial markets and affect macroeconomic stability. To eliminate unstable and speculative market conditions financial markets need to be regulated and their

Table 9.4 The findings of EGARCH analysis

American Stock Markets				European Stock Markets						
Mean Equation	The US	Mexico	Canada	Germany	The UK	France	Spain	Italy	Switzerland	Portugal
Constant	0.013*** (0.000)	0.007 (0.224)	0.012*** (0.001)	0.024*** (0.000)	0.006 (0.274)	0.003 (0.352)	−0.004 (0.662)	0.009** (0.016)	0.002 (0.349)	0.003 (0.331)
AR(1)	0.776*** (0.001)	0.282 (0.543)	0.618** (0.022)	−0.714 (0.109)	−0.662** (0.021)	−0.545 (0.461)	0.553** (0.031)	−0.775*** (0.000)	−0.721*** (0.000)	0.515*** (0.004)
MA(1)	−0.884*** (0.000)	0.385 (0.336)	−0.775*** (0.000)	0.647 (0.146)	0.774*** (0.003)	0.462 (0.344)	−0.611** (0.025)	0.942*** (0.000)	0.916*** (0.000)	−0.667*** (0.000)
Variance Equation										
Constant (ω)	1.312*** (0.000)	0.887*** (0.000)	1.110*** (0.000)	1.112*** (0.000)	−0.556** (0.023)	0.515*** (0.000)	0.554*** (0.000)	−0.652*** (0.000)	0.876*** (0.000)	0.414* (0.061)
α_1	0.889*** (0.000)	0.534*** (0.000)	0.723*** (0.002)	0.316*** (0.000)	0.255*** (0.004)	0.264** (0.035)	0.561*** (0.000)	0.371*** (0.007)	0.352*** (0.005)	0.421*** (0.000)
α_2	0.256* (0.067)			0.133* (0.079)			−0.213* (0.071)		−0.126* (0.071)	
α_3										
γ	−0.873*** (0.000)	−0.515*** (0.000)	−0.982*** (0.000)	−0.277*** (0.000)	−0.289*** (0.000)	−0.377*** (0.000)	−0.882*** (0.000)	−0.613*** (0.000)	−0.114*** (0.000)	−0.565*** (0.000)
β_1	1.776*** (0.000)	0.774*** (0.000)	0.848** (0.028)	0.714*** (0.000)	0.582*** (0.000)	0.250*** (0.001)	0.705*** (0.000)	0.474*** (0.004)	0.223*** (0.000)	0.764*** (0.000)
β_2	1.224** (0.024)	−0.117 (0.884)	0.534* (0.072)		−0.242** (0.026)					−0.117* (0.083)
β_3										−0.220 (0.359)

Diagnosis Tests	ARCH(5): 5.558 (0.385) ARCH(10): 9.312 (0.613)	ARCH(5): 4.112 (0.501) ARCH(10): 8.323 (0.707)	ARCH(5): 2.144 (0.910) ARCH(10): 6.774 (0.988)	ARCH(5): 5.252 (0.323) ARCH(10): 7.227 (0.478)	ARCH(5): 1.994 (0.744) ARCH(10): 4.800 (0.823)	ARCH(5): 1.885 (0.667) ARCH(10): 8.887 (0.812)	ARCH(5): 3.700 (0.484) ARCH(10): 7.996 (0.825)	ARCH(5): 1.557 (0.656) ARCH(10): 6.852 (0.802)	ARCH(5): 4.220 (0.323) ARCH(10): 9.114 (0.735)	ARCH(5): 2.473 (0.724) ARCH (10): 6.997 (0.883)
European Stock Markets				Asian Stock Markets						
Mean Equation	Belgium	Austria	Sweden	Japan	Australia	China	Hong Kong	S. Korea	Taiwan	Sri Lanka
Constant	0.001 (0.218)	0.010*** (0.004)	0.008** (0.011)	0.002 (0.348)	0.003 (0.340)	0.001 (0.664)	0.006* (0.053)	0.003 (0.341)	0.001 (0.246)	0.001 (0.257)
AR(1)	0.906*** (0.000)	0.656*** (0.000)	0.584*** (0.000)	0.626*** (0.000)	−0.445*** (0.000)	0.775*** (0.000)	−0.353 (0.753)	0.517*** (0.000)	0363*** (0.000)	−0.377** (0.028)
AR(2)				−0.171 (0.722)	0.224* (0.087)	0.242* (0.069)				
AR(3)						−0.156 (0.854)				
MA(1)	−1.052*** (0.000)	−0.762*** (0.000)	0.458*** (0.000)	0.447*** (0.003)	0.383*** (0.000)	0.556*** (0.000)	0.331* (0.060)	−0.464*** (0.000)	0.461*** (0.000)	−0.532*** (0.000)
MA(2)			−0.204** (0.037)		−0.235* (0.079)		−0.175 (0.241)		−0.244** (0.032)	0.684*** (0.000)
MA(3)									0.163 (0.512)	0.330 (0.156)
Variance Equation										
Constant (ω)	−0.776** (0.028)	−0.770*** (0.000)	−0.669*** (0.000)	0.239** (0.031)	0.722*** (0.000)	−0.559*** (0.000)	−0.231 (0.205)	−0.373* (0.058)	−0.550*** (0.000)	−0.224** (0.019)
α_1	0.318*** (0.000)	0.189** (0.027)	0.224*** (0.009)	0.375*** (0.000)	0.248** (0.036)	0.469*** (0.000)	0.349*** (0.009)	0.372*** (0.000)	0.449*** (0.000)	0.279** (0.021)

(continued)

Table 9.4 (continued)

European Stock Markets				Asian Stock Markets						
Mean Equation	Belgium	Austria	Sweden	Japan	Australia	China	Hong Kong	S. Korea	Taiwan	Sri Lanka
α_2						0.513*** (0.000)		−0.214* (0.062)		−0.121 (0.484)
α_3										−0.088 (0.879)
γ	−0.178*** (0.000)	−0.254*** (0.000)	−0.299*** (0.000)	−0.172** (0.013)	−0.187** (0.021)	−0.353*** (0.005)	−0.171** (0.023)	−0.236*** (0.000)	−0.252*** (0.000)	−0.127* (0.072)
β_1	0.664*** (0.000)	0.353*** (0.000)	0.587*** (0.000)	0.583*** (0.000)	0.653*** (0.000)	−0.457*** (0.000)	0.411*** (0.000)	0.279** (0.040)	0.362*** (0.007)	0.573*** (0.000)
β_2	−0.105* (0.053)		−0.334*** (0.000)	−0.119* (0.068)	−0.117* (0.084)	0.663*** (0.000)	−0.118* (0.065)		−0.113 (0.156)	
β_3					0.138* (0.075)				0.181* (0.071)	
Diagnosis tests	ARCH(5): 1.224 (0.427) ARCH (10): 5.758 (0.731)	ARCH(5): 1.676 (0.522) ARCH (10): 5.247 (0.697)	ARCH(5): 1.889 (0.487) ARCH (10): 5.075 (0.683)	ARCH (5): 5.776 (0.599) ARCH (10): 9.812 (0.893)	ARCH(5): 1.778 (0.502) ARCH (10): 5.612 (0.693)	ARCH(5): 1.663 (0.544) ARCH (10): 6.118 (0.722)	ARCH (5): 2.011 (0.474) ARCH (10): 6.881 (0.826)	ARCH(5): 2.135 (0.462) ARCH (10): 7.110 (0.851)	ARCH(5): 1.554 (0.546) ARCH (10): 5.858 (0.672)	ARCH(5): 1.686 (0.549) ARCH (10): 6.127 (0.733)

Note: ***, ** and * show the significance level at 1%, 5% and 10%, respectively. The values in parenthesis reflect the probabilities. Gaussian error distribution is considered in testing EGARCH analysis. Besides, the methods of BFGS optimization and Marquardt step procedure are taken into account. The analysis are done with a maximum of 500 iterations. The terms α, γ and β imply the shock, leverage and volatility persistence effect, respectively

Table 9.5 The findings of EGARCH analysis and spillover effects

Stock markets of fragile countries									
Mean Equation	Brazil	India	Indonesia	South Africa	Argentina	Pakistan	Egypt	Qatar	Turkey
Constant	0.001(0.226)	0.005*(0.067)	0.001(0.257)	0.001 (0.260)	0.004* (0.072)	0.001(0.232)	0.004* (0.069)	0.001 (0.306)	0.006** (0.039)
AR(1)	−0.720*** (0.000)	0.556*** (0.000)	0.613*** (0.000)	0.741*** (0.000)	0.575*** (0.000)	0.881*** (0.000)	−0.733*** (0.000)	0.718*** (0.000)	0.657** (0.027)
AR(2)		−0.199(0.127)	0.153(0.779)		0.158 (0.873)		0.142 (0.757)	−0.117 (0.774)	
AR(3)							0.101 (0.884)		
MA(1)	−0.667*** (0.000)	0.220*(0.086)	−0.912*** (0.000)	−0.703*** (0.000)	0.554*** (0.000)	−0.711*** (0.000)	0.962*** (0.000)	0.787*** (0.000)	0.353* (0.085)
MA(2)	0.112(0.354)		0.172(0.727)		0.202* (0.426)	0.220(0.417)		−0.226* (0.063)	
MA(3)			0.219(0.663)						
Variance Equation									
Constant (ω)	−0.776(0.157)	−0.117(0.926)	−0.258* (0.036)	−0.313*** (0.000)	−0.558*** (0.000)	−0.114 (0.776)	−0.664*** (0.000)	−0.886*** (0.000)	−0.556*** (0.000)
α_1	0.488***(0.000)	0.284*(0.053)	0.246*(0.059)	0.477*** (0.000)	0.545*** (0.000)	0.332*** (0.000)	0.341*** (0.000)	0.331*** (0.000)	0.659*** (0.000)
α_2					0.227 (0.327)				
α_3									
γ	−0.337*** (0.000)	−0.247* (0.062)	−0.303* (0.052)	−0.326*** (0.000)	−0.707*** (0.000)	−0.288*** (0.000)	−0.277** (0.021)	−0.273** (0.016)	−0.747*** (0.000)
β_1	0.713***(0.000)	0.351**(0.026)	0.161*(0.084)	−0.112* (0.075)	0.612*** (0.000)	0.126* (0.078)	0.401*** (0.000)	0.813*** (0.000)	0.881*** (0.000)

(continued)

Table 9.5 (continued)

Stock markets of fragile countries									
Mean Equation	Brazil	India	Indonesia	South Africa	Argentina	Pakistan	Egypt	Qatar	Turkey
β_2	−0.203*(0.062)	0.155(0.754)		0.167* (0.067)		−0.114 (0.144)	0.103 (0.887)		0.293*** (0.000)
β_3		−0.112* (0.070)		0.336*** (0.000)			−0.224** (0.029)		
Diagnosis Tests	ARCH(5): 2.882 (0.537) ARCH(10): 7.162 (0.886)	ARCH(5): 1.338 (0.446) ARCH(10): 5.762 (0.628)	ARCH(5): 1.515 (0.488) ARCH(10): 5.912 (0.650)	ARCH(5): 1.334 (0.413) ARCH(10): 5.252 (0.644)	ARCH(5): 2.775 (0.526) ARCH (10): 7.351 (0.840)	ARCH(5): 1.161 (0.396) ARCH(10): 4.995 (0.613)	ARCH(5): 1.998 (0.512) ARCH (10): 6.886 (0.783)	ARCH(5): 1.339 (0.416) ARCH (10): 5.363 (0.659)	ARCH(5): 3.551 (0.586) ARCH (10): 8.904 (0.889)
Spillover Effect									
The US	0.286***(0.000)	0.181**(0.022)	0.117**(0.031)	0.201* (0.056)	0.337*** (0.000)	0.103(0.425)	0.110 (0.420)	0.227 (0.272)	0.305*** (0.000)
Mexico	0.110*(0.067)	0.126*(0.071)	0.105(0.242)	0.155* (0.059)	0.223*** (0.006)	0.048(0.728)	0.079 (0.696)	0.109 (0.313)	0.203* (0.033)
Canada	0.177***(0.000)	0.143**(0.018)	0.121**(0.044)	0.103** (0.026)	0.205*** (0.000)	0.087(0.262)	0.140 (0.285)	0.087 (0.361)	0.259* (0.072)
Germany	0.107***(0.000)	0.092*(0.073)	0.041*(0.077)	0.116*** (0.000)	0.127*** (0.000)	0.056* (0.089)	0.087** (0.041)	0.037** (0.022)	0.218*** (0.000)
The UK	0.118***(0.000)	0.068**(0.044)	0.046*** (0.001)	0.173*** (0.000)	0.133*** (0.000)	0.018* (0.072)	0.033* (0.077)	−0.021 (0.754)	0.181*** (0.000)
France	0.153***(0.000)	0.106*** (0.000)	0.024*** (0.000)	0.153*** (0.000)	0.085** (0.061)	0.016(0.776)	0.113*** (0.000)	0.015*** (0.007)	0.272*** (0.000)
Spain	0.082*(0.062)	0.014*(0.068)	0.030*(0.068)	0.087** (0.018)	0.025 (0.661)	0.042** (0.031)	0.024 (0.557)	0.028 (0.542)	0.119* (0.057)

Italy	0.097*(0.071)	0.077*(0.063)	0.034*(0.072)	0.028* (0.056)	0.011* (0.079)	−0.001 (0.676)	0.011 (0.612)	−0.016 (0.325)	0.059* (0.063)
Switzerland	0.181***(0.004)	0.127*** (0.000)	0.100*** (0.000)	0.122*** (0.000)	0.124*** (0.000)	−0.022 (0.771)	0.121** (0.029)	0.082* (0.061)	0.193*** (0.000)
Portugal	0.026(0.554)	0.028*(0.074)	0.033(0.227)	0.018* (0.042)	0.017* (0.083)	0.011* (0.083)	0.002 (0.683)	0.005 (0.488)	0.044 (0.337)
Belgium	0.135***(0.008)	0.082*** (0.004)	0.048**(0.021)	0.084** (0.023)	0.099** (0.015)	0.026(0.881)	0.019* (0.062)	0.053 (0.193)	0.078** (0.026)
Austria	0.064**(0.017)	0.048**(0.024)	0.054*(0.083)	0.025* (0.062)	0.023 (0.338)	−0.031 (0.713)	−0.028 (0.353)	0.001 (0.816)	0.013 (0.530)
Sweden	0.079**(0.031)	0.081*** (0.003)	0.040*(0.061)	0.077* (0.076)	0.048 (0.312)	−0.004 (0.911)	0.042 (0.311)	0.054 (0.609)	0.017* (0.068)
Japan	0.137***(0.000)	0.106*** (0.000)	0.150*** (0.000)	0.064** (0.017)	0.056* (0.057)	0.027* (0.048)	0.027 (0.443)	−0.033 (0.258)	0.043** (0.026)
Australia	0.093*(0.077)	0.037**(0.038)	0.092**(0.044)	0.099** (0.028)	0.073 (0.660)	−0.013 (0.806)	0.040 (0.279)	0.051 (0.313)	0.037 (0.476)
China	0.106**(0.033)	0.099**(0.013)	0.119*** (0.000)	0.068* (0.079)	0.057* (0.071)	0.095** (0.069)	0.078 (0.527)	0.014 (0.754)	0.049* (0.061)
Hong Kong	0.007(0.883)	0.013*(0.082)	0.037(0.889)	0.037 (0.788)	0.068* (0.083)	−0.003 (0.950)	−0.004 (0.792)	0.013 (0.824)	−0.073 (0.679)
S. Korea	0.119***(0.000)	0.064**(0.028)	0.022**(0.037)	0.042* (0.067)	0.054* (0.063)	0.017* (0.087)	0.011 (0.656)	0.044 (0.775)	0.022* (0.033)
Taiwan	0.001(0.883)	0.005(0.776)	0.017*(0.048)	0.013 (0.901)	0.001 (0.939)	0.003(0.953)	0.007 (0.908)	0.005 (0.849)	0.013 (0.774)
Sri Lanka	0.003(0.905)	0.008(0.931)	0.007*(0.041)	0.007 (0.969)	0.001 (0.961)	0.011(0.913)	−0.012 (0.892)	0.002 (0.926)	0.007 (0.967)

Note: ***, ** and * show the significance level at 1%, 5% and 10%, respectively. The values in parenthesis reflect the probabilities. Gaussian error distribution is considered in testing EGARCH analysis. Besides, the methods of BFGS optimization and Marquardt step procedure are taken into account. The analysis are done with a maximum of 500 iterations. The terms α, γ and β imply the shock, leverage and volatility persistence effect, respectively

effective functioning must be guaranteed. Avoiding oligopolistic structures in the energy sector, which is one of the most important parts of the financial system, minimizing uncertainties in these markets and eliminating price fluctuations can help to make economic stability sustainable. In particular, the full transition to free market conditions following the temporary provision of regulations and restrictions to be provided by the state, which will help the elimination of cartels and oligopolies, may contribute to increased efficiency in energy sector. It should be remembered that the failure of the regulations and institutional structures provided by governments will spread to all financial and economic markets as well as energy sector, and cause these markets to have a speculative structure and make them to be short term.

References

Akıncı, M., Aktürk, E., & Yılmaz, Ö. (2013). Petrol Fiyatları İle Ekonomik Büyüme Arasındaki İlişki: OPEC ve Petrol İthalatçısı Ülkeler İçin Zaman Serisi Analizi. *Atatürk Üniversitesi Sosyal Bilimler Enstitüsü Dergisi, 17*(1), 349–361.

Arouri, M. E. H., Jouini, J., & Nguyen, D. K. (2012). On the impacts of oil price fluctuations on European equity markets: Volatility spillover and hedging effectiveness. *Energy Economics, 34*, 611–617.

Awartani, B., & Maghyereh, A. I. (2013). Dynamic spillovers between oil and stock markets in the Gulf Cooperation Council countries. *Energy Economics, 36*, 28–42.

Bouri, E. (2015). Return and volatility linkages between oil prices and the Lebanese stock market in crisis periods. *Energy, 89*, 365–371.

Burbidge, J., & Harrison, A. (1984). Testing for the effects of oil-price rises using vector autoregressions. *International Economic Review, 25*(2), 459–484.

Çağıl, G. (2012). *Enerji Sektörü ve Finansmanı*. Yalın Yayıncılık.

Chiou, J. S., & Lee, Y. H. (2009). Jump dynamics and volatility: Oil and the stock markets. *Energy, 34*, 788–796.

Darby, M. R. (1982). The price of oil and world inflation and recession. *The American Economic Review, 72*(4), 738–751.

Enders, W. (2015). *Applied econometric time series* (4th ed.). Wiley.

Ergen, I., & Rizvanoghlu, I. (2016). Asymmetric impacts of fundamentals on the natural gas futures volatility: An augmented GARCH approach. *Energy Economics, 56*, 64–74.

Fleming, J., & Ostdiek, B. (1999). The impact of energy derivatives on the crude oil market. *Energy Economics, 21*, 135–167.

Gen, J. B., Ji, Q., & Fan, Y. (2016). How regional natural gas markets have reacted to oil price shocks before and since the Shale Gas revolution: A multi-scale perspective. *Journal of Natural Gas Science and Engineering, 36*, 734–746.

Guesmi, K., & Fattoum, S. (2014). Return and volatility transmission between oil prices and oil-exporting and oil-importing countries. *Economic Modelling, 38*, 305–310.

Hammoudeh, S., Dibooglu, S., & Aleisa, E. (2004). Relationships among U.S. oil prices and oil industry equity indices. *International Review of Economics and Finance, 13*, 427–453.

Jouini, J., & Harrathi, N. (2014). Revisiting the shock and volatility transmissions among GCC stock and oil markets: A further investigation. *Economic Modelling, 38*, 486–494.

Korap, L. (2010). An Econometric Essay for the Asymmetric Volatility Content of the Portfolio Flows: EGARCH Evidence from the Turkish Economy. *Sosyal Bilimler Dergisi, 4*, 103–109.

Lin, B., & Li, J. (2015). The spillover effects across natural gas and oil markets: Based on the VEC-MGARCH framework. *Applied Energy, 155*, 229–241.

Lin, B., Wesseh, P. K., & Appiah, M. O. (2014). Oil price fluctuation, volatility spillover and the Ghanaian Equity Market: Implication for portfolio management and hedging effectiveness. *Energy Economics, 42*, 172–182.

Nelson, D. B. (1991). Conditional heteroskedasticity in asset returns: A new approach. *Econometrica, 59*, 347–370.

Oberndorfer, U. (2009). Energy prices, volatility, and the stock market: Evidence from the Eurozone. *Energy Policy, 37*, 5787–5795.

Özden, Ü. H. (2008). İMKB Bileşik 100 Endeksi Getiri Volatilitesinin Analizi. *İstanbul Ticaret Üniversitesi Sosyal Bilimler Dergisi, 7*(13), 339–350.

Regnier, E. (2007). Oil and energy price volatility. *Energy Economics, 29*, 405–427.

Sadorsky, P. (2001). Risk factors in stock returns of Canadian oil and gas companies. *Energy Economics, 23*, 17–23.

Salisu, A. A., & Fasanya, I. O. (2013). Modelling oil price volatility with structural breaks. *Energy Policy, 52*, 554–562.

Songül, H. (2010). *Otoregresif Koşullu Değişen Varyans Modelleri: Döviz Kurları Üzerine Uygulama*. Türkiye Cumhuriyet Merkez Bankası Uzmanlık Yeterlilik Tezi.

Tsuji, C. (2018). New DCC analyses of return transmission, volatility spillovers and optimal hedging among oil futures and oil equities in oil producing countries. *Applied Energy, 229*, 1202–1217.

Wang, Y., & Wu, C. (2012). Forecasting energy market volatility using GARCH models: Can multivariate models beat univariate model? *Energy Economics, 34*, 2167–2181.

Yüce Akıncı, G., & Akıncı, M. (2020). The clash of the stock markets: On the volatility dynamics and the volatility spillover effects between developed and fragile countries. In S. Evci & A. Sharma (Eds.), *Studies at the crossroads of management & economics* (pp. 303–319). IJOPEC Publication.

Zhang, B., & Wang, P. (2014). Return and volatility spillovers between China and World oil markets. *Economic Modelling, 42*, 413–420.

Chapter 10
Increasing Efficiency of Energy Saving Policies in the Future: Corporate Social Responsibility Projects

Başak Gezmen

Abstract Creating awareness and consciousness about energy saving, which are the most important topic of our era, rank among ultimate duties of media. Technological developments and new areas of the digital worlds provide advantages about awareness of consumers and taken actions of them. It has become possible to get information and be aware of every subject through the media. In this sense, the media can mobilize the masses by providing correct flow of information, forming public opinion, engaging in activities that contribute to education and performing other operations. Corporate social responsibility (CSR) projects for mass of children are more effective about creating awareness. It will have provided an achievement for limited energy correct usage of next generations in the future by teaching and gaining efficient energy consumption, saving awareness, other precautions, and correct behavior models from little ages. In this point, CSR projects draw attention. In this study, awareness aimed for mass of children and rewarded energy saving projects made by three different corporates in Turkey will be evaluated. Within these projects' scope, it will be discussed how responsibility campaigns realized by corporates on energy efficiency are effective for efficiency and creating awareness about savings.

10.1 Introduction

Today, raising on energy need and creating awareness about energy saving studies are increasing rapidly. Also, corporates make social responsibility projects for public relations' studies in order to become successful in competitive environment of the global world. Energy saving is one of the most important and sensitive topics. Furthermore, there are many of energy saving projects are organizes to form of public opinion and increasing energy saving perception. With new communication

B. Gezmen (✉)
The School of Communication, İstanbul Medipol University, İstanbul, Turkey
e-mail: bgezmen@medipol.edu.tr

S. Yüksel, H. Dinçer (eds.), *Strategic Approaches to Energy Management*, Contributions to Management Science,
https://doi.org/10.1007/978-3-030-76783-9_10

technologies' development, children are one of the most influenced mass from the media which is indispensable for us.

Children have started to satisfy their information and knowledge need through the media from little ages, and they have shown behavioral patterns by creating a role model for many things that they saw in the media. In gained correct behavioral patterns from little ages, media can use as educational and effective tool. In this point, the correct behavioral patterns can be created for children with aimed social responsibility campaigns. Moreover, environmental education makes benefit for children in order to be informed about environment. Environmental education which improves sophisticated intellection related to complicated nature of environment and the purpose of it is to solve social problems of children and gain information value and practice skills to children responsibly and effectively about quality of environment management (Mahasneh, 2017). Therefore, media can be used as effective education tool.

In this study carried out specifically for mass of children, energy saving projects and activities for creating awareness, successful social responsibility campaigns will be evaluated to their cause and consequences.

10.2 The Role of the Media on Forming Public Opinion: Reconstruction of the Agenda

Nowadays, hundreds of events occur every day both in the society we live in and in the world. With regard of these, we might not have no idea other news while we are being aware of some economics, politics, magazines, and other subjects' news. The mass communication tools are important in the determination of the societies' agenda. People are informed about the important subjects, news, and what priorities are among them thanks to the mass communication tools (Güngör, 2011). Also, the media decides what the audience-readers-listeners will talk about and what they will think of the facts with the agenda it will establish. Therewithal, the most important information resource is the media for people. Moreover, technological developments increase the power of the media as the most significant information resource. In addition to that, each individual who would like to understand what is happening in the world, has become addicted to the media (Yaylagül, 2010).

As it is known, the media is highly effective in the process of forming the relations of states most of the time. Also, the media organizations give information to society about international news and recent developments. However, it can sometimes create some problems in order to change agenda and keep it under control. The media has an active role between transmitter and receiver in mass communication and also, it has superiority in spreading, attracting attention, convincing and persuading. Thus, the media set and determine the societies' agenda because of it establishes sovereignty and acceptance on audiences (Cereci, 2012).

The theory of agenda setting is based on the media's news approach and the view of what people think, talk and how much importance they attached to these thoughts with the media's presentation methods. In line with this approach, the issues which media gives importance, are demonstrated as significant on individuals. On the other hand, topics that the media does not care and interest, are perceived by public as inadequate and insignificant issues. The most crucial point of this approach is that individuals are adequately not successful about what they will think, but they are successful to talk about what they will think (Tekinalp & Uzun, 2006). Individuals tend to know things that media interested and accept to the priority given to different issues. In essence, it refers from mass communication tools to learning functions. With regard of this, individual learns what the issue is and how it organized to the priority (McQuail & Windahl, 2010).

The basis of agenda setting theory is based on Walter Lipmann's Public Opinion study. The media creates a big sphere of influence through the images and reality patterns that people present on the perception of reality. Actually, the thoughts that individuals think to belong the outside world and their own, are semantic maps which constitutes by the media (Özçetin, 2018). Individuals are inadequate in getting information about the outside world without the media. So, these semantic maps are made mandatory in order to follow for individuals who stayed abreast of news through the media. It seemed that the knowledge and information flow of the media is related to politics. Many of things known by people reach them by second or third hand. In particular, there is evidence which most of the political information associated with political arena comes from the mass media during election period. There are discourses which are in news, opinion column and considered effective on voting decision (McCombs et al., 2009).

Bourse and Yücel (2012) emphasized that there is perspective change on mass communication tools in 70s and after. Moreover, they stated that media is not directly effective on behaviors of target groups, but it affects to shape of target groups' perspective. Hence, this situation explains that media does not form behavior, but it constitutes an intellectual agenda. In this agenda setting, some notions are ignored while some notions are prioritizing. Also, some notions are further behind. So, "media tells what we need to think, but it does not directly influence by informing what we need to think."

As historian Neumann stated in his disclosure, many of communicator, political scientist tried to define notion of public opinion, but it is not possible to treat of a definition which is age of enlightenment's notion and acceptable by everyone (Bülbül, 2001).

Today, media as mass communication tools in traditional meaning has own unique central place within the communication structure of society. It presents many inputs to the society in frame of facts and value judgments with its breaking news function which is the ultimate objective. In this point, there are other functions of the media with knowledge and information function. These are aligned as persuasion in determined issue, education, entertaining, and advertising by forming public opinion (Tokgöz, 2010).

It is reported with breaking news function of the media by connecting political, economic, social, artistic, scientific information varieties whether these are daily or not. Bülbül (2000) denoted that the function of forming public opinion is the primary duty of the media in democratic societies. Actually, media which is mentioned as industry sector in this meaning, carries out serving a public. Thus, this makes it one of the powers based on democracy. Public notion uses as a large community meaning and the media is the sighted eye and talking tong of this large community of it. Girgin (2000) dwelled on two view in forming public opinion. First viewers believe that different kind of public opinion can be learn with public opinions research. Second viewers point out that public opinion cannot completely understand with survey methods. In this context, second view supporters remarked that investigation of collective processes are required in public opinion research.

10.3 The Notion of Corporate Social Responsibility in Public Relations and Sensitivity for Environmental Problems

The notion of social responsibility for external environment refers to realize corporates economic targets without any damages to target groups' interest. In this point, corporate social responsibility (CSR) is related to organizations' taking into account all society. Individuals concern and sensibility to the environment increases with strengthening their economic, social and education level. Also, pressure groups are gathered with demand of clean and peaceful social order and increasing artistic and cultural activities. Today, many of corporations give importance these demands by taking into consideration of social responsibilities. The corporations which noticed the importance of this issue, assesses the social responsibility as strategy and politics of public relations (Biber, 2007).

The purpose of public relations' studies is to affect public opinion in an apparent subject and provide a certain public view about the subject. This is long and difficult process. Moreover, there is no expectation that spent money returns with profit quickly. Furthermore, the studies have indirectly impact on sales and profit (Onal, 2000).

One of the public relations' target is to rise corporate image and protect reputation. Therefore, the relationship between social responsibility and public relations has become clearer. With regard of these, enterprises are considered as more sensitive, environment friendly, and aiming the best service for consumers. Charity and voluntary projects within social responsibility projects made by corporates are the existence way of the enterprises with sensibility (Balta Peltekoğlu, 2016).

Bülbül (2004) explained that the notion of CSR is the result of being corporate citizen. Also, it has functions such as not to damage environment, provide secure services, and contribute to social and economic life in frame of public relations' ethics.

Ethics is among the basic principles of public relations. In social responsibility context, ethics refers to avoid from any activity which might damage to people and environment and it points out the necessity of socially responsible companies at certain limits and standards to help people. Hence, corporate studies are a purpose, not a tool (Karacan Doğan, 2018).

CSR is component of the customer satisfaction notion and it cannot ignore meeting customer need and demand. Companies do not lose customers as they meet customers' needs. So, customers who are external stakeholders and reason for the existence of companies. Today, expectations are raising because of our society consciousness level is high. Nevertheless, customers expect social responsibility with the product responsibility from firms. As Mori Company's research revealed, social responsibility is required and significant for individuals. According to the research, customers were affected by corporates which were announced to donate the charities with their part of sales revenue. Withal, 30% of the customers considered it while they were purchasing decision (Budak & Budak, 2014).

Organization culture is created by the founder or CEO's the organization. The managers of public relations cannot produce an effect if their values and ideologies are different from the organization. Also, organizational culture is influenced from large social culture and environment (Oliver, 2010).

Koçyiğit (2018) stated that advantages of social responsibility activities and increase in implementations of social responsibility with raising companies' numbers. Moreover, corporates give importance these implementations beyond their trade activities in order to emphasize their loyalty of environmental, social, and economic target. In the past, CSR seemed like additional cost, but today, it is requirement in order not to get negative reaction from society. In current competition conditions, the benefits of CSR are customer loyalty and trust, development of image and reputation, improvement of corporate identity positively and rise on all activities' efficiency. So, society and corporates utilize with better environment. Kotler and Lee (2013) emphasized that decisions in order to support social targets are aimed at meeting an obligation before the 90s and there is an approach based on pressure for looking good. In new approaches, there are tendency from obligation to strategy and corporations think different now in order to support their missions.

In previous years, there are less people than today. So, more people are using natural resources. However, the thought that next generations will use same resources, is ignored. Therefore, some future concerns occur. Hence, many of states organize some studies in order to reduce these concerns and protect next generations and their resources.

The protection and regulation of the environment is so important issue, because repairing losses and auditing them are so hard and take long time. Another significant point is that problems of border regulation and land, region, state, tourism, and environment protection cannot tackle separately. These should be investigated a whole (Tortop & Özer, 2013). Companies' social responsibility campaigns will set the agenda and increase awareness with environmental protection, environmental pollution energy saving and sustainability studies. Besides, Ülger (2003) denoted that environmental issues creates an opportunity for social responsibility project in order to show corporates themselves and there are many of CSR projects. Also, companies allocate large budgets to CSR projects. Unilever's Uluabat Living Lakes project is one of the attractive projects among these projects. Another project is Roche Firm's method which keep under control of wastes with new technologic developments. With this method, wastes are collected in the resource separately and their disposal is provided.

10.4 Energy Saving and Energy Efficiency

Energy resources are examined in two groups as fossil- based resources and renewable energy (RE) resources. RE resources renew themselves naturally faster than their depletion rate of the resources or in equal rate taken energy from the energy resources. Hydro, wind, and solar energy are RE resources. On the other hand, natural gas, oil, and coal are fossil-based energy resources (Sevim, 2019).

Fossil fuels are high carbon fuels, and they were formed by the fossilization of plants and animals lived many years ago under special conditions. Therefore, these resources cannot count as renewable. Oil, coals, and natural gas are the most commonly used resources in the world (Kademli, 2020). However, fossil-based resources have negative impact on environment. Thus, the notion of sustainable energy come up because of their running out of fossil fuel-based resources and negative impacts (Kaya & Öztürk, 2014).

The efficiency is the term of economics and it refers to relationship between inputs into the process and outputs from the process which produces goods and services (Yüceer, 2015). Today, the notions of energy efficiency and energy saving are using frequently. Moreover, these notions can use for each other because they have close relationship. However, they are two different technical term. For example, turning off the lights and throwing the wastes to recycle bin are accepted as energy saving. Furthermore, energy efficiency provides economic contribution and decreases energy consumption without hindering social welfare. In this point, it is important not to minimize the quantity and quality of the production. Actually, it is rational usage of the energy in general meaning. With regard of this, it can mention about executing energy efficiency if any of technologic process can produce with lower amount energy (Özil et al., 2012).

Energy is the most significant input and factor in order to shape human beings' material resource and provide the necessary number of services. So, cheap and high quantity of energy resources are crucial for countries and develop them positively (Özil et al., 2012).

According to climate changes and energy sustainability studies, the importance of efficient energy usage increases both in the world and in our country. These studies support to improve consciousness and create awareness about energy efficiency. Therewithal, using environment friendly methods, increasing efficiency by preventing waste in energy generation, variation and consumption politics come into prominence for our country as well. However, being sustainable in energy politics and providing supply of energy security are not easy. Otherwise, external dependency for energy supply makes the issue harder. Therefore, using energy efficiently has become more important. Energy efficiency consume energy more sufficient and economical without sacrificing from our life quality, needs and production. As common view of the world, energy saving with using efficient energy acquires with the most clean, cheapest, and fastest energy resources (Öztürk & Kaya, 2019). Moreover, Akova (2016) emphasized that it is needed to take some precautions for efficient energy consumption. These precautions are decrease energy peak, residential insulation, converting double glass of windows, preferring high class energy saving home appliances, reduction unnecessary usage of lighting, using long lifecycle highly efficient lambs, transferring to central heating in residences, reducing fuel consumption in vehicles.

10.5 Mass of Children, Media, and Creating Awareness: Social Responsibility Campaigns

Mass communication tools have crucial roles in the socialization process of individuals with family, school, peers, and other groups. An individual socializes by taking most of the behavior models and necessary information from the media throughout his/ her entire life. One of the functions the media's is education and its mass education function is important as well as the formal education. However, coming directly from the media and carried out the teacher-student relationship with educational purposes messages are not caught attention. Otherwise, receiving unnoticed and indirectly messages are effective. So, messages from the media have huge impact on children and adolescent (Hesapçıoğlu & Yılmaz, 2010).

Education refers that individuals can obtain new behaviors or changed behavior patterns by getting informed. Also, this process brings information and rationality to children. Human beings switch their behaviors by learning. In this point, learning as the large framework of the education is a whole of teaching. Teaching is defined as acquisition of information activity. Also, it makes behavior changes because human can switch behaviors by learning. In addition to that, desired behaviors can be acquired with effective learning (Sezgen, 1988).

When a child came to the world, he/she found himself/herself in a family which is a social environment. There is undisputed value of family on raising a child. The child adopts the language, religion, moral, culture, arts, tradition, norms, and values of the society. Moreover, value system of the society is so effective on formation of personality and shaping behavior and attitudes. Hence, the child exposes all this values in family at first (Yavuz, 1992).

As Yavuz (1992) referred that childhood or school age term demonstrates the child has become ready for systematic education by accomplishing appropriate level task, learning consciously, opening to research, gathering information, and increasing curiosity. This process is called as socio- cultural birth of the child.

Mass education process with the treat of meaning and shaping is the mechanism of potency and power in the society. Today, media tools such as radio, television, book, newspaper, and others have integrated to the digital worlds with technological developments. The media imbues emotion and behavior patterns to children and officiates its education function with different kind of options. Also, the simulated world of the media has countless image and pictures.

The children are the most sensitive mass of the media and they are influenced more than adults. So, huge part of the mass of children watches television and playing digital games with the digital world which creates fantastic areas and caught attention of children. As the common world political system view, television is the effective tool of politic socialization and the reason of its power for socialization is to have capacity which render into standardization and linear flow. Moreover, television conveyed common cultural norms to individuals. In this point, individuals are seemed unidirectional and passive. In modern societies, effectiveness is not worked (Özer, 2007). All critical disclosures occurred on the digital world and its tools.

The media's negative impact on mass of children is often discussed in sphere of media and digital literacy norms. Especially, this disclosure's reason is that children cannot identify what is wrong and right. So, every message from the media is accepted by children without intellectualizing, assessing, criticizing, and discussing. Thus, this non-judgmental acceptance is role model which will be adopted in the real world.

However, media uses as an effective education tool besides of all negative effects. Today, technologic, Web 2.0 and social network developments and fast internet access to our lives' changes many of thing in our lives. Once, television with its visual- auditory effectiveness, entertaining and instructive functions was seen the most powerful communication tool and it has reached many of people. Also, children and adolescents increase its education function effects. Pre-school education programs in television are aimed to support children in teaching numbers and alphabet (Kutoğlu, 2007).

In recent years, violent video games addiction of children is often on the agenda. The internet and computer games are indispensable for the children. There are different kind of games such as social effect games which prepare mass of children and adolescents to life and career. Therefore, protection mass of children from harmful games, encouraging for beneficial games and training them are important (Bilici, 2017).

Public relations' activities of institutions, ways of being in the today's digital worlds and share of messages are effective in forming public opinion and acting. One of the important points in the correct public relations' politics protects social interest. Organizations contact with different kind of masses. Further, corporations which have good relationship with their environment, seize chance to be successful. Also, a sensitive corporation should respect to society (Onal, 2000). Herein, it pays attention to the issues and creates awareness with public relations' activities.

Many of companies realize environmental sponsorship's activities. These can be CSR and various activities. For instances, a part of revenue can be allocated or savings from environment friendly packaging can be transferred into environment friendly projects (Balta Peltekoğlu, 2016).

10.6 The Importance of Social Responsibility Projects for Children in Energy Efficiency and Sample Projects

Increasing the fast flow of information creates a digital society that does not realize the content of the subject. In this fast information flow, media orientates the masses in order to become successful about which issue should be considered and how much considered. The effects of the media messages are notification and creation of awareness at first.

People try to get more information after they were aware of the news. Also, some behavior changes occur. Everyone cannot have equal chances about benefit from mass education opportunities. In this context, the importance of this function raises more. Moreover, media conveys the news which people do not know, or it creates more awareness about ignored issues because of lack of knowledge. So, prioritized media topics have become prior. Also, the media should draw attention more, corporates should increase public relations' studies and mass of children should be prompted about energy saving.

In this study, "I Protect My Energy" project by Enerjisa Energy A.Ş., "Smart Stars Energy Saving" project by Soyak Holding and "Our Energy is for Children" project by Zorlu Enerji Group are assessed.

10.6.1 Enerjisa Enerji A.Ş.: The Project of "I Protect My Energy"

"I Protect My Energy" is the first project in the sector and created by Enerjisa Enerji A.Ş. which distributes and sells electricity. The project started with the Protocol that signed with the Ministry of National Education. This project's target group is 7–10 years old children and organizes energy efficiency educations for this age group. Environment and saving consciousness and habits start in 7–10 years old children. So, this is the reason of this age group choice.

It has been aimed to create awareness and consciousness studies for children about energy efficiency since 2010. Enerjisa' the CSR project got Silver Steve Prize in 2016, prize "Responsible Consumption and Production" field in Corporate Social Responsibility Association of the Turkey in 2017. Also, they got Golden Compass and The Best Energy Management prizes in 2018. Furthermore, Enerjisa work with "Along the Life" motto as the perception of social responsibility project. Therewithal, Yetik K. Mert is the CEO of Enerjisa and he said that Enerjisa's energy efficiency target is to reach students in more schools and tell the importance of energy efficiency to next generations. Nevertheless, the crucial information was staged in theatre by professional thespians for mass of children with the project slogan "I am Protecting the Energy of the World".

The project is integrated to digital channels for reaching more students. Also, it is in the games that are waste time activity of the children. These energy efficiency games can be played from www.enerjimikoruyorum.org and mobile application of "Enerjimi Koruyorum". Further, each step teaches energy saving methods. The children build a new city to energy saver "Pozitip" kids in the game. Besides, living in this city acquires energy saving ability.

10.6.2 Soyak Holding: "Smart Stars Energy Saving"

Soyak adopts as approach of sustainable life. Soyak got social responsibility of the year "Grand Award" and in the field of environment The Best Social Responsibility Solution Award in the Pazaryeri event that its Association has been. Further, the project's ultimate goal is to create awareness, consciousness and organize activities in primary schools about energy.

The purpose of the project is to create awareness about usage of efficient energy, energy saving, and sustainability for teachers and student in 4th grade Science and Technology classes. Besides, www.akıllıyıldızlar.com gives information related to the project. There is game in the web site which tries to provide energy saving consciousness. It is aimed to training program for 2 days by professionals, scholars and teachers in determined schools.

10.6.3 Zorlu Enerji Group: "Our Energy Is for Children"

Zorlu Enerji Group's purpose is to increase energy efficiency information level and develop behavior models. It prepared by academicians. They organized activities to students about energy efficiency, saving and protection of environment in 2010. There are experiment fields with supporting interactive games. Also, there were education program "The World we Live in," "Our Life is Energy," and "Our Energy is for our Future." titled presentations. Further, there are games for children to the age groups which can play in the garden or class.

The importance of natural gas was explained in the activities. These activities started to introduce natural gas, contribution to life and environment presentations which were prepared by pedagogist. Therewithal, some implementations made with pedagogist. Interactive teaching method was preferred, and it has become both educational and entertaining with questions from children. A refrigerator magnet was gifted to children who promised to make a natural embassy for family and friends in natural gas introduction. The reason of the magnet choice as a gift is to save from the fuel by keeping temperature on 20–24 degrees.

10.7 Conclusion

Human being is a social being who is born, grows up and gains his/her personality within the society. The personality improves in childhood era. Hence, the education plays a major role in cognitive, linguistic, social and personality development of the children. It stars in the family and continues in school and social environment. According to Gebner, media is a conscious and systematic raising organ. Media confronts except family, school, and social environment while mass of children is learning and assimilating behavior patterns, being aware what happens in their environment, and assessing. Further, the ultimate goal of the media is to satisfy news and information needs of individuals. However, media has forming public opinion, socialization, creating awareness, entertaining functions. The prepared media contents for children are so significant because mass of children is vulnerable. So, studies about media and digital literacy are sensitive about it. Besides, mass of children cannot identify what is wrong and right for messages came from the media. Therefore, appropriate contents should be chosen to the age groups and given importance and sensibility for children. Moreover, educative broadcastings should be increased through children's mental and spiritual developments occurs in childhood. Today, the most important understanding is to educate while entertaining. Otherwise, mass of children gets bored so easily and cannot adapt to weighty and closed content. Hence, gaming, animation, video activities in education are more effective. Also, todays' digital world offers rich and attractive visual contents. However, the matter is to use it efficiently. In raising function of the media, the media is located as conscious and systematic project.

The importance of energy saving, and energy efficiency increases in our country as well as in the world. There are studies about it for mass of children and adults such as public service announcement and social responsibility campaigns. In this study, mass of children is investigated by considering the importance of childhood period in forming and reinforcing behavioral models. With regard of these, more successful, conscious, responsible, healthier, happier and environment adopted generations will become if saving and efficiency notions is taught in childhood. Therefore, social responsibility activities made by both the public and private sector are so important in creating awareness and action. Further, energy saving projects give information to children from little age about what energy and energy resources are, the importance of energy saving, and how energy uses efficiently. In the study, 3 different CSR projects by three different corporates were investigated and it is upheld to increase this kind of CSR project about this area. Also, it is thought that creating more awareness will bring achievements about rising conscious generations.

References

Akova, İ. (2016). *Enerji Kullanımındaki Değişimler [Changes in energy usage]*. Nobel Yayınları.

Balta Peltekoğlu, F. (2016). *Halkla İlişkiler Nedir? [What is public relations?]*. Beta Yayınları.

Biber, A. (2007). *Halkla İlişkilerde Temel Kavramlar Tanımlar Örnekler Açıklamalar Kaynaklar [Basic concepts, definitions, samples, explanations and resources in public relations]*. Nobel Yayıncılık.

Bilici, İ. (2017). *Medya Okuryazarlığı ve Eğitimi [Media literacy and education]*. Nobel Yayınları.

Bourse, M., & Yücel, H. (2012). *İletişim Bilimlerinin Serüveni [The adventures of communication sciences]*. Ayrıntı Yayınları.

Budak, G., & Budak, G. (2014). *İmaj Mühendisliği Vizyonundan Halkla İlişkiler [Public relations from the vision of image engineering]*. Nobel Yayınları.

Bülbül, A. (2000). *Genel Gazetecilik Bilgileri [Journalism guideline]*. Nobel Yayınları.

Bülbül, A. (2001). *Haberin Anatomisi ve Temel Yaklaşımlar [Anatomy of the news and fundamental approaches]*. Nobel Yayınları.

Bülbül, A. (2004). *Halkla İlişkiler [Public relations]*. Nobel Yayıncılık.

Cereci, S. (2012). *Medya Yapımları ve Yapım Teknikleri [Media production and production technics]*. Nobel Akademik Yayıncılık.

Girgin, A. (2000). *Yazılı Basında Haber ve Habercilik Etik'i [News in printed media and journalism ethics]*. İnkilap Yayınları.

Güngör, N. (2011). *İletişim Kuramlar Yaklaşımlar [Communication theories' approaches]*. Siyasal Kitabevi.

Hesapçıoğlu, M., & Yılmaz, N. (2010). *Bilgi Toplumunda Okul ve Medya: Medya Pedagojisine Bir Giriş [School and media in information society: Introduction to media pedagogy]*. Morpa Yayınları.

Kademli, M. (2020). *Temel Enerji Kaynakları [The fundamental energy resources]*. Nobel Yayınları.

Karacan Doğan, P. (2018). *Halkla İlişkiler public relations "PR"*. Gazi Kitabevi.

Kaya, D., & Öztürk, H. (2014). *Sanayide Enerji Yönetimi ve Enerji Verimliliği Uygulamalı Örneklerle [Energy management and energy efficiency in industry with practical examples]*. Umuttepe Yayınları.

Koçyiğit, M. (2018). *Halkla İlişkiler ve Kurumsal Sosyal Sorumluluk [Public relations and corporate social responsibility]*. Eğitim Yayınevi.

Kotler, P., & Lee, N. (2013). *Kurumsal Sosyal Sorumluluk [Corporate social responsibility]*. Media Cat Yayınları.

Kutoğlu, Ü. (2007). Medya Okuryzarlığı ve Çocuk Eğitimi [Media literacy and children education]. In N. Türkoğlu & M. Cinman (Eds.), *Şimşek içinde, Medya Okuryzarlığı [as cited in media literacy]* (pp. 104–117). Kalemus Yayınları.

Mahasneh, R. A. (2017). Reading social stories in the community: A promising intervention for promoting children's environmental knowledge and behavior in Jordan. *The Journal of Environmental Education, 48*, 334–346.

McCombs, M., Shaw, D., & Muratoğlu, B. (2009). Kitle Medyasının Gündem Yaratma İşlevi [The function setting agenda of mass media]. In N. Rigel & Ş. Çağlar içinde (Eds.), *Matriksi Şimşek Olan Metinler Gazetecilik [as cited in The matrix of lightning texts: Journalism]* (pp. 127–134). Anonim Yaıncılık.

McQuail, D., & Windahl, S. (2010). *İletişim Modelleri-Kitle İletişim Çalışmalarında- [Communication models in mass communication studies]*. İmge Kitabevi.

Oliver, S. (2010). *Public relations strategy*. Kopan Page.

Onal, G. (2000). *Halkla İlişkiler [Public relations]*. Türkmen Kitabevi.

Özçetin, B. (2018). *Kitle İletişim Kuramları, Kavramlar, Okullar, Modeller [Theories, concepts, schools and models of mass communication]*. İletişim Yayınları.

Özer, Ö. (2007). *Medya, Şiddet ve Toplum [Media, violence and society]*. Anadolu Üniversitesi Yayınları.

Özil, E., Şişbot, S., Özpinar, A., & Olgun, B. (2012). *Elektrik Enerjisi Teknolojileri ve Eneriji Verimliliği [Electricity energy technologies and energy efficiency]*. TESAB Yayınları.
Öztürk, H., & Kaya, D. (2019). *Enerji Yönetimi [Energy management]*. Gece Kitaplığı.
Sevim, C. (2019). *Enerji Teknolojileri ve Enerji Yatırım Projeleri [Energy technologies and energy investment projects]*. Seçkin Yayınları.
Sezgen, M. (1988). *Eğitim Psikolojisi [Educational psychology]*. Milli Eğitim Gençlik ve Spor Bakanlığı Yayınları.
Tekinalp, Ş., & Uzun, R. (2006). *İletişim Araştırma ve Kuramları [Researches and theories of communication]*. Beta Yayınları.
Tokgöz, O. (2010). *Temel Gazetecilik [The fundamental journalism]*. İmge Kitabevi.
Tortop, N., & Özer, M. (2013). *Halkla İlişkiler [Public relations]*. Nobel Yayınları.
Ülger, B. (2003). *İşletmelerde İletişim ve Halkla İlişkiler "The Big Business" [Communication and public relation in business: "The Big Business"]*. Der Yayınları.
Yavuz, K. (1992). *Çocuk Dünyası ve Gelişme [The World of children and development]*. Çocuk Vakfı Yayınları.
Yaylagül, L. (2010). *Kitle İletişim Kuramları Egemen ve Eleştirel Yaklaşımlar [Mass communication theories, sovereign and critical approaches]*. Dipnot Yayınları.
Yüceer, N. (2015). *Yapıda Çevre ve Enerji [Environment and energy in construction]*. Nobel Yayınları.

Chapter 11
Business Intelligence Application in the Natural Gas Industry: A Company Case

Kevser Şahinbaş and Bünyamin Yılmaz

Abstract Recently, information technology tools have been extensively used to accelerate the flow of data, and huge amounts of data have been generated. Therefore, the advantages of having a large amount of data are quite high, however, the increasing masses of data have caused it more complicated for businesses to make new decisions in the intense competitive environment. One of the most important stages of making strategic decisions is to process the existing raw data correctly and to produce new and meaningful information and opportunities, and to use the produced in a way to meet the needs of the market and the sector. To achieve success in a dynamic and hyper-competitive business environment, organizations need effective and timely business information to maintain their continuity. In this regard, organizations need to integrate the necessary technologies into their business processes. Thanks to a well-designed information system, organizations gain advantages in order to achieve the desired goals. Business intelligence systems provide many benefits to the organizations that use them as a type of information system. The aim of this study is to demonstrate the stages of business intelligence and information technology approaches in the natural gas sector and to evaluate and analyze their applicability. Business intelligence systems assist the organization in corporate performance management, optimizing customer relations and supporting decision making. In addition to these, it provides great convenience to decision-making mechanisms with the reports it presents to the management.

11.1 Introduction

With the rapid developments of advancing technology, it is critical to ensure the innovation for organizations to obtain values (Lin et al., 2013). Data have attracted increasing attention from organizations. Considering that data is one of the most

K. Şahinbaş (✉) · B. Yılmaz
The School of Business, İstanbul Medipol University, İstanbul, Turkey
e-mail: ksahinbas@medipol.edu.tr; bunyamin.yilmaz@std.medipol.edu.tr

S. Yüksel, H. Dinçer (eds.), *Strategic Approaches to Energy Management*, Contributions to Management Science,
https://doi.org/10.1007/978-3-030-76783-9_11

valuable assets in today's world, efforts to obtain and control data become a hot research topic in recent years. Companies and institutions take the risk of making expensive investments in technology to access and benefit from data. Since transformation of knowledge into tangible value is one of the most important problems of our day. In order for organizations to maintain their continuity, they need structures that will benefit them, improve themselves and bring them to a good position with this benefit. By these structures, it is possible to see the future better and to survive in a highly competitive environment. Information systems play a significant role in order to obtain the data that a well-structured organization will obtain from many data sources, make sense and turn it into action with the help of technology. Especially in cases where the data volume of organizations is high, support systems are used in analysis and decision-making. Business Intelligence (BI) system is one of the most successful system in this field (Negash & Gray, 2008). Many companies utilize business intelligence systems to improve business processes at different levels, and especially to improve their strategic, tactical and operational decision-making stages. Therefore, business intelligence systems, which can be preferred at all levels of management, play a significant role in increasing the productivity of the company by creating a competitive understanding (Ziora et al., 2012). Businesses need to make quick and effective decisions in order to achieve goals such as expanding their market shares and increasing their profitability, consequently managers desire to achieve success by integrating information technology solutions into their companies. For this reason, investments in business intelligence systems have increased in recent years (Ziora et al., 2012). The main purposes of using business intelligence applications are to analyze the current situation of institutions and to ensure that they can take positions on time according to possible risks. It has been developed in order to separate and store data obtained from various data sources under an order, to associate the stored data with statistical methods, and to ensure that the necessary information is reported for information users at all levels. As organizations become more and more institutional structures, they require the use of complex information systems. For this purpose, today, business intelligence platforms are widely used to obtain the required amount of information in order to be as fast as possible and to make the best decisions.

Business intelligence systems are also among the information systems that are increasingly being used today. Business intelligence systems, which appeal to many levels from employees in organizations to senior managers, aim to provide effective benefits to decision-makers, especially at the decision-making stage of managers. In an environment where environmental factors are effective, the quality of these decisions to be taken by managers who are constantly in a competitive position is very important for the organization. Managers make great efforts to reduce mistakes to be made and to minimize the risks to be taken. Sometimes a strategic decision that can be taken can put a business in a very difficult situation, sometimes it can provide positive results with the right decision. For this reason, today's business life, business intelligence systems preferred by executives and employees in the world and Turkey takes its place among information systems.

Energy has become the most important factor in economy and daily life since the eighteenth century. Energy, which started to gain a significant location in production and economy with the use of coal as an energy source in steam machines with the industrial revolution, increased its importance in the mid-nineteenth century with the development of the industry and the commercial use of oil. For many years, fossil fuels such as coal and oil have been at the forefront as energy sources but considering the rapid depletion of these resources and the damage, they cause to the bard layer, they encouraged countries to use new energy resources. Natural gas does not harm the environment as much as other fossil sources, does not generate waste materials, therefore it is a more environmentally friendly energy type compared to other sources.

We present a brief summary and overview of the surveyed research in relation to effect of business intelligence on decision making. Roldán and Leal (2003) conducted a study to reveal the effectiveness of top management information systems in decision making. In this study, decision-making success was determined as the speed of problem determination, decision-making speed and the scope of analysis presented by Leidner and Elam (1993). This study revealed that perceived satisfaction affects decision-making. Peters et al. (2014) used the speed and quality of decision making among these benefits when determining the net benefits to be gained from the use of mobile business intelligence. This study also found an effect of perceived satisfaction on net benefits. Igbaria and Tan (1997) emphasized the effect of perceived satisfaction on the quality of decision making by addressing the quality of decision making at the level of individual influence. Petter et al. stated that the decision- making structure is under net benefits and that the structures supporting decision- making such as decision-making quality, decision-making efficiency and decision- making speed are evaluated under net benefits. In this context, they reported that in all 14 studies they considered, perceived satisfaction affected net benefits. In this context, perceived satisfaction affects decision-making efficiency. Wang et al. (2007) reported in their study on hotel reservation systems that information and system quality affect perceived value, and at the same time, perceived value positively affects the intention to use. Wang (2008) reported that information quality and system quality affect perceived value and perceived value has effects on satisfaction and intention to use in their study to determine the usage behaviors of e-commerce applications.

The main purpose of this study is to demonstrate the benefits obtained from business intelligence systems and help determine the effectiveness of these systems. Today, it is inevitable that managers who are in decision-making positions will benefit from technology as much as possible in the decision-making phase. Especially in companies with large amounts of data sources and business volume, the ability to make instant and correct decisions becomes important for the continuity of companies and the effectiveness of their business processes. Therefore, the decision-making performance and decision quality of the decision maker emerge as an important issue. In this chapter, it is aimed to demonstrate the application phases of business intelligence systems for decision making process.

In this chapter, we describe the basic concepts and use of Business Intelligence for natural gas industry. We demonstrate the importance of the business intelligence applications for the decision-making process. Section 11.2 includes natural gas energy, consumption in Turkey. In Sect. 11.3 we explain the concept of Business Intelligence (BI) and the process of BI architecture. In Sect. 11.4, application of BI in a natural gas firm in Turkey is demonstrated in detailed. The all of processes are presented. Lastly, we conclude the chapter and suggest in Sect. 11.5.

11.2 Natural Gas Energy

For many years, fossil fuels such as coal and oil have been at the forefront as energy sources but considering the rapid depletion of these resources and the damage, they cause to the bard layer, they encouraged countries to use new energy resources. Natural gas does not harm the environment as much as other fossil sources, does not generate waste materials, therefore it is a more environmentally friendly energy type compared to other sources. It has a higher efficiency potential due to its gaseous state and the possibility of the combination of flammable caustic molecules. When we look in general, it is one of the rare fuels with clean, efficient and high calorific value. After electrical energy, the most efficient fuel is natural gas. Approximately 47% of Turkey's total natural gas imports from Russia are provided with amount bcm. Iran and Azerbaijan are other countries where natural gas is supplied by pipeline. In 2018, a total of 39 bcm of gas was supplied from pipelines. The remaining amount of the portion 11.3 bcm of LNG (liquefied natural gas) entered from 11 different countries to Turkey by sea tankers (https://www.tskb.com.tr/i/assets/document/pdf/dogalgaz-tr_web.pdf).

11.2.1 Natural Gas Consumption in Turkey

The natural gas consumption reached its highest value in 2017, Turkey in 2018 were 48.85 bcm of gas consumed. In 2018, natural gas consumption decreased by approximately 9% compared to the previous year, and this decline was recorded as the biggest decrease of all time. Electricity production/return-cycle power plants with a reduction of approximately 2.4 bcm and industrial facilities with a reduction of 1.4 bcm have been the consumption centers that played a major role in the contraction in consumption. Table 11.1 illustrates natural gas consumption in Turkey (https://www.tskb.com.tr/i/assets/document/pdf/dogalgaz-tr_web.pdf).

Table 11.1 Natural gas consumption in Turkey

Year	2006	2007	2008	2009	2010	2011	2012	2013	2014	2015	2016	2017	2018
Gas consumption	30	33	34	33	34	41	42	42	45	44	42	52	48

11.3 Business Intelligence

Traditional organizations have faced information overflow, weak data, incomplete information and inadequate reports due to increased data volumes. In today's world, managers can benefit from business intelligence systems in order to make effective decisions with the increasing volume of information with high acceleration (Lin et al., 2009). With the advantages of business intelligence systems, decision makers can steer organizations and increase their chances of competition. In the literature, many definitions of business intelligence have been developed by researchers over the years. Firstly, the term "business intelligence" was used by Howard Dresner in the Gartner Research in 1989 and its key concepts were explained. Accordingly, Dresner defined business intelligence as "software and solutions that provide solutions for collecting, consolidating, analyzing and accessing data that allow corporate users to make better business decisions" (Gibson et al., 2004). According to Zeng, business intelligence systems are defined as systems that collect, process and distribute data in order to minimize uncertainty (Zeng et al., 2006). It is also stated that the information obtained from these systems can be considered as auxiliary systems in strategic decision-making. Business intelligence applications generally enable businesses to query, report, and analyze data in a multidimensional way (Baars et al., 2008). In addition, business intelligence applications help companies achieve their goals within a performance management framework. In this way, it serves the desired purpose by ensuring the sustainability of companies. While achieving these goals, analyzing the business processes of companies and improving these processes are among the functions of business intelligence systems. As a result of the analysis obtained with business intelligence systems, the activity of acting can also be made possible with business intelligence. The realization of the action aimed at the intended purpose as a result of the transformation of the available data into meaningful information with effective configurations is the product of business intelligence (Golfarelli et al., 2004).

11.3.1 Business Intelligence Architecture

Business intelligence systems have an effective role in determining the decisions of the senior management by collecting and analyzing large amounts of data and creating high level reports from these data. In this context, the processes that the data passes until the business intelligence systems reach the end user side, form the architecture of business intelligence systems (Kalelkar et al., 2014). Considering the operations performed by business intelligence on data, it is seen that it has a unique structure with the capacity to obtain, access, understand, analyze and turn the data into valuable information that can turn into action. In this regard, it has the ability to collect data from sources, store data and create useful information for decision makers in today's competitive environment with the help of analytical tools. In the stage of obtaining these capabilities, business intelligence systems architectural structure is of great importance. They report that business intelligence systems consist of five main layers (Ong et al., 2011). The lowest data sources layer keeps the internal and external data sources ready for the ETL process. In the upper layer, ETL (Extract Transform Load) layer, data extraction, transformation processes, cleaning and loading stages take place. Cleaned data are stored in the data warehouse layer in a processed form to serve a specific purpose. In the end user section, the desired viewing, data mining, querying and reporting etc. on the data are obtained from the data warehouse where transactions are made. In the realization processes of these transactions, descriptive data on the real data are generated in the metadata section.

11.4 Application of Business Intelligence in a Natural Gas Firm in Turkey

In this section, we demonstrated all of processes of Business Intelligence application in a private natural gas company in Turkey. The dataset was obtained from a private natural gas company in Turkey. In Table 11.2, process of application of business intelligence is indicated. Business intelligence applications transform data into information, enable organizations to use their resources more efficiently, and increase operational efficiency. It uses analytical tools to achieve all these goals and thus combines data collection, data storage and information management.

11.4.1 Creating Dblink Between Systems

Data is extracted via dblink. As a source, the tables in the relevant databases are dblinked through the Toad program. Using the Oracle data integrator program, the models and analyzes are transferred to the data warehouse. By using the Toad program, dblinks have been created between the database and the data warehouse. Data flow between the two systems will be provided these links.

Table 11.2 Application of business intelligence in a natural gas firm

Process	Service	Transaction
1	Dblink	Oracle2Oracle link is created between OLTP system and OLAP system.
2	ODI	The created links are triggered in the ODI program.
3	DWH	Transactions in the data warehouse are performed and Fact tables are obtained.
4	RPD	JOIN conditions between tables are provided.
5	OBIEE	Data is presented to the end user by dashboards.

11.4.2 Working Principles of Dblinks Created

The dblinks created in the Toad will trigger the dblinks and retrieve data from the relevant tables, thanks to the times and procedures specified with the Oracle Data Integrator program. It will transfer the ETL date to the tables. The mentioned packages above will activate the scenarios we have created in the ETL with the triggering hours of the relevant procedures and these packages will run every night at 12.10 pm and will perform the loading process by assigning ETL date and ID to the tables in the data warehouse where we have specified the tables in the database. In this way, the data in the data warehouse will be compared with the data in the database one day.

11.4.3 Data Preparation and Storage

Data sources taken from different sources via dblinks have been transferred to a common data warehouse. It is gathered under the DWH_RAW scheme. All data have not been manipulated. These migrated data are gathered under the DWH_RAW scheme. There are all data that have not been manipulated here. The raw data transferred to Toad has been collected under the stg scheme by passing through various data cleaning processes. The data cleared under the STG scheme in Toad were collected under the FACT scheme as a summary data set in line with the demands. The dataset is ready to be modeled.

11.4.4 Data Analysis and Modeling

Relational tables that are created from the data warehouse using Oracle BI Administration program have been taken to RPD. Tables belonging to the requested reports were joined in the 3 layers and models were created. In the physical layer, the tables to be joined were imported and the data types of the tables and the data in their content were examined. In the Business Model Mapping layer, the desired joins were created between the tables and the table names in Turkish were applied. Presentation layer has been edited so that the end user can see the data. Now, we can edit the desired reports and graphics and complete our analysis over the data set we are interested in.

11.4.5 Data Presentation

In this section, we provide business intelligence analysis reports examples. To report analysis to senior manager about subscriber numbers, gas withdrawal rates, dept. amount, aging report, collection amount, work order performance, illegal natural gas notices, invoice information etc., Oracle Business Intelligence Suite Enterprise Edition (OBIEE) is used as a data presentation tool. By this program, we create the desired reports and see the efficiency of the decisions we need to take on the graphs. The dashboard of the reports is created by using the OBIEE program and the graphics created from those reports. Graphs are designed by summing the subscriber numbers table with count.

For example, one of the vital importance of the reports is daily gas usage graphs that are created using the gas withdrawal rates data from the natural gas company daily. Annual gas usage chart is generated with the gas withdrawal data of the natural gas company annually. The other crucial report is pie charts that are demonstrated in debt management charts using the company's annual business and residential debt data. In addition to that, the chart of the receivable aging report using the 2021 debt information for the year 2020 and before is given by aging charts by years. The graph of the accrued collection report prepared with the data set of the accrued debts of the natural gas company is presented in collection chart created with annual collection data chart. Besides, the graphical data presentation of the work order performance report prepared with the status codes assigned by the personnel working in the field with the work order is generated in Work Orders Performance Chart. Furthermore, the graphs of the daily, monthly and annual call numbers created by using the table that keeps calls to the company call center within a year is provided in pie chart of illegal natural gas notices coming to the notification line. The graphs indicating the reasons for the call that have been stated in call center incoming call graph is created.

These reports provide many aids to see the current situations and to achieve predictions. The benefits of business intelligence are to accelerate decision-making processes and reduce risks (Sheng, 2005). The data is in graphs, maps, tables, speed indicators, etc. using a wide variety of visual and user interactive components. In this way, it provides effective, understandable and easy data analysis. The decision-makers can access the analysis whenever and wherever they want via mobile, tablet or computer, thus increasing the efficiency of the decision-making process. Besides, it provides achieving predictions and predictions for the future using statistical methods and accessing to reliable corporate information through a single environment and increasing business continuity and efficiency institutionally (Azvine et al., 2006; Watson & Wixom, 2007; Ranjan, 2009).

In Table 11.3, meter reading, number of notice and payment types are demonstrated.

In Table 11.4, consumption and collection amount are illustrated. Thanks to these tables, business intelligence graphics are created.

Table 11.3 Bill information for the natural gas firm

Date	Meter reading	Number of bill issued	Number of notice	Company payments	PTT payments	Number of bank payments	Number of other payments	Total number of payments
202,001	314,404	301,647	91	2846	40,988	239,258	18,708	301,800
202,002	315,157	302,133	107	3013	39,148	233,034	18,493	293,688
202,003	313,341	303,292	145	3452	34,323	233,568	16,003	287,346
202,004	306,444	297,289	150	2279	4710	258,522	10,871	276,382
202,005	301,705	294,812	39	1806	5661	240,866	9121	257,454
202,006	311,494	301,374	8	5474	18,015	241,563	10,517	275,569
202,007	319,711	307,723	377	6127	19,566	239,526	9970	275,189
202,008	316,170	303,783	68	5702	19,001	238,973	9287	272,963
202,009	322,877	311,270	51	6212	20,265	240,537	8594	275,608
202,010	321,413	309,185	10	4994	20,146	248,401	8711	282,252
202,011	322,074	311,353	19	3476	19,043	255,850	8655	287,024
202,012	325,783	315,755	92	3014	21,735	268,649	9176	302,574
Total amount	3,790,573	3,659,616	1157	48,395	262,601	2,938,747	138,106	3,387,849

Table 11.4 Consumption and collection amount in the natural gas firm

Consumption SM3	Accruement amount	Number of accruement (Number of invoice)	Sector	Collection amount	Number of collection	Collection rate (%)
50.030.443	90.287.592	100.157	Business	82.503.712	93.845	91.38%
65.119.582	120.983.825	347	Industry	120.983.825	344	99.49%
336.117.229	611.690.46	3.519.667	Hosuing	573.356.459	3.372.805	93.73%

11.5 Conclusion and Discussion

Factors such as the increase in data volume, changing customer needs, and the obligation to make instant decisions create the necessity of providing better performance in the business world. In today's challenging business life, information is one of the most valuable assets that organizations have. Organizations that obtain information and use it in the best way take great steps in achieving success. Organizations need to use the data they have in the best way and make sense of them in order to ensure their continuity. In this way, organizations that can evaluate the data they have in the best way and produce valuable information and respond quickly to business processes can gain advantages and get ahead. In order to gain benefits from well-obtained and structured data, it is becoming necessary to benefit from information technologies. Companies that are successful today are companies that can make the best use of their data and make effective decisions. For a company to achieve optimum profits, organizations need to apply business intelligence reports and graphics while taking vital decisions about the company. Business Intelligence is a set of theories, methodologies, processes, architectures and technologies that transform raw data into meaningful and useful information for business purposes. Business Intelligence manages large amounts of information to help identify and develop new opportunities. It is used to develop new opportunities and effective strategies. Senior managers need to be people who know data mining and its approaches. They know that companies use the summary data they obtain from various source systems for business intelligence reports as well as for data mining applications. For this reason, a manager who has mastered the entire data science process knows very well how the company's resources will be 6 months later, or an energy company that makes a yes-no survey to its subscribers can demonstrate how many people will say yes and how many people will say no in advance with data mining applications. The main subject of this study is to explain the reporting systems required by the Energy Markets Supervisory Board and to indicate its effect on decision support mechanisms. However, these reports indicate graphics for past and current data and need to be analyzed. Data mining departments need to be established in companies, and applications could be implemented using data that has passed through the entire data science process in existing data warehouses. In this way, data warehouses will go beyond serving the historical reporting needs and will help all future data mining applications. First, we need a data warehouse where we model the data. In order for company to transfer data to this data warehouse, bridges are created with dblink between various source systems and the data warehouse. Dblink is completely free technology. If a larger project is to be managed, this data transfer can also be performed with Golden Gate. Scenarios were created with Oracle Data Integrator program to trigger and run the dblinks, and data transfer is performed when the conditions of these scenarios are met. After the data transfer is provided, the data are kept in 3 separate categories as Raw, Stg and Fact. In these 3 categories, the data are prepared completely in accordance with the purpose and after the Fact category, the data is modeled and transferred to the OBIE program. We briefly summarize the reports made in this program and the added value they have provided to the company; the daily report of the subscriber numbers, the purpose of which customers work with the company and the annual report by the Energy Markets Supervisory Board. The number of subscribers has a vital importance in terms of providing this when requested. Another crucial issue is that daily gas usage reports are very important in terms of calculating monthly and annual consumption. Thanks to this report, the company can see how many kWh of natural gas it will need to purchase next year. In the debt management report, on the other hand, the dwellings where the debt holders are invoiced to the firm decide whether the company is a business or a house, and accordingly, it provides some payment facilities for subscriber groups with high debt. Furthermore, the receivable aging report by years, on the other hand, are the parameters showing how the company is in terms of collecting its receivables compared to previous years. The accrual collection report demonstrates how much of the invoiced receivables of the company accrued within the current year and month. By this report, as the accrual amount is much higher than the collection amount in 2019, the company has facilitated its subscribers to pay debt by credit card. The work orders performance report, on the other hand, enables the personnel working in the field to be evaluated on the service and the company provides a more efficient service to its subscribers. In addition to that the incoming call reports of the call center are reports that are used to determine the speed of response to denunciations and other complaints. As a result, by analyzing the reports shown in the study more broadly, we can develop data mining applications in the direction we want and determine how many call calls will arrive next year. Because the data of where the company has infrastructure problems, and in which districts the complaints are received more quickly are stored in bulk in the databases. Reports are created in accordance with the purpose by processing and modeling these data. Therefore, it is concluded that revealing the benefits of business intelligence systems is of great importance in this sense.

References

Azvine, B., Cui, Z., Nauck, D. D., & Majeed, B. (2006). Real time business intelligence for the adaptive enterprise. In *E-commerce technology, 2006. The 8th IEEE International Conference on and Enterprise Computing, E-Commerce, and E-Services*, p.29.

Baars, H., Kemper, H. G., Lasi, H., & Siegel, M. (2008). Combining rfid technology and business intelligence for supply chain optimization scenarios for retail logistics. In *Proceedings of the 41st Annual Hawaii International Conference on System Sciences (HICSS 2008)*, pp. 73–73.

Gibson, M., Arnott, D., Jagielska, I., & Melbourne, A. (2004). Evaluating the intangible benefits of business intelligence: Review & research agenda. In *Proceedings of the 2004 IFIP International Conference on Decision Support Systems (DSS2004): Decision Support in an Uncertain and Complex World*, 295–305.

Golfarelli, M., Rizzi, S., & Cella, I. (2004). Beyond data warehousing: What's next in business intelligence? In *Proceedings of the 7th ACM international workshop on data warehousing and OLAP*, pp 1–6.

Igbaria, M., & Tan, M. (1997). The consequences of information technology acceptance on subsequent individual performance. *Information & Management, 32*(3), 113–121.

Kalelkar, M., Churi, P., & Kalelkar, D. (2014). Implementation of model-view-controller architecture pattern for business intelligence architecture. *International Journal of Computer Applications, 102*(12), 975–8887.

Leidner, D. E., & Elam, J. J. (1993). Executive information systems: Their impact on executive decision making. *Journal of Management Information Systems, 10*(3), 139–155.

Lin, H. E., McDonough, E. F., Lin, S. J., & Lin, C. Y. Y. (2013). Managing the exploitation/exploration paradox: The role of a learning capability and innovation ambidexterity. *J. Prod. Innov. Manage, 30*(2), 262–278.

Lin, Y. H., Tsai, K. M., Shiang, W. J., Kuo, T. C., & Tsai, C. H. (2009). Research on using anp to establish a performance assessment model for business intelligence systems. *Expert Systems with Applications, 36*(2), 4135–4146.

Negash, S., & Gray, P. (2008). *Business intelligence handbook on decision support systems 2.* Springer.

Ong, I. L., Siew, P. H., & Wong, S. F. (2011). *A five-layered business intelligence architecture.* Communications of the IBIMA.

Peters, T., Popovič, A., Isik, O., & Weigand, H. (2014). *The role of mobile bi capabilities in mobile BI success.*

Ranjan, J. (2009). Business intelligence: Concepts, components, techniques and benefits. *Journal of Theoretical and Applied Information Technology, 9*(1), 60–70.

Roldán, J. L., & Leal, A. (2003). *A validation test of an adaptation of the delone and mclean's model in the spanish eis field, in critical reflections on information systems: A systemic approach* (pp. 66–84). IGI Global.

Sheng, Z. (2005). *Measuring index system and method of organizational intelligence.* School of Economics University of Jinan.

Wang, Y. S. (2008). Assessing e-commerce systems success: A respecification and validation of the delone and mclean model of is success. *Information Systems Journal, 18*, 529–557.

Wang, Y. S., Wang, H. Y., & Shee, D. Y. (2007). Measuring e-learning systems success in an organi- zational context: Scale development and validation. *Computers in Human Behavior, 23* (4), 1792–1808.

Watson, H. J., & Wixom, B. H. (2007). The current state of business intelligence. *Computer, 40*(9), 96–99.

Zeng, L., Xu, L., Shi, Z., Wang, M., & Wu, W. (2006). Techniques, process, and enterprise solutions of business intelligence, in 2006 IEEE international conference on systems. *Man and Cybernetics, 6*, 4722–4726.

Ziora, L., et al. (2012). The concept of real-time business intelligence. Review of practical applications. *Informatyka Ekonomiczna, 23*, 67–75.

Chapter 12
Environmental Impact of Economic Globalization and Renewable Energy Consumption in the OPEC Countries

Ibrahim Nandom Yakubu, Ayhan Kapusuzoglu, and Nildag Basak Ceylan

Abstract This study seeks to investigate the effect of economic globalization and renewable energy consumption on environmental quality in the Organization of the Petroleum Exporting Countries (OPEC) over the period 1990–2017. In doing so, we disentangle the economic globalization index into de facto and de jure components to examine their impact on environmental quality. Applying the Pooled Mean Group (PMG) estimator, the long-run estimates show that economic globalization in its overall, de facto, and de jure forms reduce environmental quality. Renewable energy consumption is documented to have a favourable effect on the quality of the environment. The study further finds an inverted U-shaped relationship between real per capita income and environmental pollution, confirming the Environmental Kuznets Curve (EKC) hypothesis in the OPEC bloc. Given these findings, it is crucial for policymakers in the OPEC member countries to review and implement policies to mitigate the negative environmental consequences of economic globalization. Also, environmental sustainability can be achieved by adequately investing in renewable energy.

12.1 Introduction

Globally, climate change has been on the rise due to activities such as excessive fossil fuel usage, agriculture, mining and construction, and many other human events. From these occurrences, human life is not only harmed but the environment also deteriorates as anthropogenic greenhouse gases (GHG) are emitted. Owing to the growth of global economies coupled with an upsurging population in some countries, there has been an escalation in the emission of anthropogenic GHG and

I. N. Yakubu
Graduate School of Social Sciences, Ankara Yildirim Beyazit University, Ankara, Turkey

A. Kapusuzoglu (✉) · N. B. Ceylan
Faculty of Business, Ankara Yildirim Beyazit University, Ankara, Turkey
e-mail: akapusuzoglu@ybu.edu.tr; nbceylan@ybu.edu.tr

S. Yüksel, H. Dinçer (eds.), *Strategic Approaches to Energy Management*, Contributions to Management Science,
https://doi.org/10.1007/978-3-030-76783-9_12

this presents a great concern for nations. International treaties and agreements including the Kyoto Protocol, the United Nations Framework Convention on Climate Change, and the Paris Agreement have keenly emphasized on the need for environmental sustainability urging for measures that will aid in reducing the atmospheric concentration of anthropogenic GHG emissions. Given this scenery, research on the factors influencing environmental degradation has drawn global attention. The work of Grossman and Krueger (1991) on the link between income and environmental degradation resulted in the formulation of the Environmental Kuznets Curve (EKC) hypothesis. Per this theory, environmental pollution is caused by an increase in income during the initial stages of development. However, once income hits a certain level, more increases in income decrease environmental pollution. While some studies have validated this hypothesis (Shahbaz et al., 2016; Khan & Ullah, 2019; Phong, 2019), others have presented contrasting results (Ulucak et al., 2020; Onifade et al., 2021).

In the wake of the EKC hypothesis, the link between environmental degradation and various facets of the economy and society has been studied extensively. The environmental consequence of globalization is one of the pivotal areas that has gained research attention.

The champions of globalization opine that lower environmental degradation is associated with higher globalization levels. According to them, due to globalization firms are subjected to strict environmental regulations which reduce their pollution levels (Christmann & Taylor, 2001). Conversely, globalization opponents argue that environmental quality is reduced by higher globalization levels. They posit that globalization leads to more firms expanding their operations and increasing their production levels. The expansion in production activities is associated with higher demand for energy which increases energy consumption and thus contributes to environmental depletion.

Several research works have looked at the effects of globalization including its subcomponents (social, political, and economic globalization) on environmental degradation at individual country and cross-country levels (Acheampong et al., 2019; Khan & Ullah, 2019; Zafar et al., 2019; Saint Akadiri et al., 2020; Ulucak et al., 2020; Aluko et al., 2021). However, findings from these studies remain unsettled. This study, therefore, seeks to contribute to the ongoing discussion on the globalization-environment link by examining how environmental degradation is impacted by economic globalization in the Organization of the Petroleum Exporting Countries (OPEC). The study is relevant given the fact that the OPEC bloc is among the biggest energy Alliances in the world. Due to globalization (precisely economic globalization), the OPEC states have been regular suppliers of petroleum to various consuming countries, and most countries in the bloc rely mainly on crude oil revenue for supporting economic growth.

Aside from the economic benefit in the form of petrodollar received by OPEC member countries for oil exports facilitated by economic globalization, it is also imperative to find out whether these countries maintain a balance between achieving their target economic benefit from globalization and ensuring environmental sustainability. Hence, the purpose of this chapter.

The study also aims to appraise the influence of renewable energy usage on environmental pollution. Renewable energy utilization and production have intensified in recent times and become one of the most significant factors that improve the growth of the economy. In general, the renewable energy share of the overall energy consumption is trivial in relation to fossil fuels, which take the majority of the total energy consumption. For instance, from the total world energy consumption in 2017, approximately 20.3% constituted renewable energy. The total renewables are expected to reach 45% by the year 2040. Adopting renewable energy particularly in industrialized sectors will not only lessen the need for fossil fuels but will also increase the quality of the environment by abating GHG emissions. In line with this, we investigate whether the use of renewable energy has any notable impact on environmental quality in the OPEC region.

The paper will contribute to the extant literature in three ways. First, the study will probably serve as the pioneering attempt to look at the environmental effect of economic globalization in the OPEC bloc. Second, rather than employing the overall economic globalization index, we disentangle this index into de facto and de jure indicators, thereby departing from earlier empirical studies. In the environmental economics literature, existing studies have generally employed the overall globalization indices to examine their impact on environmental quality. However, the overall indices for the various forms of globalization consist of de facto and de jure measures which need to be distinguished as their respective impact differs. In essence, Gygli et al. (2019) explain that the de facto measure of globalization indicates real flows and activities while the de jure globalization reflects the institutions and policies facilitating the flows and activities. From these definitions, it can be deduced that the de facto indicator reveals actual globalization whiles the de jure factor shows the intention to globalize (Aluko et al., 2021). The use of the de facto and de jure measures of globalization will provide a more comprehensive view on which aspect of the overall economic globalization index exerts more explanatory power on CO_2 emission. The final contribution of the paper is that we add to the few studies that have investigated the influence of energy transition on the environment.

The remaining of the chapter is laid out as follows: Section 12.2 contains a summary of the literature, while Sect. 12.3 focuses specifically on methodology and data. We address our findings in Sect. 12.4, and the conclusion and recommendations are highlighted in Sect. 12.5.

12.2 Literature Review

12.2.1 Theoretical Arguments

As noted by some theoretical presumptions, globalization is a critical factor for environmental quality. For instance, the "pollution haven hypothesis" is among the theoretical frameworks that elucidate the connection between globalization and the environment. According to this hypothesis, firms tend to locate in countries with laxer environmental policies in an attempt to dodge strict environmental regulatory costs coupled with higher energy prices. By operating in such places, firms expand their market and customer base while maintaining a lower cost of production. Also, these lax environments which are mostly developing economies in terms of environmental policies in essence turn out to be a haven for environmental polluters. In proportion to the "pollution haven hypothesis", the "race to the bottom" theory postulates that less developed countries wittingly relax their environmental norms to increase their income levels through the expansion of activities by foreign entities in the host nations. In contrast to this supposition, the "race to the top" hypothesis infer that countries in the developing world do not need to relax environmental regulatory standards to draw investors' attention. The theory emphasizes that under the right conditions, competitions, creativity, development, and environmental quality would always be encouraged (Copeland & Taylor, 2004).

12.2.2 Review of Empirical Studies

Environmental quality issues have attracted much research attention recently owing to the need of preserving the ecosystem and economic activities. In the literature, numerous studies have analyzed how different factors influence environmental quality. For our case, we seek to look at the influence of both economic globalization and the use of renewable energy on the environment. With these objectives, the empirical literature is discussed in the contexts of economic globalization-environmental quality relationship and the effect of renewable energy consumption on the environment.

12.2.2.1 The Impact of Economic Globalization on Environmental Quality

Due to the environmental effects of globalization, there has been growing literature on the influence of the various globalization components on the environment. The influence of economic globalization is noteworthy among the globalization sub-indices. We discuss below some of the studies linking economic globalization and the environment.

Using countries in the OECD and some selected non-OECD countries, Bu et al. (2016) assessed the connection between globalization and climate change over the period 1990–2009. Employing the fixed effects and the two-stage least squares (2SLS) methods, they revealed that all the globalization indices including economic globalization contribute positively to CO_2 emission growth.

Lv and Xu (2018) applied the mean group (MG) and the augmented mean group (AMG) estimators to determine whether economic globalization enhances environmental quality. Using data from 15 selected emerging economies spanning

1970–2012, the study found that economic globalization reduces CO_2 emission irrespective of the technique employed.

In Saudi Arabia, Xu et al. (2018) investigated how financial development influences CO_2 emissions through the moderating role of globalization. Employing the (ARDL) technique on annual time series data covering the period 1971–2016, the study revealed that in both short- and long-run periods, economic globalization positively and significantly drives CO_2 emission.

In light of the EKC hypothesis, Khan and Ullah (2019) investigated the connection between globalization and CO_2 emissions in Pakistan with data spanning from 1975 to 2014. Employing the sub-indices of globalization and the ARDL framework as an estimation technique, the results showed an overall significant long-term effect of globalization on pollution. For the sub-measures, all the indices including economic globalization positively influence CO_2 emission. The study further noted that the EKC premise is valid in Pakistan in the presence of globalization.

Phong (2019) studied the effect of globalization and the development of the financial sector on environmental degradation in the ASEAN-5 nations using annual panel data spanning 1971–2014. Applying the fixed and random effects techniques, the author evidenced that globalization tends to raise CO_2 emission, with economic globalization exerting more impact on CO_2 emission.

Employing data of countries in the North American Free Trade (NAFTA) extending from 1990–2015, Kalayci and Hayaloglu (2019) looked into the effects of economic globalization on CO_2 emissions in the NAFTA region, as well as whether the EKC hypothesis stands true. Deploying the fixed and random effects methods, the results evidenced that economic globalization positively influences CO_2 emission. The study also established the existence of the EKC theory in NAFTA.

In analyzing the effect of the globalization dimensions on the environment, Destek (2020) considered panel data of 12 nations in the Central and Eastern European Countries for the years 1995–2015. The results of the augmented mean group (AMG) approach revealed that rising CO_2 emissions are associated with an improved level of economic globalization. The study, in addition, validated the EKC supposition in the sampled nations.

Suki et al. (2020) scrutinized the influence of globalization on environmental sustainability in Malaysia in view of the EKC hypothesis. Using ecological footprint for gauging environmental degradation, the authors employed the quantile ARDL technique on quarterly data which spanned from 1970 to 2018. The findings demonstrated that, in the long run, globalization in overall and its economic index increase environmental degradation. The study further confirmed the applicability of the EKC hypothesis in the Malaysian economy.

Ulucak et al. (2020) explored the impact of financial globalization (a sub-component of economic globalization) on environmental degradation in emerging economies for the years 1974–2016. Applying different panel techniques including the mean group (MG), pooled mean group (PMG), and dynamic OLS (DOLS), the authors reported that financial globalization reduces environmental

Table 12.1 Summary of the empirical literature (economic globalization)

Author(s)	Country(s)	Period	Estimation technique(s)	Result
Bu et al. (2016)	OECD and non-OECD countries	1990–2009	Fixed effects, two-stage least squares (2SLS)	Positive
Lv and Xu (2018)	Emerging economies	1970–2012	Mean group, augmented mean group	Negative
Xu et al. (2018)	Saudi Arabia	1971–2016	Autoregressive distributed lag (ARDL)	Positive
Khan and Ullah (2019)	Pakistan	1975–2014	Autoregressive distributed lag (ARDL)	Positive
Phong (2019)	ASEAN-5	1971–2014	Fixed effects, random effects	Positive
Kalayci and Hayaloglu (2019)	NAFTA countries	1990–2015	Fixed effects, random effects	Positive
Destek (2020)	Central and eastern European countries	1995–2015	Augmented mean group	Positive
Suki et al. (2020)	Malaysia	1970–2018	Quantile ARDL	Positive
Ulucak et al. (2020)	Emerging economies	1974–2016	Mean group, pooled mean group, dynamic OLS	Negative
Yameogo et al. (2020)	Sub-Saharan Africa	2002–2017	System GMM	Negative
Aluko et al. (2021)	Industrialized countries	1991–2016	Augmented mean group	Negative

degradation, hence contributes to improving a sustainable environment. For the selected economies, the EKC hypothesis was rejected.

Yameogo et al. (2020) in the context of Sub-Saharan Africa investigated how economic globalization affects environmental quality via institutional quality. With data covering the period 2002–2017, the researchers relied on the system generalized method of moments (GMM) estimation technique in the analysis. The findings indicated that economic globalization on its own and through its interaction with institutional quality negatively influence environmental degradation.

Aluko et al. (2021) incorporated globalization into the STIRPAT model to explain its impact on the environment in some selected industrialized economies. Applying the augmented mean group (AMG) technique on panel data covering the period 1991–2016, the findings established that economic globalization in overall and its de facto form lessens environmental degradation.

From the discussion above, given the disparate findings, it is worth noting that no consensus has been reached on the effect of economic globalization on environmental pollution. Thus, we contribute to the ongoing discussions on the economic globalization-environmental quality nexus.

The empirical studies that demonstrate the effect of economic globalization on environmental pollution are summarized in Table 12.1.

12.2.2.2 The Impact of Renewable Energy Consumption on Environmental Quality

The effect of renewable energy use on CO_2 emissions has also been investigated. For instance, employing panel data over the period 1977–2010 and comprising of 17 countries in the OECD, Bilgili et al. (2016) assessed the influence of renewable energy usage on CO_2 emissions within the framework of the EKC. Deploying the FMOLS and the DOLS methods, the findings indicated a negative relationship between renewable energy use and CO_2 emissions. The EKC hypothesis was found to be valid.

Using data from North African nations for the years 1980–2011, Jebli and Youssef (2017) examined the impact of renewable energy consumption and other sets of variables on CO_2 emissions. Employing panel cointegration techniques such as the FMOLS and DOLS, the long-term estimates showed that renewable energy usage is positively correlated with CO_2 emissions.

In analyzing the EKC hypothesis in China, Dong et al. (2018) considered the link between natural gas and renewable energy and CO_2 emission. Using data covering from 1965 to 2016, the authors considered the ARDL estimator to unearth the short- and long-run association between the variables. The empirical results established that natural gas and the use of renewable energy reduces the emissions of CO_2 in China.

Employing data of 25 developing economies which cover from 1996 to 2012, Hu et al. (2018) examined how renewable energy usage affects CO_2 emission. Using the FMOLS and DOLS approaches, the authors found that CO_2 emission rises with increasing demand for renewable energy.

Riti et al. (2018) examined whether renewable energy consumption promotes environmental quality in China using annual data spanning 1971–2013. Relying on the ARDL model as an estimation technique, the researchers found that using renewable energy mitigates emissions levels in the long run.

Using the FMOLS and the DOLS techniques with data over the period 1980–2014, the effect of renewable energy use on CO_2 emissions in developing economies within the Asian continent was studied by Hasnisah et al. (2019). The findings showed a negative insignificant connection between renewable energy usage and CO_2 emission. The EKC hypothesis was also established in the selected Asian countries, according to the analysis.

Adams and Nsiah (2019) looked at the effect of renewable energy on CO_2 emissions in a total of 28 Sub-Saharan African countries from 1980 to 2014. Using the FMOLS and the system GMM techniques, the results have shown that the use of non-renewable and renewable energy induces CO_2 emissions positively though the effect is insignificant for renewable energy.

Alola et al. (2019) studied the influence of renewable energy use and migration on the level of CO_2 emission in the largest economies in the European Union for the period 1990–2016. In the data analysis, the authors applied the FMOLS and DOLS.

The results indicated that the consumption of renewable energy is a vital pollution abatement measure in the sampled economies.

The effects of renewable energy and other factors on CO_2 emissions over the period 2000–2013 have been examined in the BRICS nations by Cheng et al. (2019). Findings from their panel OLS and quantile regression portrayed that in the BRICS, utilizing renewable energy helps to reduce CO_2 emissions.

Salahuddin et al. (2020) empirically analyzed renewable energy's importance in maintaining environmental quality in Sub-Saharan Africa with data covering from 1984 to 2016. Relying on the mean group (MG) approach and common correlated effects mean group (CCEMG) technique, they found that increasing usage of renewable energy boosts environmental sustainability.

Applying the ARDL method, Sahoo and Sahoo (2020) assessed the impact of both renewable and non-renewable energy consumption on emissions in India for the years 1965–2018. The findings revealed that the various sources of non-renewable energy consumption spur CO_2 emission. For renewable energy consumption sources, while hydro energy consumption showed a direct albeit insignificant influence on CO_2 emission, nuclear energy consumption exhibited an inverse and insignificant effect on CO_2 emission.

Table 12.2 summarizes empirical studies that demonstrate how renewable energy use impacts pollution.

12.3 Research Methodology

12.3.1 Data and Variables

To accomplish the objective of the study, a total of 11 countries in the OPEC member states with data covering the period 1990–2017 are considered. The sampled countries presented in alphabetical order include Algeria, Angola, Congo, Gabon, the Islamic Republic of Iran, Iraq, Libya, Nigeria, Saudi Arabia, the United Arab Emirates, and Venezuela. Data are obtained from the World Bank's World Development Indicators and the KOF Globalization Indices by Gygli et al. (2019) which advanced the dataset produced by Dreher (2006). The indices are scaled from 0 to 100, where 0 is the lowest and the highest value is 100. For our measure of environmental degradation, the study uses carbon dioxide (CO_2) emission in kilotons (kt). Economic globalization is proxied by the KOF economic globalization index in its overall, de facto, and de jure forms. Renewable energy consumption (percentage of the final energy used) measures renewable energy consumption. To check the existence of the Environmental Kuznets Curve (EKC), the analysis controls for the effect of real per capita income (constant 2010 US Dollars) and its square.

Table 12.2 Summary of the empirical literature (renewable energy)

Author(s)	Country(s)	Period	Estimation technique (s)	Result
Bilgili et al. (2016)	OECD countries	1977–2010	FMOLS, DOLS	Negative
Jebli and Youssef (2017)	North Africa	1980–2011	FMOLS, DOLS	Positive
Dong et al. (2018)	China	1965–2016	ARDL	Negative
Hu et al. (2018)	Developing countries	1996–2012	FMOLS, DOLS	Positive
Riti et al. (2018)	China	1971–2013	ARDL	Negative
Hasnisah et al. (2019)	Asia	1980–2014	FMOLS, DOLS	Negative
Adams and Nsiah (2019)	Sub-Saharan Africa	1980–2014	FMOLS, GMM	Positive
Alola et al. (2019)	European Union largest economies	1990–2016	FMOLS, DOLS	Negative
Cheng et al. (2019)	BRICS	2003–2013	Panel OLS	Negative
Salahuddin et al. (2020)	Sub-Saharan Africa	1984–2016	MG, CCEMG	Negative
Sahoo and Sahoo (2020)	India	1965–2018	ARDL	Positive/negative depending on the measure of renewable energy consumption.

12.3.2 Model and Analytical Approach

To analyze the influence of economic globalization and renewable energy consumption and the control factors on carbon dioxide (CO_2) emission (the dependent variable), the basic empirical model is specified as:

$$ln\, CO_{2it} = \alpha_0 + \beta_1\, ln\, EGO_{it} + \beta_2\, ln\, REN_{it} + \beta_3\, ln\, Y_{it} + \beta_4\, ln\, Y^2_{it} + \varepsilon_{it} \quad (12.1)$$

where CO_2, EGO, REN denote carbon dioxide emission, economic globalization, and renewable energy consumption in a specific country i at period t respectively. Y designates real per capita income for the respective countries. All the variables in the model have been set to natural logarithms.

This study makes use of the Pesaran et al. (1999) panel ARDL model. The mean group (MG) estimation technique and the pooled mean group (PMG) estimator are two separate estimators used in the panel ARDL approach. The advantage of the panel ARDL method is that irrespective of the stationarity order of the variables, the approach is applicable. We use the Pooled Mean Group (PMG) as our estimation approach. With this technique, we evaluate the long- and short-run impact of the independent factors on environmental quality. In the general form, the equation for the PMG estimator is specified as:

$$Y_{it} = \sum_{j=1}^{p} \delta_i Y_{it-j} + \sum_{j=0}^{q} \lambda_{ij} X_{it-1} + \mu_i + \varepsilon_{it} \tag{12.2}$$

where Y_{it} represents the dependent variable. The vectors of the explanatory factors are denoted by X_{it-1} and λ_{ij} is the coefficient matrix. The unit-specific effects are connoted by μ_i and the error term is signified by ε_{it}. The optimum lags are symbolized by p and q. Following the error correction model, Eq. (12.2) can be augmented as follows:

$$\Delta Y_{it} = \vartheta_i \{ Y_{i(t-1)} - \varnothing_i X_{it} \} + \sum_{j=1}^{p-1} \pi_{ij} \Delta Y_{i(t-j)} + \sum_{j=0}^{q-1} \lambda_{ij} \Delta X_{i(t-1)} + \mu_i + \varepsilon_{it} \tag{12.3}$$

where ϕi is the vector for the long-run estimates. $\{Y_{i(t-1)} - \phi_i X_{it}\}$ signifies the error correction term. The short-run coefficients are represented by πij and λij.

12.4 Empirical Findings

12.4.1 Descriptive Statistics

The summary statistics of the variables (not in logarithm form) are provided in Table 12.3. The sampled countries on average are less economically globalized in overall given the mean score of 48.433, though the average de facto economic globalization score is greater than 50. It is observed that all the series show positive skewness aside from the square of per capita income. CO_2 emission and economic globalization in its overall and de jure forms have high kurtosis values (greater than 3), hence show a leptokurtic distribution. Our variables are not distributed normally, as shown by the Jarque-Bera test's respective probability values.

12.4.2 Panel Unit Root Test

Before testing for cointegration, it is crucial to inspect the statistical properties of the variables through a unit root test (Bekun et al., 2019). The unit root test is conducted to ascertain the order in which the variables are integrated. To reduce biases in the

Table 12.3 Descriptive statistics

Variables	CO_2	EG	EGdf	EGdj	REN	Y	Y^2
Mean	134128.6	48.43295	52.56686	44.01137	29.30418	185925.2	121.6902
Maximum	649480.7	87.00417	90.56947	85.61817	88.83185	690068.8	140.1594
Minimum	575.7190	16.64840	15.58910	16.22024	0.005977	0.00000	96.30681
Std. dev.	154440.3	13.77980	17.60383	15.90372	35.46847	9563.489	12.41555
Skewness	1.700814	0.623252	0.024785	1.021124	0.594676	0.766129	−0.571052
Kurtosis	5.220940	3.673515	2.215187	3.257954	1.514463	2.702277	2.091729
Jarque-Bera	205.6079	25.00879	7.704083	52.78988	45.11634	30.35412	26.52824
Probability	0.000000	0.000004	0.021236	0.000000	0.000000	0.000000	0.000002
Observations	299	299	299	299	299	299	299

Table 12.4 Panel unit root test results

Levels			
Variables	IPS	Fisher-ADF	Fisher-PP
ln CO_2	0.684 (0.742)	13.510 (0.918)	18.604 (0.670)
ln EG	−0.761 (0.223)	28.481 (0.160)	33.130 (0.060)
ln EGdf	−1.537 (0.062)	36.144 (0.029)	48.849 (0.001)
ln EGdj	−0.859 (0.195)	22.162 (0.450)	19.768 (0.598)
ln REN	−1.255 (0.105)	31.085 (0.094)	43.891 (0.004)
ln Y	3.563 (1.000)	6.117 (1.000)	8.231 (0.997)
ln Y^2	3.665 (1.000)	5.777 (1.000)	8.043 (0.997)
First difference			
ln CO_2	−10.814 (0.000)	144.098 (0.000)	242.306 (0.000)
ln EG	−7.203 (0.000)	92.466 (0.000)	181.894 (0.000)
ln EGdf	−7.728 (0.000)	99.497 (0.000)	196.081 (0.000)
ln EGdj	−8.689 (0.000)	113.439 (0.000)	161.550 (0.000)
ln REN	−9.407 (0.000)	123.423 (0.000)	199.061 (0.000)
ln Y	−6.831 (0.000)	89.205 (0.000)	176.612 (0.000)
ln Y^2	−6.793 (0.000)	88.697 (0.000)	171.868 (0.000)

Note: p-values are in ()

estimation, it is critical to eliminate all variables showing second-order integration. The summary of the panel unit root test results from the IPS by Im et al. (2003), Fisher Augmented Dickey and Fuller (Fisher ADF) and Phillips and Perron (1988) tests are outlined in Table 12.4. We note from the results that, for the IPS test, de facto economic globalization (EGdf) is stationary at level. For the Fisher-ADF test, only the de facto economic globalization (EGdf) and renewable energy consumption show stationarity at levels. The Fisher-PP results report economic globalization, de facto economic globalization, and renewable energy consumption to be stationary at level. The test results further depict that at first difference, all the factors are stationary. We, therefore, conclude that our variables do not require further differencing to be stationary.

12.4.3 Panel Cointegration Test

After establishing that most of the factors are stationary at first difference, the cointegration analysis is performed to ascertain how the variables are correlated in the long-term. In doing so, the study employs the panel cointegration approach by Pedroni (1999). In order to establish that cointegration exists, Pedroni (1999) suggested seven test statistics. The first four statistics tests are computed as within dimension. The rest of the three statistics are based on between dimension estimation. The null hypothesis of the Pedroni (1999) test is that there is no cointegration against the alternative hypothesis that the factors are cointegrated. Based on the

Table 12.5 Panel cointegration test results

Alternative hypothesis: common AR coefficients (within-dimension)				
	Statistic	Prob.	Weighted Statistic	Prob.
Panel v-statistic	−0.324	0.627	−0.942	0.827
Panel rho-statistic	−0.562	0.287	−0.418	0.338
Panel PP-statistic	−3.357	0.000	−3.080	0.001
Panel ADF-statistic	−2.068	0.019	−2.741	0.003
Alternative hypothesis: Individual AR coefficinets (between-dimension)				
	Statistic	Prob.		
Group rho-statistic	0.808	0.791		
Group PP-statistic	−2.716	0.003		
Group ADF-statistic	−2.062	0.020		

Table 12.6 Pooled mean group results

Long-run estimates			
Variables	Overall	De facto	De jure
ln EG	1.040*** (0.000)	0.478*** (0.005)	0.408*** (0.000)
ln REN	−1.675*** (0.000)	−0.230** (0.058)	−0.107 (0.142)
ln Y	−1.264 (0.616)	4.904** (0.015)	6.495*** (0.000)
ln Y^2	0.098 (0.388)	−0.174* (0.057)	−0.246*** (0.002)
Short-run estimates			
ECT	−0.224** (0.026)	−0.302*** (0.003)	−0.300*** (0.000)
ln EG	−0.245 (0.602)	−0.530** (0.043)	0.052 (0.805)
ln REN	−0.284430 (0.459)	−0.404 (0.3108)	−0.411 (0.261)
ln Y	−3.827630 (0.918)	−8.947(0.721)	1.815 (0.945)
ln Y^2	0.346930 (0.837)	0.511 (0.6630)	0.012 (0.992)
C	1.614052** (0.038)	−8.711*** (0.003)	−11.269*** (0.000)

Notes: p-values are in ()
*, **, and *** denote the level significance at 10%, 5%, and 1% respectively

significance level of some of the test statistics, our results in Table 12.5 present evidence of cointegration among the variables, and thus rejecting the null hypothesis.

12.4.4 *Estimation Results*

After confirming that there is cointegration amid the variables, the long-run coefficients are then estimated using the ARDL Pooled Mean Group (PMG) approach. From the results in Table 12.6, overall economic globalization has a positive and significant effect on CO_2 emissions, according to long-run estimation. Specifically, CO_2 emission increases by 1.04% as overall economic globalization improves by a percentage. This result conforms with prior studies (Khan & Ullah, 2019; Phong,

2019; Destek, 2020) but controverts Suki et al. (2020) and Yameogo et al. (2020). A similar finding holds for the de facto and de jure economic globalization indicators. The results in overall suggest that as OPEC countries are more engaged in oil-exporting activities facilitated by economic globalization, environmental pollution increases in the long-run. However, economic globalization, with its overall and de facto measures, tend to reduce CO_2 emission in the short term, though the coefficient of the overall economic globalization is insignificant.

The results report that renewable energy utilization reduces CO_2 emission in the long-term, hence contributing to environmental quality. This signifies that the energy transition policies put in place by the countries in the OPEC are laudable. The finding is in line with recent empirical studies that documented the significance of renewable energy consumption in reducing environmental pollution (see Cheng et al., 2019; Alola et al., 2019; Salahuddin et al., 2020) but inconsistent with Hu et al. (2018) and Adams and Nsiah (2019). For the short-run estimates, renewable energy utilization maintains its negative influence on CO_2 emission though statistically insignificant.

In addition, the link between real per capita income and CO_2 emission in the long-run is positive and significant when economic globalization is proxied by the de facto and de jure measures. Correspondingly, the impact of the square of real per capita income on CO_2 emission is negative and significant with the de facto and de jure indicators. The findings point to an inverted U-shaped relationship between real per capita income and CO_2 emissions, confirming the existence of the Environmental Kuznets Curve (EKC) hypothesis in OPEC countries. The finding agrees with most studies that verified the presence of the EKC hypothesis (Shahbaz et al., 2016; Khan & Ullah, 2019; Phong, 2019). In the short-run, however, the EKC hypothesis does not hold.

Finally, the error correction term (ECT) has a negative coefficient, which is statistically significant. From the estimates, the coefficient of the ECT suggests that at the speed of 22.4% annually (for the overall economic globalization model), the short-run disequilibrium should adjust to reach the long-run equilibrium state.

12.5 Conclusion and Policy Implications

In this study, we looked at the environmental effect of economic globalization and renewable energy usage in the OPEC bloc. We analyzed how the various components of economic globalization affect environmental quality. The paper employed the Pooled Mean Group (PMG) technique on panel data covering from 1990 to 2017. We also tested for the presence of the EKC hypothesis by using the real per capita income and its square. The findings established that economic globalization in its overall, de facto, and de jure forms positively and significantly influence pollution in the long-term. This signifies that the growing level of environmental pollution in the OPEC is driven by an increase in trade levels and financialization processes. The study also discovered that using renewable energy helps to maintain environmental sustainability, though the impact is insignificant in the short term. From the findings, the coefficients of real per capita income and its square in the long-run provided evidence that the EKC hypothesis exists in the OPEC nations.

Given the findings of this study, it is crucial for policymakers in the OPEC member countries to review and implement policies to mitigate the negative environmental consequences of economic globalization. They could perhaps impose stringent environmental regulations on firms, particularly foreign firms. For instance, firms seeking to do business or engage in exporting activities must have environmentally friendly technologies that reduce pollution risks. Firms in the OPEC countries should be encouraged to increase their reliance on renewable energy sources. Likewise, environmental sustainability can be achieved by adequately investing in renewable energy.

References

Acheampong, A. O., Adams, S., & Boateng, E. (2019). Do globalization and renewable energy contribute to carbon emissions mitigation in sub-Saharan Africa? *Science of the Total Environment, 677*, 436–446.

Adams, S., & Nsiah, C. (2019). Reducing carbon dioxide emissions; does renewable energy matter? *Science of the Total Environment, 693*, 133288.

Alola, A. A., Yalçiner, K., Alola, U. V., & Saint Akadiri, S. (2019). The role of renewable energy, immigration and real income in environmental sustainability target. Evidence from Europe largest states. *Science of The Total Environment, 674*, 307–315.

Aluko, O. A., Opoku, E. E. O., & Ibrahim, M. (2021). Investigating the environmental effect of globalization: Insights from selected industrialized countries. *Journal of Environmental Management, 281*, 111892.

Bekun, F. V., Alola, A. A., & Sarkodie, S. A. (2019). Toward a sustainable environment: Nexus between CO_2 emissions, resource rent, renewable and nonrenewable energy in 16-EU countries. *Science of the Total Environment, 657*, 1023–1029.

Bilgili, F., Koçak, E., & Bulut, Ü. (2016). The dynamic impact of renewable energy consumption on CO_2 emissions: A revisited environmental Kuznets curve approach. *Renewable and Sustainable Energy Reviews, 54*, 838–845.

Bu, M., Lin, C. T., & Zhang, B. (2016). Globalization and climate change: new empirical panel data evidence. *Journal of Economic Surveys, 30*(3), 577–595.

Cheng, C., Ren, X., Wang, Z., & Yan, C. (2019). Heterogeneous impacts of renewable energy and environmental patents on CO_2 emission-evidence from the BRIICS. *Science of the Total Environment, 668*, 1328–1338.

Christmann, P., & Taylor, G. (2001). Globalization and the environment: Determinants of firm self-regulation in China. *Journal of International Business Studies, 32*(3), 439–458.

Copeland, B. R., & Taylor, M. S. (2004). Trade, growth, and the environment. *Journal of Economic Literature, 42*(1), 7–71.

Destek, M. A. (2020). Investigation on the role of economic, social, and political globalization on environment: Evidence from CEECs. *Environmental Science and Pollution Research, 27*(27), 33601–33614.

Dong, K., Sun, R., & Dong, X. (2018). CO_2 emissions, natural gas and renewables, economic growth: Assessing the evidence from China. *Science of the Total Environment, 640*, 293–302.

Dreher, A. (2006). Does globalization affect growth? Evidence from a new index of globalization. *Applied Economics, 38*(10), 1091–1110.

Grossman, G. M., & Krueger, A. B. (1991). Environmental impacts of a north American free trade agreement (no. w3914). *National Bureau of economic Research.*

Gygli, S., Haelg, F., Potrafke, N., & Sturm, J. E. (2019). The KOF globalization index–revisited. *The Review of International Organizations, 14*(3), 543–574.

Hasnisah, A., Azlina, A. A., & Che, C. M. I. (2019). The impact of renewable energy consumption on carbon dioxide emissions: Empirical evidence from developing countries in Asia. *International Journal of Energy Economics and Policy, 9*(3), 135.

Hu, H., Xie, N., Fang, D., & Zhang, X. (2018). The role of renewable energy consumption and commercial services trade in carbon dioxide reduction: Evidence from 25 developing countries. *Applied Energy, 211*, 1229–1244.

Im, K. S., Pesaran, M. H., & Shin, Y. (2003). Testing for unit roots in heterogeneous panels. *Journal of Econometrics, 115*(1), 53–74.

Jebli, M. B., & Youssef, S. B. (2017). The role of renewable energy and agriculture in reducing CO_2 emissions: Evidence for North Africa countries. *Ecological Indicators, 74*, 295–301.

Kalayci, C., & Hayaloglu, P. (2019). The impact of economic globalization on CO_2 emissions: The case of NAFTA countries. *International Journal of Energy Economics and Policy, 9*(1), 356–360.

Khan, D., & Ullah, A. (2019). Testing the relationship between globalization and carbon dioxide emissions in Pakistan: Does environmental Kuznets curve exist? *Environmental Science and Pollution Research, 26*(15), 15194–15208.

Lv, Z., & Xu, T. (2018). Is economic globalization good or bad for the environmental quality? New evidence from dynamic heterogeneous panel models. *Technological Forecasting and Social Change, 137*, 340–343.

Onifade, S. T., Alola, A. A., Erdoğan, S., & Acet, H. (2021). Environmental aspect of energy transition and urbanization in the OPEC member states. *Environmental Science and Pollution Research*, 1–12.

Pedroni, P. (1999). Critical values for cointegration tests in heterogeneous panels with multiple regressors. *Oxford Bulletin of Economics and Statistics, 61*(S1), 653–670.

Pesaran, M. H., Shin, Y., & Smith, R. P. (1999). Pooled mean group estimation of dynamic heterogeneous panels. *Journal of the American Statistical Association, 94*(446), 621–634.

Phillips, P. C., & Perron, P. (1988). Testing for a unit root in time series regression. *Biometrika, 75* (2), 335–346.

Phong, L. H. (2019). Globalization, financial development, and environmental degradation in the presence of environmental Kuznets curve: Evidence from ASEAN-5 countries. *International Journal of Energy Economics and Policy, 9*(2), 40–50.

Riti, J. S., Song, D., Shu, Y., Kamah, M., & Atabani, A. A. (2018). Does renewable energy ensure environmental quality in favour of economic growth? Empirical evidence from China's renewable development. *Quality & Quantity, 52*(5), 2007–2030.

Sahoo, M., & Sahoo, J. (2020). Effects of renewable and non-renewable energy consumption on CO_2 emissions in India: Empirical evidence from disaggregated data analysis. *Journal of Public Affairs*, e2307.

Saint Akadiri, S., Alola, A. A., Bekun, F. V., & Etokakpan, M. U. (2020). Does electricity consumption and globalization increase pollutant emissions? Implications for environmental sustainability target for China. *Environmental Science and Pollution Research, 27*(20), 25450–25460.

Salahuddin, M., Habib, M. A., Al-Mulali, U., Ozturk, I., Marshall, M., & Ali, M. I. (2020). Renewable energy and environmental quality: A second-generation panel evidence from the sub-Saharan Africa (SSA) countries. *Environmental Research, 191*, 110094.

Shahbaz, M., Solarin, S. A., & Ozturk, I. (2016). Environmental Kuznets curve hypothesis and the role of globalization in selected African countries. *Ecological Indicators, 67*, 623–636.

Suki, N. M., Sharif, A., Afshan, S., & Suki, N. M. (2020). Revisiting the environmental Kuznets curve in Malaysia: The role of globalization in sustainable environment. *Journal of Cleaner Production, 264*, 121669.

Ulucak, Z. Ş., İlkay, S. Ç., Özcan, B., & Gedikli, A. (2020). Financial globalization and environmental degradation nexus: Evidence from emerging economies. *Resources Policy, 67*, 101698.

Xu, Z., Baloch, M. A., Meng, F., Zhang, J., & Mahmood, Z. (2018). Nexus between financial development and CO_2 emissions in Saudi Arabia: Analyzing the role of globalization. *Environmental Science and Pollution Research, 25*(28), 28378–28390.

Yameogo, C. E., Omojolaibi, J. A., & Dauda, R. O. (2020). Economic globalisation, institutions and environmental quality in sub-Saharan Africa. *Research in Globalization, 3*, 100035.

Zafar, M. W., Saud, S., & Hou, F. (2019). The impact of globalization and financial development on environmental quality: Evidence from selected countries in the Organization for Economic co-operation and Development (OECD). *Environmental Science and Pollution Research, 26* (13), 13246–13262.

Chapter 13
Energy Policy Recommendations for ASEAN Countries: Empirical Evidence from the Bootstrap Panel Granger Causality Analysis

Zafer ADALI, Özge KORKMAZ, and Orkun ÇELİK

Abstract In social sciences, especially in the economic literature, energy has been one of the most important topics. Energy seems to be dominant on the political and economic agendas because of nearly all economic activities linked to energy. Hence, all economies have endeavored to detect and implement policies to increase their efficiency and mitigate energy's detrimental effects on the earth. This study's principal mission is to check the connection between economic growth, non-renewable and renewable energy consumption in ASEAN-5 countries over 1990–2014 to recommend energy-saving policies. Within this view, bootstrap panel Granger causality test is applied. The model results indicate that economic growth causes renewable energy consumption in the Philippines and economic growth induces non-renewable energy consumption in Indonesia and Malaysia. A unidirectional causality relationship operating from renewable energy consumption to economic growth is approved for Brunei Darussalam. Finally, a unidirectional causality relationship working from renewable energy consumption to Brunei Darussalam's economic growth is confirmed.

Z. ADALI (✉)
Department of Economy, Artvin Coruh University, Artvin, Turkey
e-mail: zaferadali@artvin.edu.tr

Ö. KORKMAZ
Department of Management Information Technology, Malatya Turgut Özal University, Malatya, Turkey

O. ÇELİK
Department of Management and Organization, Gümüşhane University, Gümüşhane, Turkey
e-mail: ocelik@gumushane.edu.tr

S. Yüksel, H. Dinçer (eds.), *Strategic Approaches to Energy Management*, Contributions to Management Science,
https://doi.org/10.1007/978-3-030-76783-9_13

13.1 Introduction

Modern times provide a vast dilemma for countries. Rapid population rates, fluctuating energy prices, environmental degradation, high required economic growth rate for higher standards of living, sustainable development policies, and a lively environment for their future generation provide a can of worms for the policymakers. Economic growth seems to be a prime objective for the governments among these objectives, but they are aware of required energy policies to sustain their economic goals. However, the current economic structure regarding the link among energy and economic growth should be noticed to implement sound policies. Therefore, the theoretical framework and empirical results related to the association between energy and economic development become a pioneer area for the countries' political and economic agendas.

ASEAN has endeavored to increase their economic activities, provide essential and high needs for the galloping population and urbanization. That is why sound policies influencing current and future economic and social structure appear to be irreplaceable and efficient, and environmentally friendly policies have come to the forward to increase its geographical, economic, and political figure in the present and future. Within this scope, we investigate the nexus between various kind of EC[1] consisting of RE and NONRE, and EG by using bootstrap panel Granger causality test improved by Kónya (2006) on data covering the period 1990–2014 only for Brunei Darussalam, Indonesia, Malaysia, the Philippines, and Singapore due to the available and uninterrupted variables database. Established and implement energy policies require the nexus between EC and EG to sustain and not block current economic activities. We believe that the analysis results will provide recommendations to policymakers, supervisory, and authorities regarding this concept within this context.

13.2 Theoretical Framework for the Nexus Between Energy and Economic Development

The nexus between EC and EG has been received massive awareness since the seminal study conducted by Kraft and Kraft (1978). There have also been theoretical striving studies to detect the nexus between EC and EG. In the literature, four principal hypotheses consist of conservation hypothesis (CH), growth hypothesis (GH), feedback hypothesis (FH), and neutrality hypothesis (NH) to clarify the relationship between EC and economic improvement. The first hypothesis is the

[1] Hereafter, energy consumption, non-renewable energy consumption, renewable energy consumption, economic growth, non-renewable energy sources, renewable energy sources, non-renewable energy, renewable energy denote EC, NONREC, REC, EG, NONRES, RES, NONRE, and RE respectively.

CH, in which there is a unidirectional causality link running from economic development to EC. This system implies that EG stimulates EC as energy demand for various sectors operating in the economy such as manufacturing, construction, and transformation rises because of an increase in economic activities. In light of the CH, energy-saving implications can have no harmful impact on EG. The second hypothesis is the GH, in which there is a unidirectional causality connection moving from EC to EG. This argument argues that EG is conditioned on EC because energy plays a vital role in production as serving as a direct input and indirectly as a complement to labor and capital inputs. According to the GH, policies and measures purposed energy saving can have a detrimental influence on GDP. The third hypothesis is the FH, in which there is a bi-directional causality association between EC and economic development. This implication refers that economic development and EC cause each other. Concerning this case, policy makers should consider the feedback by conducting and proposed implementations to reduce environmental degradation. A performance aimed at shifting from less efficient energy resources to more efficient and environmentally friendly may set a boosting rather than a barrier to economic improvement.

Finally, the end hypothesis is the NH, in which the lack of a causal nexus among EC and EG exists. This hypothesis is that production and economic structure are based on shifting towards services associated with energy-intensive activities. In other words, the cost and function of energy is a tiny dimension of the economic system (Vlahinić-Dizdarević & Žiković, 2010). The policies aimed to decrease environmental degradation do not impact EG, and the fluctuation of economic activities may not influence EC (Saatci & Dumrul, 2013).

13.3 Various Kinds of Energy Sources and their Impacts on the Economy and Environment

Energy has been accepted as an essential part of society regardless of the primitive community and the contemporary society. Various kinds of resources, starting from wood to nuclear energy, have been used to demand human needs regarding human history (Zhou et al., 2021; Liu et al., 2021). Today, there are two significant types of energy resources used for heating, as an input for the production, transformation, etc. One energy source is RES involving wind, solar, hydro, biomass, geothermal, biodiesel, and tidal waves, which have an endless supply because of being recycled or replaced (Zhong et al., 2020; Li et al., 2020). Other energy sources are NONRES involving oil, gas, and cola, a limited resource. The demand for the requisite energy regarding the current system and world population will increasingly continue day by day in developed and developing countries (Li et al., 2021). It is reported that the global energy demand could boost to five times of current energy demand. All countries in the world have increasingly experienced an increase in energy demand, possible exhaustion in NONRES fluctuation of NONRE prices, the detrimental effects of the non-renewable on the earth climate change, CO_2 emissions, health, land, and deforestation. These experiences lead countries to search for alternative and less-harmful, more efficient, and sustainable energy sources (Ali et al., 2017). This section provides alternative energy sources' general features and empirical studies to explain their environmental and economic development effects.

Today industrialization level, the higher standard of living, accessing electricity, and accessibility result from NONRES, but energy provision becomes a global issue resulting in a calamity involving wars, occupation, global climate changes, natural habitats, and living creatures' health. NONRES do not include equal distribution globally, and some countries and their unbalanced distribution make some countries foreign dependent and a significant weight on the economies. NONRES have a limited lifespan which will not suffice future energy demand across the world (Kaya, 2020). Meeting future requirements and preventing environmental issues make RES one of the essential solutions. Within this content, every county, regardless of energy sources imported or exported, developed and developing have been within the process to promote and adopt policies for switching over their NONRES towards RES (Tiwari, 2011). The detecting policies for future energy processes, the effects of NONRES, and understanding of the current accessibility of RES impact on the economy and the environment stimulate massive attention in the literature. Many studies used different econometric approaches, time-variant, and countries' scope confirmed that RES plays an essential role in decreasing CO_2 emissions (Dong et al., 2018; Zoundi, 2017; Bilgili et al., 2016; Raza & Shah, 2018; Dogan & Seker, 2016).

In the theoretical framework, the purpose of studies investigating the relationship between NONRES and CO_2 emission is linked to NONREC in economic activity. When NONRE, which is accepted as detrimental energy sources deteriorating the environment, is the primary source used in economic activity, the energy policies reducing CO_2 emission are likely to dampen economic performance. The policymakers should diversify energy sources such as RE. On the other hand, the total economy is less dependent on NONRE. The policymakers can implement policies related to diversifying energy sources and reduce NONREC to create a more sustainable environment.

13.4 Renewable and Non-Renewable Energy in the ASEAN Countries

In 1967, in order to encourage regional economic, social and cultural co-operation, the ASEAN was established by Indonesia, Malaysia, the Philippines, Singapore, and Thailand (Luukkanen & Kaivo-oja, 2002: 282). Later, the association launched initiatives to increase regionalism by the participation of Brunei (1984), Vietnam (1995), Laos and Myanmar (1997), and Cambodia (1999) (Maizland & Albert, 2020).

One of the most significant initiatives of these countries is energy, RE, and global warming. Especially in the fields regarding RE and energy efficiency, they are well-positioned to take advantage of Clean Development Mechanism (CDM) projects by the Kyoto Protocol, which came into force on 16 February 2005 (Lidula et al., 2007). In 2008, the association pioneered in debate related to mitigation in CO_2 emissions across the world via talks on the Kyoto protocol and the post-Kyoto talks in Bali (Lean & Smyth, 2010).

Although all these advancements, Asia-Pacific Energy Research Center (APERC) foresaw that CO_2 emission would be quadruple in Southeast Asia countries by 2030 if they do not take sufficient and required measures (Lee et al., 2013). The electricity production sector is the primary driver of CO_2 emissions in the ASEAN countries. The sector bases exceedingly on fossil fuels as energy sources (Ang & Goh, 2016). According to the IRENA report (2016), the ASEAN countries have no adequate fossil fuel resources to meet their growing energy demand by 2025, so the share of imported fossil fuel will increase. Also, this case will create significant energy security results. The report IEA (2020) reveals that Indonesia (26%), Vietnam (22%), Thailand (19%), and Malaysia (15%) are the four largest countries by electricity consumption in the ASEAN. Furthermore, the association has a target of integrating 23% RE by 2025. Historically, fossil fuels (such as coal in Indonesia, the Philippines and Malaysia, and oil and gas in Vietnam, Thailand, and Singapore) have dominated the energy mix of the ASEAN countries; nevertheless, this domination has made the energy sector which is the biggest diffuser in the region of greenhouse gases (GHG). Therefore, these countries are also the most significant contributor to global warming (Sandu et al., 2019).

In addition, the ASEAN countries are also rich in RE, especially such as solar, wind, hydro, and geothermal. However, they are unequally shared throughout the region; for instance, Laos's hydro resource is Southeast Asia's battery. The existence's plentiful potential for utility-scale in the land-based wind and solar PV development in many countries such as Thailand and Vietnam, at a range of expected generation costs (Suryadi, 2020).

In the future, targets on RE of the countries in the region are to; (a) grow the capacity of RE to 2.080 MW by 2020 and 4.000 MW until 2030 for Malaysia, (b) enhance the share of new and RE in primary energy supply to reach 23% until 2025 and 31% until 2050 for Indonesia, (c) obtain 10% of electricity generation from RE until 2035 for Brunei, (d) advance solar photovoltaic potential to 350 MW until 2020 for Singapore, (e) raise the growth of REC to 30% until 2036 for Thailand, (f) establish nuclear power plants of 2 GW until 2036 for the Philippines, (g) increase capacity of hydropower to 2.241 MW until 2020 for Cambodia, (h) enhance the share of non-hydro RE-based power generation capacity to 12.5% until 2025 and 21% until 2030, (i) augment the share of non-hydro RE-based power generation capacity to 12.5% until 2025 and 21% until 2030 for Vietnam. Lastly, Lao aims to achieve a 30% share of renewables in the primary energy supply until 2025 (Erdiwansyah et al., 2019: 1098). To reach 100% RE supply of the ASEAN countries, Vidinopoulos et al. (2020) demonstrate that there is a reliance that an energy system decarbonization pathway in these countries is possible. Nevertheless, they show that present policies and actions are insufficient for accomplishment any decarbonization level until 2050.

In sum, the ASEAN is a prominent association with significant power in EG in the region. Although countries in the association have different energy resources and potentials, they should collaborate to determine energy policies.

13.5 Literature Review

In this section, we present a couple of studies based on the hypothesis based on the EC and EG for ASEAN countries. From these studies, Fatai et al. (2004) found the NH for Indonesia, Thailand, the Philippines, New Zealand, India, and Australia by applying Toda and Yamamoto approach on data covering 1961–1990. In other saying, the appearance of the causal connection between EC and EG does not exist. Moreover, Chiou-Wei et al. (2008) employed the VAR model to investigate the causal connection between EC and EG in selected Asian countries. Empirical evidence posed that unidirectional causality operating from EG to EC is valid for the Philippines and Singapore. In contrast, the growth hypothesis is revealed for Taiwan, Hong Kong, Malaysia, and Indonesia.

Azam et al. (2015) made a study to investigate the nexus between EC and EG in the ASEAN-5 countries involving Indonesia, Malaysia, Thailand, Singapore, and the Philippines by using Granger causality analysis on yearly data covering 1980–2012. According to the causality result, there is no causal connection between variables for Indonesia. In the case of Malaysia, the causality link moving from economic to EC is found. The results achieved from the analysis emphasize that EC seems to have an essential and long-term relationship to EG for almost ASEAN-5 countries. Yıldırım et al. (2014) studied the connection between EC and EG for Thailand, Malaysia, Singapore, and the Philippines using panel data causality and time-series causality test 1971–2009. According to the result, the CH is endorsed for the Philippines, Indonesia, and Malaysia. Besides, a bi-directional causality is concluded for Thailand. In the case of Singapore, the NH is approved. Chang et al. (2013) tested the hypothesis based on the nexus EC and EG in 12 Asian countries by employing the bootstrap panel Granger causality analysis for 1970–2010. The NH is favored for Indonesia, Malaysia, Taiwan, and Singapore, while the Philippines' growth hypothesis is approved. In addition, the presence of the FH is accepted for Thailand and Vietnam. Kyophilavong et al. (2015) also found the FH for Thailand.

Furthermore, Vo and Le (2019) endeavored to confirm the EKC by concentrating on the CO_2 emissions, RE, population growth, and EG nexus in the ASEAN countries for the 1971–2014 period. They intended to investigate the EKC hypothesis by focusing on each country using time series techniques. According to the results of Granger causality, cointegration, and FMOLS-DOLS analysis, they concluded a long-run association between variables in Indonesia, Myanmar, and

Malaysia. In contrast, the long-run relation does not appear in Thailand and the Philippines. Besides, the causality relationship between variables does not find regarding Malaysia, Thailand, and the Philippines. Contrarily, there is a unidirectional causality working from EG to REC, and EG leads to CO_2 emission and EC in Indonesia. On the other hand, Myanmar experienced a unidirectional causality operating from EG, EC, and population to RE adoption. Ghazouani et al. (2020) used a bootstrap ARDL approach to review REC and EG's connection in Asia-Pacific countries during 1980–2017. The results of the Granger causality test approved the FH for Malaysia, Indonesia, and Thailand.

13.6 Data and Methodology

This section examines the nexus between energy sources and economic improvement in ASEAN countries to determine efficient energy policies to decrease environmental degradation and sustain economic welfare. Within this purpose, we collected data consisting of GDP (GDP per capita constant 2010 US$), REC (Renewable energy consumption % of total final energy consumption), and NONREC (Non-renewable energy consumption % of total final energy consumption). All data are achieved from the World Bank (2021) database for 1990–2014 only for Brunei Darussalam, Indonesia, Malaysia, the Philippines, and Singapore because of the available and uninterrupted variables database. In this study, the causality connection between variables is examined in the light of the following models.

$$GDP = f(REC) \tag{13.1}$$

$$REC = f(GDP) \tag{13.2}$$

$$GDP = f(NONREC) \tag{13.3}$$

$$NONREC = f(GDP) \tag{13.4}$$

$$REC = f(NONREC) \tag{13.5}$$

$$NONREC = f(REC) \tag{13.6}$$

Regarding T (time dimension) and N (cross-section dimension), Panel Seemingly Unrelated Regression (SUR) model is recommended to apply as T is greater than N. Within this scope, the connection between variables was first analyzed using the SUR model, but the SUR model results were statistically insignificant. Therefore, the study aims to investigate the causality connection between variables by applying the bootstrap panel Granger causality test based on the SUR model.

At the bootstrap panel Granger causality test improved by Kónya (2006), the series can be stationary or not, whereas the assumption of homogeneity should be valid. The bootstrap critical values of test are associated with Wald tests. The bootstrap panel causal model is delivered below:

$$w_{1t} = \delta_{11} + \sum_{l=1}^{p_1} \mu_{11l} w_{1t-l} + \sum_{l=1}^{p_1} \rho_{11l} z_{1t-l} + \varepsilon_{11t}$$

$$w_{Nt} = \delta_{1N} + \sum_{l=1}^{p_1} \mu_{1Nl} w_{Nt-l} + \sum_{l=1}^{p_1} \rho_{Nl} z_{Nt-l} + \varepsilon_{1Nt} \tag{13.7}$$

$$z_{1t} = \delta_{21} + \sum_{l=1}^{p_2} \mu_{21l} w_{1t-l} + \sum_{l=1}^{p_2} \rho_{22l} z_{2t-l} + \varepsilon_{22t}$$

$$z_{Nt} = \delta_{2N} + \sum_{l=1}^{p_2} \mu_{2Nl} w_{Nt-l} + \sum_{l=1}^{p_2} \rho_{2Nl} z_{Nt-l} + \varepsilon_{2Nt} \tag{13.8}$$

In the equations, t and i display the time dimension (t = 1,......, T), and cross-sectional dimension (i = 1,......,N), respectively. Moreover, δ, μ, ρ present common factors and l to the lag. ε demonstrate white-noise errors terms. The causality test results represents per each cross-section (Kónya, 2006). The causal connection can be found that no causality, uni-directional causality (w causes to z or z causes to w), bi-directional causality (See in more details Kónya, 2006).

However, the top necessity for using the bootstrap panel Granger causality test is the heterogeneity of slope coefficients. The Slope Homogeneity Test improved by Pesaran and Yamagata (2008) is used to ascertain whether the slope coefficients are heterogeneous or not. According to the test has two distinct test statistics to inquire hypotheses (Pesaran & Yamagata, 2008):

$$H_0 : \beta_i = \beta_i$$

$$H_0 : \beta_i = \beta_j$$

Whereas the null hypothesis shows the slope coefficients are homogeneous and alternative hypotheses are heterogeneous. $\widetilde{\Delta}$ and $\widetilde{\Delta}_{adj}$ use for large and small samples when it investigates the valid assumption of homogeneity, respectively. If the probability value is less than 5%, it is decided that the homogeneity assumption is valid (See in more details Pesaran & Yamagata, 2008).

13.7 Findings

Firstly, cross-section dependence was initially examined, and the outcomes are presented in Table 13.1. As known, firstly, the heterogeneity of slope coefficients should be used. For this, by Slope Homogeneity Test developed by Pesaran and

Table 13.1 Tests of Cross-section Dependence and Homogeneity

Models	LM	CDLM	CD	LM adj	$\widetilde{\Delta}$	$\widetilde{\Delta}_{adj}$
GDP=f(REC)	175.6870*** (0.0000)	37.0490*** (0.0000)	12.8700*** (0.0000)	5.9030*** (0.0000)	8.8470*** (0.0000)	9.4310*** (0.0000)
REC=f(GDP)	53.8830*** (0.0000)	9.8130*** (0.0000)	0.6010 (0.2740)	5.2490*** (0.0000)	15.1700*** (0.0000)	16.1710*** (0.0000)
GDP=f (NONREC)	176.6810*** (0.0000)	37.2710*** (0.0000)	13.1080*** (0.0000)	1.7600** (0.0390)	15.8540*** (0.0000)	16.9010*** (0.0000)
NONREC=f (GDP)	18.6320** (0.0450)	1.9300** (0.0270)	0.0370 (0.4850)	5.6250*** (0.0000)	14.3640*** (0.0000)	15.3120*** (0.0000)
REC=f (NONREC)	52.3220*** (0.0000)	9.4630*** (0.0000)	4.9160*** (0.0000)	4.8750*** (0.0000)	14.7930*** (0.0000)	15.7690*** (0.0000)
NONREC=f (REC)	33.9670*** (0.0000)	5.3590*** (0.0000)	−0.7460 (0.2280)	4.4370*** (0.0000)	14.4360*** (0.0000)	15.3890*** (0.0000)

Notes: (1) ***, ** and * show the significance level at 1%, 5% and 10%, respectively. (2) $\widetilde{\Delta}$ and $\widetilde{\Delta}_{adj}$ show the homogenity tests. (3) $H_0{:}\beta_{i=}\beta_i$; $H_1{:}\beta_{i=}\beta_j$; for homogenity test. (4) LM Breusch and Pagan (1980), CDLM (Pesaran, 2004), CD (Pesaran, 2004), LMadj (Pesaran et al., 2008a, 2008b) are tests of cross-section dependence

Yamagata (2008) is used. The consequences of homogeneity and cross-section dependence tests are displayed in Table 13.1.

According to Table 13.1, the null hypotheses of the slope coefficient implying homogeneity and no dependence are rejected at 1% level. Hence, it is understood that the slope coefficients are heterogeneous, and it is determined that cross-sectional dependency does not exist. Then, the causality test results are exhibited in Table 13.2.

As examining the causality analysis result, it is observed that the unidirectional causality relationship moving EG to REC exists in only the Philippines. Besides, the results attest to the presence of a bi-directional causality connection between REC and NONREC for the Philippines. Nevertheless, the REC causes the EG just for Brunei Darussalam. Regarding Indonesia and Malaysia, EG leads to NONREC.

13.8 Conclusion and Policy Recommendation

Human being current and future life leads to a considerable dilemma and discussion for policymakers and ordinary people. Economic well-being, a higher standard of life, and reaching high economic development and sustainability generate pressure to increase in economic activities. Furthermore, devastating human activities associated with economic greed destroy and harm the earth; maybe there will be no livable environment for future generations. Therefore, non-governmental foundations, government agencies, and even green political parties have created a consensus that governments and firms should change their current economic and energy framework towards RES. Almost all economic activities are associated with energy,

Table 13.2 The Results of Boostrap Panel Granger Causality Test

Null Hypothesis: *GDP* ↛ *REC*	Brunei Darussalam	Indonesia	Malaysia	The Philippines	Singapore
Wald-statistics	0.3550 (0.6400)	0.0640 (0.9110)	1.2440 (0.6600)	8.1390* (0.0670)	0.6350 (0.3200)
Cv 1%	4.5380	5.7230	14.8420	10.6620	6.3710
Cv 5%	3.3490	3.7960	9.0510	8.6510	3.0300
Cv 10%	2.6550	2.8960	7.1130	7.2620	1.9540
Null hypothesis: *REC* ↛ *GDP*	Brunei Darussalam	Indonesia	Malaysia	The Philippines	Singapore
Wald-statistics	2.6700* (0.053)	0.1660 (0.8190)	0.6420 (0.8730)	0.0010 (0.9980)	1.3500 (0.2490)
Cv 1%	5.0100	5.8710	11.6070	5.9300	6.7740
Cv 5%	2.7140	3.6540	8.0610	4.0350	3.7970
Cv 10%	1.9410	2.7840	6.4890	3.3220	2.7840
Null hypothesis: *GDP* ↛ *NONREC*	Brunei Darussalam	Indonesia	Malaysia	The Philippines	Singapore
Wald-statistics	0.3470 (0.9640)	6.1700*** (0.0010)	6.4320*** (0.0100)	0.9500 (0.7380)	4.7690 (0.5910)
Cv 1%	17.0670	3.3570	6.4060	6.0940	2.7150
Cv 5%	11.9720	2.0610	3.9270	4.1640	3.4760
Cv 10%	9.3460	1.3770	2.7150	3.4760	10.1300
Null hypothesis: *NONREC* ↛ *GDP*	Brunei Darussalam	Indonesia	Malaysia	The Philippines	Singapore
Wald-statistics	4.3350 (0.5010)	3.4920 (0.2030)	0.1120 (0.8950)	0.8970 (0.3040)	1.4040 (0.9620)
Cv 1%	19.7070	8.3910	6.4540	4.5550	12.2380
Cv 5%	14.0250	5.4350	4.3930	3.1340	9.8020
Cv 10%	10.8290	4.4730	3.3690	2.1430	8.3060
Null hypothesis: *REC* ↛ *NONREC*	Brunei Darussalam	Indonesia	Malaysia	The Philippines	Singapore
Wald-statistics	0.7270 (0.2320)	0.1450 (0.8340)	4.4150 (0.1770)	2.4790* (0.0970)	0.7420 (0.4450)
Cv 1%	3.8880	5.8200	9.4750	5.1970	10.1140
Cv 5%	2.2870	3.3570	6.4280	3.3630	5.8290
Cv 10%	1.5090	2.5890	5.3000	2.4050	3.8410
Null hypothesis:*NONREC* ↛ *REC*	Brunei Darussalam	Indonesia	Malaysia	The Philippines	Singapore
Wald-statistics	0.7040 (0.2690)	1.1550 (0.9030)	2.2720 (0.4840)	23.4270*** (0.0000)	6.1360 (0.3560)
Cv 1%	4.9490	12.8840	12.3660	14.5460	20.7320
Cv 5%	2.4700	8.9930	8.0470	11.5760	14.2630
Cv 10%	1.6810	7.0350	6.2430	10.2610	11.8260

Note: (1)***, ** and * show the significance level at 1%, 5% and 10%, respectively. (2) The paranthesis shows the probability values. (3) Cv shows the boostrap critical values.

and hence it is required to detect the connection between energy and economic activities to mitigate the effects of energy on environmental degradation and sustain EG.

Parallel to this context, we aim to review the connection between EG, REC and NONREC in ASEAN countries. In this study, Brunei Darussalam, Indonesia, Malaysia, the Philippines, and Singapore are to be selected, and the period covers 1990–2014. Bootstrap panel Granger causality test is applied to review the nexus between EC and EG in ASEAN countries. With the causality test, we achieve finding causality result separately for each country. According to the results, it was observed that the conservation hypothesis exists for the Philippines; in other words, changes in REC result from a change in EG. According to this hypothesis, energy-saving policies do not harm EG because EG stimulates EC as energy demand for several sectors working in the economy. In addition to this, there is also a bi-directional causality relationship between REC and NONREC for the Philippines. According to above mentioned two causality results, it can be implied that typically, fossil fuels such as coal have dominated energy sources in the Philippines. However, the Philippine's economic structures are based on services or less-energy-dependent sectors because an increase in EG causes REC. It is recommended to the Philippines to increase RES and shifting funds and pieces of equipment used for the NONRES to RE investment. Another conservation hypothesis is detected for Indonesia and Malaysia, but there is a uni-directional causality relationship moving from EG to NONREC, unlike the Philippines. Like the Philippines, fossil fuel energy sources also have prime energy sources in Malaysia and Indonesia, but the countries are based on manufacturing and more-energy dependent sectors.

As a result of this finding, Indonesia and Malaysia should focus on energy-saving policies to change in current economic energy structure and invest in research and development activities to establish RE used types of equipment. This change in economic and energy policies will play an important role in Malaysia and Indonesia in the long-run because NONRES are limited and unsustained energy sources, and NONRES have detrimental effects on human-being, animals, forest, and other natural country' endowments. Besides, the growth hypothesis is found for Brunei Darussalam; in other words, there is a unidirectional causality relationship between REC and EG. As a result of this conclusion, Brunei Darussalam should increase funds for investment in RE sectors, and it becomes an indispensable object on the government's agenda.

References

Ali, S., Anwar, S., & Nasreen, S. (2017). Renewable and non-renewable energy and its impact on environmental quality in south Asian countries. *Forman Journal of Economic Studies, 13*, 177–194.

Ang, B. W., & Goh, T. (2016). Carbon intensity of electricity in ASEAN: Drivers, performance and outlook. *Energy Policy, 98*, 170–179.

Azam, M., Khan, A. Q., Bakhtyar, B., & Emirullah, C. (2015). The causal relationship between energy consumption and economic growth in the ASEAN-5 countries. *Renewable and Sustainable Energy Reviews, 47*, 732–745.

Bilgili, F., Koçak, E., & Bulut, Ü. (2016). The dynamic impact of renewable energy consumption on CO_2 emissions: A revisited environmental Kuznets curve approach. *Renewable and Sustainable Energy Reviews, 54*, 838–845.

Breusch, T. S., & Pagan, A. R. (1980). The Lagrange multiplier test and its applications to model specification in econometrics. *The Review of Economics Studies, 47*(1), 239–253.

Chang, T., Chu, H. P., & Chen, W. Y. (2013). Energy consumption and economic growth in 12 Asian countries: Panel data analysis. *Applied Economics Letters, 20*(3), 282–287.

Chiou-Wei, S. Z., Chen, C. F., & Zhu, Z. (2008). Economic growth and energy consumption revisited—Evidence from linear and nonlinear granger causality. *Energy Economics, 30*(6), 3063–3076.

Dogan, E., & Seker, F. (2016). Determinants of CO_2 emissions in the European Union: The role of renewable and non-renewable energy. *Renewable Energy, 94*, 429–439.

Dong, K., Sun, R., Li, H., & Liao, H. (2018). Does natural gas consumption mitigate CO_2 emissions: Testing the environmental Kuznets curve hypothesis for 14 Asia-Pacific countries? *Renewable and Sustainable Energy Reviews, 94*, 419–429.

Erdiwansyah, R. M., Sani, M. S. M., & Sudhakar, K. (2019). Renewable energy in Southeast Asia: Policies and recommendations. *Science of the Total Environment, 670*, 1095–1102.

Fatai, K., Oxley, L., & Scrimgeour, F. G. (2004). Modelling the causal relationship between energy consumption and GDP in New Zealand, Australia, India, Indonesia, the Philippines and Thailand. *Mathematics and Computers in Simulation, 64*(3–4), 431–445.

Ghazouani, T., Boukhatem, J., & Sam, C. Y. (2020). Causal interactions between trade openness, renewable electricity consumption, and economic growth in Asia-Pacific countries: Fresh evidence from a bootstrap ARDL approach. *Renewable and Sustainable Energy Reviews, 133*, 110094.

IEA. (2020). *Electricity market report December 2020*. https://www.iea.org/reports/electricity-market-report-december-2020/2020-regional-focus-southeast-asia. Accessed 05 March 2021.

IRENA. (2016). *Renewable energy outlook for ASEAN*. https://www.irena.org/publications/2016/Oct/Renewable-Energy-Outlook-for-ASEAN. Accessed 05 March 2021.

Kaya, M. (2020). Evaluation of the existing solar energy and rainwater potential in the Total roof area of buildings: Izmit District example. *Advances in Materials Science and Engineering, 2020*, 1–16.

Kónya, L. (2006). Exports and growth: Granger causality analysis on OECD countries with a panel data approach. *Economic Modelling, 23*(6), 978–992.

Kraft, J., & Kraft, A. (1978). On the relationship between energy and GNP. *The Journal of Energy and Development, 3*, 401–403.

Kyophilavong, P., Shahbaz, M., Anwar, S., & Masood, S. (2015). The energy-growth nexus in Thailand: Does trade openness boost up energy consumption? *Renewable and Sustainable Energy Reviews, 46*, 265–274.

Lean, H. H., & Smyth, R. (2010). CO_2 emissions, electricity consumption and output in ASEAN. *Applied Energy, 87*(6), 1858–1864.

Lee, Z. H., Sethupathi, S., Lee, K. T., Bhatia, S., & Mohamed, A. R. (2013). An overview on global warming in Southeast Asia: CO_2 emission status, efforts done, and barriers. *Renewable and Sustainable Energy Reviews, 28*, 71–81.

Li, Y. X., Wu, Z. X., Dinçer, H., Kalkavan, H., & Yüksel, S. (2021). Analyzing TRIZ-based strategic priorities of customer expectations for renewable energy investments with interval type-2 fuzzy modeling. *Energy Reports, 7*, 95–108.

Li, X., Zhu, S., Yüksel, S., Dinçer, H., & Ubay, G. G. (2020). Kano-based mapping of innovation strategies for renewable energy alternatives using hybrid interval type-2 fuzzy decision-making approach. *Energy, 211*, 118679.

Lidula, N. W. A., Mithulananthan, N., Ongsakul, W., Widjaya, C., & Henson, R. (2007). ASEAN towards clean and sustainable energy: Potentials, utilization and barriers. *Renewable Energy, 32* (9), 1441–1452.

Liu, J., Lv, J., Dinçer, H., Yüksel, S., & Karakuş, H. (2021). Selection of renewable energy alternatives for green Blockchain investments: A hybrid IT2-based fuzzy modelling. *Archives of Computational Methods in Engineering, 28*(1), pp. 1–15.

Luukkanen, J., & Kaivo-oja, J. (2002). ASEAN tigers and sustainability of energy use decomposition analysis of energy and CO_2 efficiency dynamics. *Energy Policy, 30*(4), 281–292.

Maizland, L, & Albert, E. (2020). *What is ASEAN?* https://www.cfr.org/backgrounder/what-asean. Accessed 05 March 2021.

Pesaran, M. H. (2004). General diagnostic tests for cross section dependence in panels. *Cambridge Working Papers in Economics no. 435*, University of Cambridge.

Pesaran, M. H., Ullah, A., & Yamagata, T. (2008a). A bias-adjusted LM test of error cross-section independence. *The Econometrics Journal, 11*(1), 105–127.

Pesaran, M. H., Ullah, A., & Yamagata, T. (2008b). A bias-adjusted LM test of error Crosssection Independence. *The Econometrics Journal, 11*(1), 105–127.

Pesaran, M. H., & Yamagata, T. (2008). Testing slope homogeneity in large panels. *Journal of Econometrics, 142*(1), 50–93.

Raza, S. A., & Shah, N. (2018). Testing environmental Kuznets curve hypothesis in G7 countries: The role of renewable energy consumption and trade. *Environmental Science and Pollution Research, 25*(27), 26965–26977.

Saatci, M., & Dumrul, Y. (2013). The relationship between energy consumption and economic growth: Evidence from a structural break analysis for Turkey. *International Journal of Energy Economics and Policy, 3*(1), 20.

Sandu, S., Yang, M., Mahlia, T. M. I., Wongsapai, W., Ong, H. C., Putra, N., & Rahman, S. M. (2019). Energy-related CO_2 emissions growth in ASEAN countries: Trends, drivers and policy implications. *Energies, 12*(24), 4650.

Suryadi, B. (2020). *Energy priorities in high-performing ASEAN.* https://www.energyforgrowth.org/memo/energy-priorities-in-the-asean-region/. Accessed 05 March 2021.

Tiwari, A. K. (2011). Comparative performance of renewable and nonrenewable energy source on economic growth and CO_2 emissions of Europe and Eurasian countries: A PVAR approach. *Economics Bulletin, 31*(3), 2356–2372.

Vidinopoulos, A., Whale, J., & Hutfilter, U. F. (2020). Assessing the technical potential of ASEAN countries to achieve 100% renewable energy supply. *Sustainable Energy Technologies and Assessments, 42*, 100878.

Vlahinić-Dizdarević, N., & Žiković, S. (2010). The role of energy in economic growth: The case of Croatia. *Zbornik radova Ekonomskog fakulteta u Rijeci: časopis za ekonomsku teoriju i praksu, 28*(1), 35–60.

Vo, D. H., & Le, Q. T. T. (2019). CO_2 emissions, energy consumption, and economic growth: New evidence in the ASEAN countries. *Journal of Risk and Financial Management, 12*(3), 145.

World Bank. (2021). https://data.worldbank.org/. Accessed 28 February 2021.

Yıldırım, E., Aslan, A., & Ozturk, I. (2014). Energy consumption and GDP in ASEAN countries: Bootstrap-corrected panel and time series causality tests. *The Singapore Economic Review, 59* (02), 1450010.

Zhong, J., Hu, X., Yüksel, S., Dinçer, H., & Ubay, G. G. (2020). Analyzing the investments strategies for renewable energies based on multi-criteria decision model. *IEEE Access, 8*, 118818–118840.

Zhou, P., Luo, J., Cheng, F., Yüksel, S., & Dinçer, H. (2021). Analysis of risk priorities for renewable energy investment projects using a hybrid IT2 hesitant fuzzy decision-making approach with alpha cuts. *Energy, 224*(4), 120184.

Zoundi, Z. (2017). CO_2 emissions, renewable energy and the environmental Kuznets curve, a panel cointegration approach. *Renewable and Sustainable Energy Reviews, 72*, 1067–1075.

Chapter 14
Strategic and Management Thinking: An Examining of Sustainable Energy Management

Zafer Adiguzel

Abstract In general, energy management has emerged as a vital concept to not only address the increasingly serious energy problems but to reduce the consumption, cost, and environmental impact of scarce and valuable energy resources. Because energy is a cost element for both states and companies, the importance of energy management stems from the global need for energy savings at all scales. Among the objectives of energy management are to conserve resources and cost savings while reducing environmental impacts and providing uninterrupted access for energy users. However, with the adoption of a strategically sustainable management approach, together with energy management, future-oriented investments are needed. In the study, energy, strategy and sustainability concepts are examined.

14.1 Introduction

Due to the effect of technological developments within the last century, production using human and animal energy has mostly been replaced by industrial production. In fact, energy has been one of the most important factors in the production of industrial products or services to improve the quality of life. Yet rapidly increasing energy consumption has caused the depletion of natural resources, bringing environmental problems that threaten the improving quality of life. The fact that a significant amount of energy production is based on the use of resources that are both limited and harmful to the environment, in addition to constantly increasing energy demands that cannot be sufficiently met, raises the issue of the efficient use of energy. Furthermore, businesses aim to continue their activities in a less costly manner, causing the concepts of energy efficiency and energy saving to be more urgently discussed. As a result, the importance of energy management practices that aim to use energy more efficiently and increase energy efficiency continues to

Z. Adiguzel (✉)
Medipol Business School, İstanbul Medipol University, İstanbul, Turkey
e-mail: zadiguzel@medipol.edu.tr

S. Yüksel, H. Dinçer (eds.), *Strategic Approaches to Energy Management*, Contributions to Management Science,
https://doi.org/10.1007/978-3-030-76783-9_14

increase. The concept of energy is of primary importance in businesses targeting clear profit. After the industrial revolution, the sustainability of enterprise work in the manufacturing sector became dependent on the use of existing energy resources, moving towards mechanization. Today, energy demand is growing faster than energy production. Especially, the high energy consumption based on fossil fuels has triggered threats, such as global warming and environmental pollution, while the rapid consumption of scarce resources has caused an increase in energy costs (Martins et al., 2019). From this point of view, it can be argued that the concept of the energy problem is actually due both to the limited number of resources from which energy is obtained and from environmental pollution. Investigating whether energy management systems, especially those more recent government programs within the framework of energy policies, are applied by generational companies show that more studies should be done on this subject. Not surprisingly, the fiercely competitive conditions that businesses have to struggle with, and at the same time, the increase in the sensitivity to protect the livable environment, have forced businesses to rethink their goals. Today, the general aim is to try and prevent enterprises from harming nature and polluting the environment while producing products/services through both independent organizations and government institutions, such as ministries. As a result, while certain incentives, advantages, and rewards are brought to enterprises that protect the environment, deterrent measures are taken against the businesses that harm the environment. Of course, in order to meet energy needs, it is necessary to consider the procedures applied by the countries importing energy. For this reason, it is important for each country to continue its current business activities with less energy consumption. In line with all these reasons, energy management is applied in order to protect natural resources and to increase the profitability of the enterprises by reducing energy costs and is a concept that cannot be ignored by enterprises that want to survive. In the study, basic concepts related to Energy Management, Strategy, and Sustainability were examined and subjected to a detailed literature review.

14.2 The Concept of Energy

The use of energy is a must in order for physical actions to take place and a transformation to occur. The movement of an entity with material content on different points, and the physical or chemical transformation of an entity are actions, and energy must be used for these actions to take place. The use of energy, whose origins have been present since the existence of humanity, is an indispensable input for the continuance of many things we see in our daily life to continue performing within their functions. Without energy, the situation of the world and humanity would be very different from what it is now. For this reason, issues such as energy production, energy resources, and energy distribution are always at the top of the issues that humanity deals with and discusses. Suffice to say, the limitation of energy resources and the continuous increase in the world population have brought major problems in the distribution of energy in the world, oftentimes leading to great wars. The energy need of all developed or developing economies is of critical importance in terms of its strategic potential. In determining the value of energy, the presence of and the need for an economic approach is key (Heffron & McCauley, 2017). Indeed, the need for energy amongst the ever-increasing global demands increases the importance of energy, which is limited in nature. Equally so, the emergence of the need for large energy use within industrial production has made the energy issue even more critical. Due to the limited fossil resources and the resulting damage to nature, the trend towards renewable energy sources is increasing. However, the initial investments of these resources create high costs for many economic units. Therefore, reducing these costs will result in the use of more renewable energy resources (Twidell & Weir, 2015).

With the discovery that energy can transform into other forms, human life has seen great development, leaving the foundations of modern civilization (Gielen et al., 2019). Energy, which increases the quality of life among modern people and contributes to economic and social development, is of vital importance in meeting both basic and complex needs (Bergasse & Paczynski, 2013). Energy, which is so important for people and life, is an important parameter in various fields from economy to sociology, physics to environmental science, and politics to health (Azam, 2020). According to the data of the World Bank, approximately one fourth of the people living in the world cannot benefit from electricity (World Bank, 2018). Based on this data, multiple questions regarding communication and transportation arise: "why hasn't energy use become widespread around the world and "is it too late to address this problem?" come to mind. As an answer to this question, it is stated that "The scientific description of energy was made quite late" (Smil, 2018). Shortly after the scientific definition of energy, the Law of Conservation of Energy took its place within literature (Farhodovna & Abdishukurovna, 2021). But, unfortunately, awareness of the importance of conserving limited energy reserves in the universe came late. Even still, innovations in non-traditional resources that emerged with developing technology promise to fulfill our responsibility towards future generations.

14.3 Importance of Energy Concept in Management Thinking

When talking about energy efficiency, many concepts are used, starting with the word energy. Haider (2020) stated that the world will need 60% more energy in 2030 compared to 2008, and in fact, 80% of this energy demand will be met from fossil fuels while the biggest increase in demand among fossil resources will be experienced in natural gas. For this reason, countries need to invest in renewable energy resources, take advantage of the resources they own, and reduce their dependence on foreign sources. Energy efficiency means ensuring that the "end use" (heating, cooling, lighting, etc.) of energy is kept at the same quality and level with lower energy input level (Beggs, 2010). Today's developments and changes involve the evaluation of the concept of productivity, the favorite term of the industrial age, with a new perspective. Productivity should now be perceived as a business culture and a lifestyle (Liu et al., 2019). In this context, the importance of productivity in management is an undeniable fact. Minimizing costs is of vital importance in increasing productivity during the production phase. The fact that important costs in the production sector are energy costs reveals the importance of energy in the production sector. Indeed, the more effective the use and management of energy, the more productivity there will be.

The fact that energy needs continue to increase also impacts the costs and the supply required to meet the demand. This demand, in turn, has a polluting effect on the environment, leading to developments that contribute to the adoption of an energy-oriented management approach. These developments can be classified as legal obligations imposed by countries, technical factors, economic factors, social factors and global factors.

14.4 Definition of Energy Management

The concept of energy management is not new to human history, but has been an important part of human life for centuries. Energy management is carried out to ensure the efficient use of energy resources and energy, including training, study, measurement, monitoring, planning and implementation activities (Capehart et al., 2020). Energy management is similar to other management activities and if planning, coordination and control mechanisms are independent from each other, functions that will remain ineffective come together and form a whole. Energy Management is a disciplined work that is structured and organized in line with the more efficient and rational use of energy resources and energy without sacrificing the quality of the product, safety (or other environmental conditions), and without reducing production. As civilization has developed, the importance of the efficient and rational use of energy has become evident due to multiple reasons, such as increasing energy needs, decreasing natural resources, and the global climate change threat. Energy management focuses on the efficient and rational use of energy to reduce consumption and negative effects on the environment without affecting performance. In the broadest terms, energy management is the efficient and effective use of energy in order to strengthen the competitive position of an enterprise and maximize profits (minimizing costs) (Thollander & Palm, 2012). Energy management helps to minimize energy costs, increase the quality of life, and reduce environmental impacts with the efficient and sensitive use of energy (Lee & Cheng, 2016). In order to meet environmental and economic needs after the supply and conversion of energy, energy management is needed for the distribution, use, organization, and coordination of the resulting energy (Thollander & Ottosson, 2010). In other words, the use of energy according to the principle of efficiency, the examination of many interrelated situations in energy management, and the expression of a holistic perspective are examined within the concept of energy management. However, the common point in these definitions is to meet the current energy requirement of the enterprise in the most efficient way without any restrictions and without any reduction in production. Undoubtedly, energy savings that would be achieved by decreasing production is not a desired result for companies. Rather, companies aim to increase profitability as a result of the efficient use of energy, while at the same time exhibiting a sensitive management behavior that minimizes the negative effects on the environment. Thanks to energy management, an organization can find the opportunity to improve in various aspects such as the quality and safety of energy systems.

14.5 The Concept of Strategy

In management science, strategy shows the ways a business can achieve its purpose (Freedman, 2015). In particular, the strategies that businesses follow in the market when involved with competitors is what first led to the emergence of a discipline called strategic planning and then strategic management (Contreras Sierra, 2013). While strategic management was known and applied by multinational companies, large holdings, and companies in the private sector even before the 1980s, today many enterprises are implementing strategic management (Peng, 2021). Even in very small enterprises, the determination of policies can be done by people who know or predict all possibilities and conjunctures (Gray, 2015). Businesses need a body that determines the basic strategies in order to realize the policies that have the characteristics of purpose (Teece, 2010).

There are methods or processes that we can describe as a set of strategic tools that companies use today to ensure energy efficiency. These methods and processes provide companies with a set of principles in providing energy efficiency. Some of these strategic tools can be listed as cleaner production, life cycle assessment, environmental design, green supply chain management and strategic environmental assessment.

14.6 Strategic Environmental Assessment (SEA)

Environmental pollution, which emerged with the industrial revolution, has ceased to be point-based in time, and its cumulative effects have become an issue to be taken into account. In addition, the fact that some environmental problems that have emerged affect the whole world offer a common future concern and the need for a holistic approach to these problems (Fischer, 2010). SEA is important for the participation of environmental values in the decision-making process. Although the SEA should be a guide, it should also be transparent in decision-making since the SEA process is a participatory process (Lobos & Partidario, 2014). It is stated that the need for the evaluation of policies, plans, and programs emerged as a continuation of the Environmental Impact Assessment, first put into practice in 1969 with the National Environmental Protection Law in the United States (Therivel, 2012). SEA can be defined as the process of evaluating the effects of policies, plans, and programs on the environment in accordance with the inclusive approach. The Environmental Impact Assessment, which includes only project-based applications, is a tool for eliminating the deficiencies of combating environmental problems (Therivel & Paridario, 2013). In fact, the SEA has a very important role in the integration of the environment into strategic decision-making mechanisms (Therivel et al., 2013).

14.7 The Importance of Energy in Business

One of the important issues in terms of energy in businesses is the cost of energy. Increasing energy costs and competitive conditions have made the efficient use of resources an extremely important issue. With energy saving efforts, the energy costs of enterprises will be optimized and their competitive power will be increased (Burger & Weinmann, 2012). Businesses that do not take energy costs into consideration and do not implement energy management are more likely to be out of the competition. For this reason, every business must have sustainable energy management, and energy requirements in companies may differ depending on various factors (Hellström et al., 2015). Heating, cooling, lighting, the operation of computers, machines of various sizes and functions, and transportation logistics are all possible with the use of energy. Particularly in large industrial enterprises, the cost of energy reach can be quite high due multiple factors: the area to be heated or cooled, the number of lighting units and computers, and the machines demands. For this reason, it is necessary to determine the energy consumption in the enterprises. For this purpose, energy monitoring systems are rapidly becoming widespread in industrial enterprises in calculating energy consumption costs. Monitoring data, such as voltage fluctuations, energy consumption, power factors, and changes in frequency and current have become important factors for businesses in order to measure energy quality. Different energy consumption values during the day have also been an important factor in calculating production and consumption costs for the enterprise (Baker, 2019). In the studies, the data obtained (as a result of monitoring the instantaneous and previous energy usage of the enterprises via computers) are continuously recorded and the opportunity to make energy analysis of the previous period has been provided. In this way, the energy consumption of the business can be determined in the desired time interval. As well, opportunities such as reducing the cost of the energy used to the enterprise can be provided by shifting the busy working hours to times when energy is cheaper (Bryant et al., 2018).

14.8 Sustainability

Most of the concepts widely accepted today, with regard to sustainability, stem from the 1987 United Nations' World Environment and Development Commission's Our Common Future Report (also called Brundtland Report). "Sustainability is an economic development activity that meets today's needs without compromising the ability of future generations to meet their own needs." (Wced, 1987). The 1987 report of the Commission defined sustainability as three equal parts, with elements including environment, economy, and equity, all of which provided the basis for understanding and elaborating on sustainability (Caradonna, 2014).

Sustainability means being productive in a way that supports future generations so that people and nature can create and sustain the conditions in which they can exist (Clayton & Radcliffe, 2018). The simple and basic principle on which sustainability is based on the idea that everything needed to live and survive in prosperity is found in nature. Sustainability reveals a concept that requires change in our overall approach and style without compromising the quality of life. The basis of this idea is to modify or eliminate the consumerism lifestyle and to target climate management, social responsibility, and economic solutions that are in global solidarity (Gibson et al., 2013).

In the widespread use of sustainability practices, it has the feature of being a very comprehensive, multidimensional, and multidisciplinary concept. Indeed, sustainability's importance increases in developmental studies as its critical attitudes take all issues and solutions into account. In order to understand the concept of sustainable development, it is necessary to examine the individual sectors because after analyzing what sustainable development projects can be applied in the world, the right investments can be made incorporating the right decisions (Bonini & Gorner, 2011). In order for the lifestyle in modern industrial societies to develop with basic social change, it is necessary to switch to a sustainable and environmentally friendly energy management system. However, in order to ensure sustainability, social and ethical rules should be given importance, particularly considering that it is not possible to solve energy problems only with an economic approach but also social values and a healthy lifestyle (Jenkins, 2013).

Sustainability can be seen through the lens of long-term planning and activities of a particular organization with the ultimate goal of delivering services and products in a sustainable manner (Blowfield, 2013). While sustainable organizations try to have the most positive economic and social impact, they also need to avoid adversely affecting the environment (Wu & Zhi, 2016).

14.9 Sustainable Energy Management

Sustainable energy can be defined as meeting the energy needs of enterprises without damaging their ability to meet their goals in the future and consists of two main elements: the use of renewable energy resources and the more efficient use of energy (Cumo et al., 2012). For this reason, the efficient use of energy has become one of the most important issues for businesses in terms of sustainability. If energy is used without proper planning, problems arise that threaten the environment and public health. In order to prevent this situation, decisions should be taken for the efficient use of energy by both ensuring the encouragement of sustainable energy management the correct use of natural resources, and minimizing environmental problems (Bansal & DesJardine, 2014).

Providing an energy system that is adequate, safe, and at the same time protects the environment, is one of the most important factors affecting the prosperity, competitive position, and development of an economy. Clearly, energy management is both an important subsystem and problem of sustainable development because the increased energy consumption has a negative effect on environmental subsystems while also having a positive effect on the economic subsystem values.

All elements of modern management science also affect the concept of sustainable energy management. In sustainable energy management, where long-term planning is a necessity, there is also a need for the integration of energy policies that must be implemented in the sustainable development process. Understanding and achieving the determined targets is important for sustainable energy management. However, thanks to the correct use of energy resources, the positive effects of sustainable energy management will be at traceable, measurable and controllable levels (Radovanovic et al., 2012). If the energy management is done correctly, it is possible to be successful in areas such as industrial development, agriculture, urban and rural development, which are critical for sustainable development.

Sustainable energy management and related green engineering designs and production processes affect every aspect of our lives (Oh & Son, 2016). The reason for designing sustainable energy management is clear: a need for countries to adopt an acceptable management approach both geographically and in terms of nature and economy both ensures the determination of the most appropriate criteria in the management of energy production and energy consumption system, and secondarily, provides the necessary control in terms of how the targets set in energy management can be achieved. For this reason, sustainable energy management should not be seen as a model that only determines the goals related to the development of the needed energy (Stritih et al., 2015). Policy makers around the world take a holistic approach to the issue of sustainable energy, and it depends on their well-read evaluation of this complex structure.

The basic concept of sustainable energy development can generally be summarized as follows (Afgan et al., 1998; Anderson et al., 1999; Jefferson, 2006):

- The principle of protecting existing non-renewable resources.
- The principle of using renewable energy resources.

- Energy efficiency principles.
- The principle of intergenerational justice.
- The principle of harmonization of economic development and energy consumption.
- The principle of payment of damages (caused by the use of excessive and unsuitable energy sources).
- The principle of measurability and the need for sustainable strategic management.
- The principle of promotion and training.

The basic principles of sustainability are very important as they enable the planning and implementation of energy management. Energy system management will definitely bring challenges, but energy transitions that can be achieved by adapting innovative policy arrangements and new business models will play an important role in determining the success of societies around the world in accessing safe, sustainable, and equal energy.

As work in energy management applications mature, the success of an energy management program increases. It should be clearly stated in the beginning that top management has a firm commitment that at least one person is appointed as the "Energy Manager" in accordance with the structure of the business. Due to their positive outputs (including reduced costs and providing a green image), in addition to the obstacles to the implementation of the energy management program, the existence of motivating and supportive factors that enable energy -efficient measures to be taken- should be well understood. At this stage, it is important that an organization that decides to implement the energy management program carefully implement the initiation-planning, analysis-audit, and implementation-continuous evaluation stages, respectively, in order to achieve the highest efficiency.

Humanity has the capacity, creativity, technologies and resources to create a better world. However, the absence of appropriate institutions, coordination duties, political will, and governance structures complicates this task. Clearly, developing and implementing energy management systems within globally defined sustainability criteria must be seen as an obligation.

14.10 Discussion

The need for energy manifests itself in every small or large business, from providing requirements such as lighting, heating, and transportation to operating large mass production machines and bringing giant furnaces to very high temperatures. Energy is obtained by using numerous different sources, especially solid and liquid fuels, natural gas and electricity, and energy consumption, especially high in industrial enterprises and one of the main costs. A business that pays more to generate energy will find it more difficult to use its limited capital available in other areas necessary for the operation or development of the business. In other words, energy, which is one of the basic inputs in businesses, directly affects the profitability of the business.

Energy, mostly imported natural gas and coal in Turkey, has emerged as a significant problem because of the high cost of its operation. In addition to operating profitability, the use of the aforementioned fuels to provide necessary energy causes environmental pollution, thereby causing serious harm to human health. This situation imposes sanctions on businesses through certain legal obligations, and also affects customer satisfaction, particularly in today's world where nature and human health have gained a greater focus within society.

As well, it is important to implement procedures in energy management to strategically carry out activities on energy in accordance with the procedures in energy management. Energy management standards also guide the adaptation of the culture towards continuous improvement of management practices. It is important to examine the energy consumption patterns distributed across industries around the world to understand the future of global energy use. In order to implement and finance energy management systems in industries with the largest share in energy consumption, supportive policies and institutional mechanisms that require forward thinking and continuous investment are required. And, in order to create a major energy system transformation, companies must increase their innovative mechanisms and widely implement these systems.

14.11 Conclusion

So that companies reduce their greenhouse gas emissions and protect the environment, they generally need to apply a systematic approach to both improving their performance in energy management and managing energy savings. In order for companies to keep their corporate reputation at a high level, they must fulfill their legal obligations, ensure customer satisfaction, never compromise on quality and of course, make environmentally friendly energy management a priority. (Petrecca, 2012). Sustainable energy, defined by renewable energies and energy efficiency, should be a vital part of any energy strategy and contain a large savings potential (Müller-Steinhagen & Nitsch, 2005). Therefore, businesses are encouraged to manage systematically and more harmoniously in energy-related matters to continuously improve their energy efficiency.

Indeed, energy management has become increasingly popular over the past 50 years. Before the 1970s, energy management was mostly unheard of, except for a few exceptional industries that were only energy-intensive (Introna et al., 2014). However, it can be stated that the concept of energy management has become a much more important and popular concept today. When it comes to energy management, it is the sum of all measurements and activities including organizational and technical processes that are carried out in a planned company or organization to minimize energy consumption (Schippers & Hogenes, 2011). In other words, the concept of energy management can be defined as the process of optimizing energy consumption. It has been argued that many sciences, such as architecture, engineering, management, and finance can work together for greater energy management

(Droste-Franke et al., 2012). The criteria specified by Kannan and Boie (2003) are important for the implementation of energy management. These criteria include the following:

- In energy management, the top management, that is, the decision-making mechanism, should be involved in this business.
- An effective and sustainable energy management is not possible without the direct request and effort of senior management.
- After the decision-making mechanism that will make the necessary decisions, take responsibilities and make investments, and approve the energy management, the first step is to check the current energy status of the enterprise.
- Before any regulation is made, all units from general to specific are examined in the most detailed way, such as which type of energy is consumed in the enterprise, how much energy is used, which units are used to perform which operations, etc.
- Later, without reducing work efficiency in these units, energy saving measures are considered, and losses, if any, are determined. At this stage, it is necessary to demonstrate the technical and economical applicability of the savings to be achieved, since the ways to increase energy efficiency can be made primarily in line with physical and technical conditions and will also reveal the cost element.
- The next step is the implementation of the measurements and the planned solutions. In the last stage, it is examined whether what has been done as a result of the applications made worked as planned, and whether the expected results have been achieved. In addition, at this stage, energy budgeting for the next year, i.e. consumption and cost estimation is made.

Finally, the fact that energy management stages typically run in cycles must be considered. In other words, applications repeat each other continuously in a cycle, and as a result, energy management must be continuous and sustainable.

References

Afgan, N. H., Al Gobaisi, D., Carvalho, M. G., & Cumo, M. (1998). Sustainable energy development. *Renewable and Sustainable Energy Reviews, 2*(3), 235–286.

Anderson, T., Doig, A., Rees, D., & Khennas, S. (1999). *Rural energy services: A handbook for sustainable energy development*. Intermediate Technology Publications Ltd (ITP).

Azam, M. (2020). Energy and economic growth in developing Asian economies. *Journal of the Asia Pacific Economy, 25*(3), 447–471.

Baker, G. (2019). *The energy equation: Unlocking the hidden power of energy in business*. John Wiley & Sons.

Bansal, P., & DesJardine, M. R. (2014). Business sustainability: It is about time. *Strategic Organization, 12*(1), 70–78.

Beggs, C. (2010). *Energy: Management, supply and conservation*. Routledge.

Bergasse, E., & Paczynski, W. (2013). *The relationship between energy and economic and social development in the southern Mediterranean* (no. 27). MEDPRO technical report.

Blowfield, M. (2013). *Business and sustainability*. Oxford University Press.

Bonini, S., & Gorner, S. (2011). *The business of sustainability*. McKinsey & Company.
Bryant, S. T., Straker, K., & Wrigley, C. (2018). The typologies of power: Energy utility business models in an increasingly renewable sector. *Journal of Cleaner Production, 195*, 1032–1046.
Burger, C., & Weinmann, J. (2012). *The decentralized energy revolution: Business strategies for a new paradigm*. Springer.
Capehart, B. L., Turner, W. C., & Kennedy, W. J. (2020). *Guide to energy management*. Crc Press.
Caradonna, J. L. (2014). *Sustainability: A history*. Oxford University Press.
Clayton, T., & Radcliffe, N. (2018). *Sustainability: A systems approach*. Routledge.
Contreras Sierra, E. R. (2013). The concept of strategy as a basis for strategic planning. *Pensamiento & gestión, 35*, 152–181.
Cumo, F., Garcia, D. A., Calcagnini, L., Rosa, F., & Sferra, A. S. (2012). Urban policies and sustainable energy management. *Sustainable Cities and Society, 4*, 29–34.
Droste-Franke, B., Paal, B., Rehtanz, C., Sauer, D. U., Schneider, J. P., Schreurs, M., & Ziesemer, T. (2012). *Balancing renewable electricity: energy storage, demand side management, and network extension from an interdisciplinary perspective* (Vol. 40). Springer Science & Business Media.
Farhodovna, A. M., & Abdishukurovna, E. M. (2021). Application of the law of conservation of energy in economics. *The American Journal of Applied sciences, 3*(01), 93–103.
Fischer, T. B. (2010). *The theory and practice of strategic environmental assessment: Towards a more systematic approach*. Routledge.
Freedman, L. (2015). *Strategy: A history*. Oxford University Press.
Gibson, B., Hassan, S., & Tansey, J. (2013). *Sustainability assessment: Criteria and processes*. Routledge.
Gielen, D., Boshell, F., Saygin, D., Bazilian, M. D., Wagner, N., & Gorini, R. (2019). The role of renewable energy in the global energy transformation. *Energy Strategy Reviews, 24*, 38–50.
Gray, C. S. (2015). *The future of strategy*. John Wiley & Sons.
Haider, W. H. (2020). Estimates of total oil & gas reserves in the world, future of oil and gas companies and smart investments by E & P companies in renewable energy sources for future energy needs. In *International Petroleum Technology Conference* (pp. 1–16). OnePetro. https://doi.org/10.2523/IPTC-19729-MS
Heffron, R. J., & McCauley, D. (2017). The concept of energy justice across the disciplines. *Energy Policy, 105*, 658–667.
Hellström, M., Tsvetkova, A., Gustafsson, M., & Wikström, K. (2015). Collaboration mechanisms for business models in distributed energy ecosystems. *Journal of Cleaner Production, 102*, 226–236.
Introna, V., Cesarotti, V., Benedetti, M., Biagiotti, S., & Rotunno, R. (2014). Energy management maturity model: An organizational tool to foster the continuous reduction of energy consumption in companies. *Journal of Cleaner Production, 83*, 108–117.
Jefferson, M. (2006). Sustainable energy development: Performance and prospects. *Renewable Energy, 31*(5), 571–582.
Jenkins, W. (2013). *The future of ethics: Sustainability, social justice, and religious creativity*. Georgetown University Press.
Kannan, R., & Boie, W. (2003). Energy management practices in SME—Case study of a bakery in Germany. *Energy Conversion and Management, 44*(6), 945–959.
Lee, D., & Cheng, C. C. (2016). Energy savings by energy management systems: A review. *Renewable and Sustainable Energy Reviews, 56*, 760–777.
Liu, Y., Wei, T., & Park, D. (2019). Macroeconomic impacts of energy productivity: A general equilibrium perspective. *Energy Efficiency, 12*(7), 1857–1872.
Lobos, V., & Partidario, M. (2014). Theory versus practice in strategic environmental assessment (SEA). *Environmental Impact Assessment Review, 48*, 34–46.
Martins, F., Felgueiras, C., Smitkova, M., & Caetano, N. (2019). Analysis of fossil fuel energy consumption and environmental impacts in European countries. *Energies, 12*(6), 1–11. https://doi.org/10.3390/en12060964

Müller-Steinhagen, H., & Nitsch, J. (2005). The contribution of renewable energies to a sustainable energy economy. *Process Safety and Environmental Protection, 83*(4), 285–297.

Oh, E., & Son, S. Y. (2016). Toward dynamic energy management for green manufacturing systems. *IEEE Communications Magazine, 54*(10), 74–79.

Peng, M. W. (2021). *Global strategy*. Cengage Learning.

Petrecca, G. (2012). Industrial energy management: Principles and applications: Principles and applications. Springer Science & Business Media.

Radovanovic, M., Popov, S., & Dodic, S. (2012). *Sustainable energy management: Planning, implementation, control and strategy*. Academic Press.

Schippers, M. C., & Hogenes, R. (2011). Energy management of people in organizations: A review and research agenda. *Journal of Business and Psychology, 26*(2), 193–203.

Smil, V. (2018). *Energy and civilization: A history*. MIT Press.

Stritih, U., Paksoy, H., Turgut, B., Osterman, E., Evliya, H., & Butala, V. (2015). Sustainable energy management. *Management of Environmental Quality: An International Journal., 26*(5), 764–790. https://doi.org/10.1108/MEQ-06-2013-0063

Teece, D. J. (2010). Business models, business strategy and innovation. *Long Range Planning, 43* (2–3), 172–194.

Therivel, R. (2012). *Strategic environmental assessment in action*. Routledge.

Therivel, R., & Paridario, M. R. (2013). *The practice of strategic environmental assessment*. Routledge.

Therivel, R., Wilson, E., Heaney, D., & Thompson, S. (2013). *Strategic environmental assessment*. Routledge.

Thollander, P., & Ottosson, M. (2010). Energy management practices in Swedish energy-intensive industries. *Journal of Cleaner Production, 18*(12), 1125–1133.

Thollander, P., & Palm, J. (2012). Improving energy efficiency in industrial energy systems: An interdisciplinary perspective on barriers, energy audits, energy management, policies, and programs. Springer Science & Business Media.

Twidell, J., & Weir, T. (2015). *Renewable energy resources*. Routledge.

Wced, S. W. S. (1987). World commission on environment and development. *Our Common Future, 17*(1), 1–91.

World Bank. (2018). The World Bank annual report 2018. The World Bank.

Wu, X., & Zhi, Q. (2016). Impact of shared economy on urban sustainability: From the perspective of social, economic, and environmental sustainability. *Energy Procedia, 104*, 191–196.

Chapter 15
Urbanization, Fossil Fuel Consumption and Carbon Dioxide Emission in Ghana: The STIRPAT Model Approach

Ibrahim Nandom Yakubu, Ayhan Kapusuzoglu, and Nildag Basak Ceylan

Abstract Relying on the Stochastic Impacts by Regression on Population, Affluence and Technology (STIRPAT) model, this chapter examines the effect of urbanization and fossil fuel consumption on CO_2 emission in Ghana. The study applies the autoregressive distributed lag (ARDL) estimator to investigate the short- and long-run influence of the variables of interest on CO_2 emission with data spanning from 1971 to 2017. The findings indicate that urbanization exerts a negative significant influence on CO_2 emission in both long- and short-term periods, thus confirming the validity of the compact city theory in Ghana. While affluence significantly reduces CO_2 emission in the short-term, technology contributes to environmental pollution regardless of the period. The results further reveal that although fossil fuel consumption triggers pollution in the short-run, it impedes CO_2 emission in the long-run. The findings suggest the need to espouse policies that will assist in urban development. Besides, adopting environmentally friendly technologies and increasing the dependence on renewable energy sources are imperative to enhancing environmental quality in Ghana.

15.1 Introduction

Climate change is one of the biggest threats facing this modern era. Among the causes, carbon dioxide (CO_2) emission has been the most profound. Climate change as posited by Belaïd and Zrelli (2016) is a multifaceted issue emanating from the dynamic interactions between the environment and energy, as well as economic activities. This interaction between energy and climate change thus shows that energy is a prerequisite for production, and hence for stimulating the growth of the

I. N. Yakubu
Graduate School of Social Sciences, Ankara Yildirim Beyazit University, Ankara, Turkey

A. Kapusuzoglu (✉) · N. B. Ceylan
Faculty of Business, Ankara Yildirim Beyazit University, Ankara, Turkey
e-mail: akapusuzoglu@ybu.edu.tr; nbceylan@ybu.edu.tr

S. Yüksel, H. Dinçer (eds.), *Strategic Approaches to Energy Management*,
Contributions to Management Science,
https://doi.org/10.1007/978-3-030-76783-9_15

economy. However, in order to achieve the desired growth, overreliance on energy may pose some significant cost as energy consumption largely contributes to greenhouse gas (GHG) emission, specifically carbon dioxide that degrades environmental quality. Given this, researchers have extensively investigated how CO_2 emission is influenced by energy consumption (Apergis & Ozturk, 2015; Behera & Dash, 2017; Dogan & Aslan, 2017; Usman et al., 2020). In relation to the theory of the Environmental Kuznets Curve (EKC), several of these studies have yielded intriguing results regarding the energy-CO_2 emissions relationship.

Arguably, earlier studies disregarded the significance of demographic factors on the environment, which to a large extent influence energy usage and environmental depletion. Following the seminal work of Jones (1989) which studied the interaction between urbanization and energy utilization, researchers have noted the impact of demographic indicators such as population growth on the environment. The important role of urbanization cannot be overstated. The pattern of energy demand and most pollutant emissions is significantly impacted by urbanization (Raheem & Ogebe, 2017). The International Energy Agency (2008) reported that urbanization accounts for about 70% of emissions related to energy consumption, and this is projected to reach 76% by the year 2030. Recent studies have noted urbanization as one of the major drivers of CO_2 emission (Shahbaz et al., 2015; Zhang et al., 2015; Sheng & Guo, 2016; Raheem & Ogebe, 2017; Yi et al., 2017).

This research aims to look into the effects of urban population growth on CO_2 emission in Ghana within the context of the Stochastic Impacts by Regression on Population, Affluence and Technology (STIRPAT) model. This is necessary as the country has experienced growth in urban population coupled with industrial activities in recent years. Urbanization may exercise a direct or negative influence on CO_2 emission. For instance, urbanization can aid reduce emissions as urban dwellers may rely on efficient energy consumption mechanisms due to the dearth of available energy resources. On the other hand, urbanization can propel carbon dioxide emissions. Due to large-scale industrial activities and hefty vehicular traffics in urban areas, fumes emissions are prevalent which harms the environment.

In addition to urbanization, we investigate the influence of fossil fuel consumption on CO_2 emission. In Ghana, electricity and fossil fuel are significant contributors to the country's economic development (Kwakwa, 2012). Despite the emphasis on transition to cleaner energy, there has been a decline in renewable energy utilization in the country and fossil fuel use continues to increase and takes a larger portion of the overall energy consumption mix in Ghana (Abokyi et al., 2019).

On the link amid urbanization, energy consumption and CO_2 emission, this study is quite similar to the work of Kwakwa and Alhassan (2018). However, our paper is distinct in three ways. First, we apply the STIRPAT model where we explore the influence of urban population growth on CO_2 emission in the midst of affluence and technology. Though this approach has been applied in different countries, no study has considered that in Ghana. Second, rather than employing total energy consumption as in the case of most studies, we detach fossil fuel consumption to examine its impact on CO_2 emission given its significance in the energy mix in Ghana. Finally, we also differ from earlier studies in Ghana in that we aim to look at the short- and long-term influence of urbanization and fossil fuel utilization on CO_2 emissions.

The following is how the rest of the chapter is organized. Section 15.2 provides a summary of the literature on the connection between urbanization, fossil fuel use, and emissions, while Sect. 15.3 discusses methodology and data. In Sect. 15.4, we elaborate on our empirical results before concluding with suggestions in Sect. 15.5.

15.2 Literature Review

In this section, we review the studies on the relationship between urbanization and CO_2 emission, as well as the relationship between fossil fuel consumption and CO_2 emission.

15.2.1 Urbanization and CO_2 Emission Nexus

Theoretically, the urban transition theory relates that in the urban vicinities, environmental challenges depend on the affluence level (Marcotullio & Lee, 2003). This suggests that countries in the low-income bracket are faced with environmental problems that are localized and immediate. However, environmental problems in high-income economies are considered to be more intense as they are global challenges and maybe intergenerational with a severe threat to the ecosystems. Essentially, environmental pollution surges as urban areas develop and transit to high-income areas from low-income cities. However, once a high-income status is achieved, pollution levels will be reduced as more environmental safety policies will be instituted and innovation in terms of technology increases (Adams et al., 2016). According to the compact city theory, as urban areas become condensed due to the expansion of existing urban cities instead of development in the exurbs, there will be more clustering in activities as well as floor space. As a result, commuting to the workplace and access to certain services becomes convenient, hence CO_2 emitted through transportation lessens (Sadorsky, 2014; Yi et al., 2017). In a nutshell, the compact city theory elaborates on the advantages of urban population growth by arguing that it causes less harm to the environment.

Empirically, there are numerous studies examining the influence of urbanization on CO_2 emission though the conclusions are significantly divergent. For example, Shahbaz et al. examined the effect of urbanization on CO_2 emission in Malaysia using the STIRPAT model. Employing quarterly data for the years 1970Q1 to 2011Q4, the ARDL analysis showed that urban growth and CO_2 emission possess a U-shaped relationship, suggesting that urbanization lessens CO_2 emission to a certain level and then adversely affects the environment by increasing CO_2 emission. Similarly, Bekhet and Dash empirically investigated the effect of growth in urban population on the level of CO_2 emission in Malaysia during the period 1971–2015.

Applying the vector error correction model (VECM), the authors discovered that in the long-run urbanization and CO_2 emission exhibit a U-shaped relationship.

In India, Franco et al. (2017) analyzed the link between urbanization, energy usage and CO_2 emission by employing census data on urbanization covering from 1901 to 2011. The results suggested that urbanization motivates CO_2 emission.

In a cross-country study, Zhang et al. (2017) assessed the effect of urbanization on the emission of carbon dioxide with data spanning from 1961 to 2011. By applying the fixed effects model, the authors revealed a positive significant influence of urbanization on CO_2 emission. They noted that a U-shaped link exists between the two variables.

Using less developed economies, McGee and York (2018) examined the urbanization- CO_2 emission relationship with panel data which spanned from 1960 to 2010. Using the generalized least squares estimation method, the results portrayed that CO_2 emission is substantially reduced when urbanization declines.

Applying the ARDL approach with cointegration techniques (Fully Modified OLS and Canonical Cointegration Regression), Pata (2018) investigated how urbanization and other variables impact CO_2 emission in Turkey for the period 1974–2014. Within the ECK hypothesis, the results presented a direct influence of urbanization on CO2 emission.

Shuai et al. (2018) analyzed the main factors influencing CO_2 emission levels in China using the STIRPAT model. Among the factors determining CO_2 emission, urbanization rate was noted as the most significant factor inhibiting the amount of CO_2 emission.

Salim et al. (2019) employed data of 13 developing Asian countries covering the period 1980–2010 to study the effect of urbanization on pollutant emission. Employing techniques for instance the pooled mean group (PMG) and augmented mean group (AMG) with the ARDL method to check for robustness, the findings generally indicated that CO_2 emission is directly influenced by urbanization.

In Indonesia, Ahmed et al. (2019) examined the correlation between urbanization and the amount of CO_2 emission for the period 1971–2014. Applying the ARDL estimator, the findings evidenced a U-shaped influence of urban development on environmental pollution.

Exploring how urbanization influences CO_2 emission in Pakistan, Ali et al. (2019) invoked the ARDL approach on time series data extending from 1972 to 2014. The findings established that urbanization drives CO_2 emission in both the short- and long-term.

In a city-level analysis in China, Chen et al. (2019) looked into the influence of urbanization on CO_2 emission with data covering 188 cities which spanned from 2005 to 2013. The results reported that in the western part of the country, there exists a U-shaped impact of urbanization on CO_2 emission.

Lv and Xu (2019) assessed the impact of urbanization and openness to trade on CO_2 emission using data from middle-income countries extending from 1992 to 2012. Applying the pooled mean group estimation technique, the study revealed that

CO_2 emission is negatively and significantly influenced by urbanization in the short- and long-term.

Sarwar and Alsaggaf (2019) analyzed the role of urbanization on CO_2 emission in thirty provinces in China over the period 1998 to 2016. Applying the mean group and system GMM techniques, the results noted that urbanization positively and significantly affects CO_2 emission regardless of the estimation technique employed.

Khoshnevis and Golestani (2019) investigated the correlation between urbanization, economic growth, and CO_2 emission in Asian countries. Relying on the pooled mean group method and employing data for the years 1980 to 2014, the research documented that urbanization promotes more CO_2 emission in the sampled countries.

Zhou et al. (2019) investigated the influence of urbanization on the emission of carbon dioxide in China using Yangtze River Delta as a case study and with data spanning from 1992 to 2013. Dividing urbanization into sub-categories and applying the fixed effects technique, the results depicted that urbanization has a varying effect on CO_2 emission depending on the sub-system of urbanization. For instance, while economic urbanization and spatial urbanization positively influence CO_2 emission, social urbanization and pollution were established to be negatively correlated.

Abbasi et al. (2020) applied the fully modified OLS and the VECM techniques to research the influence of urbanization alongside energy consumption on CO_2 emission in 8 countries from Asia for the period 1982–2017. The results found that CO_2 emission increases with urban development.

Akorede and Afroz (2020) tested the relationship that exists between emissions of carbon dioxide and urbanization in the Nigerian economy over the period 1970–2017. Using the ARDL technique, the results found an inverse albeit significant impact of urbanization on CO_2 emission.

Applying the fixed effects model, Anwar et al. (2020) looked at the factors that influence CO_2 emissions in the far Eastern countries from 1980 to 2017. The authors revealed that the emissions level in the selected countries is positively and significantly motivated by urbanization.

Using the Driscoll-Kraay standard errors estimator, Musah et al. (2020) analyzed the nexus between CO_2 emission and urbanization in West African nations over the period 1990–2018. The study established that CO_2 emission is significantly and positively impacted by urban development.

Recently, Khan and Su (2021) empirically investigated the urbanization-CO_2 emission link in the newly industrialized countries over the period 1991–2019. Using the panel threshold regression analysis, the findings showed that when urbanization is below the threshold value, it impacts positively on CO_2 emission. On the other hand, when urbanization exceeds the threshold value, it tends to lower pollution.

In the case of Ghana, Kwakwa and Alhassan (2018) adopted the fully modified OLS technique with data covering from 1971 to 2013 to examine the impact of

Table 15.1 Summary of the empirical literature (urbanization and CO_2 emissions)

Author(s)	Country/Region	Period	Estimation technique(s)	Result
Shahbaz et al.	Malaysia	1970–2011	Autoregressive distributed lag (ARDL)	Negative/ positive
Bekhet and Dash	Malaysia	1971–2015	Vector error correction model (VECM)	Negative/ positive
Zhang et al. (2017)	Cross country study	1961–2011	Fixed effects	Negative/ positive
McGee and York (2018)	Developing economies	1960–2010	Generalized least squares	Negative
Pata (2018)	Turkey	1974–2014	ARDL, fully modified, canonical cointegration regression	Positive
Salim et al.	Asian countries	1980–2010	Pooled mean group, augmented mean group, ARDL	Positive
Kwakwa and Alhassan (2018)	Ghana	1971–2013	Fully modified OLS	Positive
Ahmed et al. (2019)	Indonesia	1971–2014	ARDL	Negative/ positive
Ali et al. (2019)	Pakistan	1972–2014	ARDL	Positive
Lv and Xu (2019)	Middle-income countries	1992–2012	Pooled mean group	Negative
Sarwar and Alsaggaf (2019)	China	1998–2016	Mean group, system GMM	Positive
Khoshnevis and Golestani (2019)	Asian countries	1980–2014	Pooled mean group	Positive
Zhou et al. (2019)	China	1992–2013	Fixed effects	Positive/ negative
Abbasi et al. (2020)	Asian countries	1982–2017	Fully modified OLS and VECM	Positive
Akorede and Afroz (2020)	Nigeria	1970–2017	ARDL	Negative
Anwar et al. (2020)	Far eastern countries	1980–2017	Fixed effects	Positive
Musah et al. (2020)	West Africa	1990–2018	Driscoll-Kraay standard errors	Positive
Khan and Su (2021)	Newly industrialized countries	1991–2019	Panel threshold regression	Negative/ positive

urbanization and energy on CO_2 emission. Controlling for other factors, the findings evidenced a direct influence of urbanization on CO_2 emission.

Table 15.1 shows the nexus between urbanization and CO_2 emissions as discussed in the literature.

15.2.2 The Relationship Between Fossil Fuel Consumption and CO_2 Emission

The effect of aggregate energy consumption on CO_2 emissions has been lengthily studied in the literature (see Wang et al., 2016; Mirzaei & Bekri, 2017; Zaman & Abd-el Moemen, 2017; Baloch & Suad, 2018; Sarkodie & Strezov, 2019; Khan et al., 2020; Osobajo et al., 2020). However, the independent effect of fossil fuel consumption on CO_2 emission has been less examined. Among the scanty studies, Rafindadi (2014) analyzed the link between fossil fuel consumption and pollution in the Asia-Pacific for the years 1975 to 2012. Using the least squares techniques, the results found that fossil fuel usage positively and significantly affects pollution levels in the region.

Behera and Dash (2017) investigated the impact of energy utilization and other sets of factors on CO_2 emission in the South and Eastern Asian countries over the period 1980–2012. Employing the FMOLS and the DOLS methods, a positive effect of fossil fuel usage on CO_2 emission was established for countries in the middle and high-income categories. Kwakwa and Alhassan (2018) reported a positive influence of fossil fuel use on CO_2 emission in Ghana.

Ibrahiem and Hanafy (2020) assessed the dynamic relationship between fossil fuel consumption and environmental degradation in Egypt over the period 1971–2014. In establishing the long-run relationship between the variables, the authors applied the FMOLS and the DOLS techniques. The results showed that fossil fuel consumption contributes to environmental pollution.

In examining the relation between CO_2 emission and fossil fuel energy, Naseem et al. (2020) applied the ARDL technique on data spanning from 1969 to 2018. They revealed that fossil fuel energy reduces environmental quality in Pakistan.

Applying the pooled mean group technique, Onifade et al. analyzed the impact of energy transition in the OPEC countries for the years 1990 to 2014. The findings evidenced that the consumption of fossil fuel significantly increases environmental degradation.

Xia and Wang (2020) looked at the impact of fossil fuels on CO_2 emissions in China from 1965 to 2016. The ARDL results indicated that a surge in the consumption of fossil fuels intensifies CO_2 emissions.

In Table 15.2, we provide a summary of the scanty literature on the influence of fossil fuel utilization on CO_2 emission.

15.3 Research Methodology

15.3.1 Theoretical Framework and Model Construction

The Stochastic Impacts by Regression on Population, Affluence and Technology (STIRPAT) model serves as the framework for this study. Dietz and Rosa (1994)

Table 15.2 Summary of the empirical literature (fossil fuel consumption and CO_2 emission)

Author(s)	Country/Region	Period	Estimation technique (s)	Result
Rafindadi et al. (2014)	Asia-Pacific	1975–2012	Least squares techniques	Positive
Behera and Dash (2017)	South and eastern Asia	1980–2012	FMOLS and DOLS	Positive
Ibrahiem and Hanafy (2020)	Egypt	1971–2014	FMOLS and DOLS	Positive
Naseem et al. (2020)	Pakistan	1969–2018	ARDL	Positive
Onifade et al.	OPEC countries	1990–2014	Pooled mean group	Positive
Xia and Wang (2020)	China	1965–2016	ARDL	Positive

introduced the STIRPAT model which has been adopted by several researchers to observe the link between some selected economic factors and the environment. The STIRPAT model is the advanced form of the IPAT equation which describes how population, affluence, and technology impact the environment. Given the IPAT's drawback of not allowing hypothesis testing, Dietz and Rosa (1994) reshaped the IPAT model into a stochastic equation. The STIRPAT model resolves this shortcoming by allowing the inclusion of other variables into the model that can potentially explain environmental quality. The STIRPAT model is basically expressed as:

$$I_t = aP_t^b A_t^c T_t^d \varepsilon_t \tag{15.1}$$

where I denotes CO_2 emission (a pollution proxy). P, A, and T symbolize population, affluence, and technology for time t respectively and ε is the error term.

To evaluate the influence of fossil fuel utilization on CO_2 emission, eq. (15.1) is expanded to include fossil fuel (FOF). The population proxy, in this case, is urban population (U). By transmuting the variables into natural logarithm, the model is restated as:

$$ln\,CO_{2t} = \alpha_0 + \alpha_1\, ln\,U_t + \alpha_2\, ln\,A_t + \alpha_3\, ln\,T_t + \alpha_4\, ln\,FOF_t + \varepsilon_t \tag{15.2}$$

In Eq. (15.2), similar to Musah et al. (2020), the study measures CO_2 emission by employing the volume of emissions in metric tons per capita. In keeping with Abbasi et al. (2020), urbanization (U) is proxied by annual percentage growth in urban population. We use GDP per capita growth to define affluence (A). Technology is measure by the interaction term of industry sector value-added and services sector value-added. This measure is drawn from the work of Shahbaz et al. (2017). Fossil fuel consumption (FOF) is the quantity of fossil fuel utilized (% of total energy consumption). Data spanning 1971–2017 for all the variables are sourced from World Bank's World Development Indicators.

15.3.2 Analytical Approach

We seek to assess how the explanatory factors affect the emissions of CO_2 in the short- and long-term. In this direction, the study employs the ARDL estimator by Pesaran et al. (2001). In comparison to other techniques, the ARDL technique is simple to use and offers accurate long-run estimates. It is best used in studies with a limited sample size (Yakubu, 2020). To perform the long and short-run estimations, the ARDL model in conformity with Pesaran et al. (2001) is specified as:

$$
\begin{aligned}
\ln CO_{2t} = {} & \alpha_0 + \sum\nolimits_{i=0}^{n} \alpha_{1i} \Delta \ln CO_{2t-1} + \sum\nolimits_{i=0}^{n} \alpha_{2i} \Delta \ln U_{t-1} \\
& + \sum\nolimits_{i=0}^{n} \alpha_{3i} \Delta \ln A_{t-1} + \sum\nolimits_{i=0}^{n} \alpha_{4i} \Delta \ln T_{t-1} \\
& + \sum\nolimits_{i=0}^{n} \alpha_{5i} \Delta \ln FOF_{t-1} + \delta_1 \ln CO_{2t-1} + \delta_2 \ln U_{t-1} + \delta_3 \ln A_{t-1} \\
& + \delta_4 \ln T_{t-1} + \delta_5 \ln FOF_{t-1} + \rho ECT_{t-1} + \varepsilon_t
\end{aligned}
\tag{15.3}
$$

where Δ represents the difference operator. $\alpha 1$–$\alpha 5$ symbolizes short-run estimates. $\delta 1$–$\delta 5$ are the long-run parameters and the lag length is indicated by n. The error correction term is signified by ECT with ρ connoting its coefficient.

The study carries out the bound-testing for long-run relationships. The null hypothesis (H0: $\delta 1 = \delta 2 = \delta 3 = \delta 4 = \delta 5 = 0$) of non-existence of long-run association amid the factors in contrast to the alternative hypothesis (H1: $\delta 1 \neq \delta 2 \neq \delta 3 \neq \delta 4 \neq \delta 5$) which specifies cointegration is tested. The results of the F-statistics are compared to the critical bounds to determine if cointegration exists. When the upper critical bound is less than the F-statistics, cointegration is assumed.

15.4 Empirical Findings

15.4.1 Descriptive Statistics and Correlation Analysis

Table 15.3 displays the descriptive statistics as well as the correlation matrix results. From the results, the variables are positively skewed, with the exception of urbanization and affluence, which are skewed to the left. Our variables are normally distributed at 5%, according to the Jarque-Bera probability values, with the exception of affluence. The standard deviation of affluence (as measured by GDP per capita growth) is the highest, indicating greater variability. The correlation matrix indicates that the variables are free from multicollinearity as they are weakly associated. Furthermore, the variance inflation factor (VIF) and tolerance values show that there are no issues with multicollinearity.

Table 15.3 Descriptive statistics and correlation matrix

Variables	$\ln CO_2$	lnU	lnA	lnT	lnFOF
Mean	−0.478	0.608	0.288	2.862	1.446
Maximum	−0.233	0.709	1.054	3.135	1.721
Minimum	−0.682	0.412	−1.614	2.569	1.180
Std. dev.	0.122	0.070	0.468	0.133	0.188
Skewness	0.558	−0.779	−1.812	0.329	0.241
Kurtosis	2.526	3.755	8.456	2.968	1.497
Jarque-Bera	2.330	4.748	67.926	0.686	3.943
Probability	0.312	0.093	0.000	0.710	0.139
Observations	38	38	38	38	38
Correlation and multicollinearity analysis					
Coefficients		lnU	lnA	lnT	lnFOF
lnU		1.000			
lnA		−0.059	1.000		
lnT		−0.2079	0.181	1.000	
lnFOF		−0.489	0.324	0.839	1.000
VIF		1.69	1.21	4.42	6.09
Tolerance		0.592	0.825	0.226	0.164

Table 15.4 Unit root test results

	ADF		PP	
Variables	t-stats.	Decision	t-stats.	Decision
$\ln CO_2$	−9.212***	I(1)	−12.888***	I(1)
lnU	−3.025**	I(1)	−3.025**	I(1)
lnA	−8.114***	I(0)	−7.275***	I(0)
lnT	−4.636***	I(1)	−4.346***	I(1)
lnFOF	−9.381***	I(1)	−9.547***	I(1)

Notes: **, and *** denote the level significance at 5% and 1% respectively

15.4.2 Unit Root Test Results

The order in which the variables are integrated is first appraised prior to performing the ARDL. In the ARDL analysis, variables must be integrated at level I(0) or at the first difference I(1). If any of the series is found to be integrated at second difference (I(2)), then the ARDL technique becomes inappropriate. In this study, we rely on the Augmented Dickey-Fuller (ADF) and the Phillips-Perron (PP) unit root tests to check the integration order of our variables. Table 15.4 shows that the ADF and PP results for all variables indicate integration at I(I), with the exception of affluence, which indicates integration at I(0). Since the variables are integrated in a mixed order, and none of them is stationary at second difference (I(2)), we proceed with the estimation of the ARDL model.

15.4.3 Bound Test Results

The bounds-testing results in Table 15.5 indicate that the upper critical bound value is less than the approximated F-statistics value (4.203) at 5% significance, suggesting that our variables are cointegrated. As a result, we can estimate the long-term impact of our explanatory variables on carbon dioxide emission. The lag order determination is based on the Akaike Information Criterion (AIC).

15.4.4 Long-Run and Short-Run Results

Table 15.6 outlines the long- and short-run impact of the variables on CO_2 emission. From the results, urbanization exerts a significant negative influence on CO_2 emission in the long-term. Precisely, a percentage growth in urban population contributes significantly to mitigating environmental degradation by lessening CO_2 emission approximately by 1.691%. The result shows that despite the growth in migration to

Table 15.5 Bounds test results

	Value	Level	Level	1st difference
F-Statistics	4.203**	10%	2.2	3.09
k	4	5%	2.56	3.49
		1%	3.29	4.37

Note: ** denotes the level significance at 5%

Table 15.6 Long- and short-run results

ARDL (3, 2, 2, 1, 3)				
Long-run estimates				
Variable	Coefficient	Std. error	t-stats.	Prob.
$\ln U_t$	−1.691	0.423	−4.001	0.002***
$\ln A_t$	0.042	0.035	1.192	0.255
$\ln T_t$	0.581	0.138	4.217	0.001***
$\ln FOF_t$	−0.270	0.116	−2.318	0.037**
C	−0.716	0.596	−1.200	0.252
Short-run estimates				
$\ln CO_{2t-2}$	0.195	0.113	1.718	0.110
$\ln U_{t-1}$	−0.626	0.418	−1.499	0.158
$\ln A_{t-1}$	−0.036	0.017	−2.112	0.055*
$\ln T_t$	0.542	0.143	3.784	0.002**
$\ln FOF_{t-2}$	0.273	0.133	2.050	0.061*
ECT_{t-1}	−1.599	0.271	−5.909	0.000***
R^2	0.841			
D.Watson	1.664			
F-stats.	26.685		0.000	

Notes: *, **, and *** denote the level significance at 10%, 5% and 1% respectively

the urban areas and industrial concentration in most urban centers in Ghana, the environmental policies put in place to curb pollution are laudable. For example, the Partnership for Action on Green Economy is one of the key measures to ensuring a green economy in Ghana through the adoption of appropriate policies to reduce environmental pollution from industrial activities. It can be inferred from the results that the pollution abatement impact of urbanization advocated by the compact city theory is valid in Ghana. The finding conforms with studies such as Lv and Xu (2019) and Akorede and Afroz (2020) but inconsistent with most earlier studies that documented a positive influence of urbanization on CO_2 emission (see Kwakwa & Alhassan, 2018; Pata, 2018; Ali et al., 2019; Abbasi et al., 2020; Musah et al., 2020). In the short-run, urbanization still exhibits a decreasing effect on CO_2 emission though the impact is not significant.

The effect of affluence (which measures wealth) in the long-run is positive albeit insignificant. This suggests that affluence in the long-term has no significant influence on CO_2 emission. In the short-term, however, affluence negatively and significantly influences CO_2 emission. This indicates that a percentage decline in GDP per capita will reduce environmental pollution by 0.036%. The implication is that as wealth decreases in the short-term, demand for goods and services also declines, and thus production level reduces. As a result, energy consumption declines and likewise less pressure on natural resources. The negative effect of affluence on CO_2 emission goes contrary to prior studies (Zaidi et al., 2019).

In both periods, technology and CO_2 emission are positively and significantly correlated confirming the finding of Lv and Xu (2019). Our results show that technology has the greatest influence on the rate of CO_2 emission. Environmental pollution rises by 0.581% and 0.542% in the long- and short-term respectively with every 1% increase in technology. The finding implies that in the quest to enhance domestic output through the adoption of sophisticated technologies, energy consumption increases which raises environmental pollution (Shahbaz et al., 2017).

Fossil fuel consumption in the short-term exerts a positive significant influence on CO_2 emission. As fossil fuel utilization increases by a percentage, the emission level rises by 0.273% in the short-run. This finding can be based on the argument that fossil fuel is a 'dirty' form of energy that increases CO_2 emission (Kwakwa & Alhassan, 2018). The result of a positive relationship is congruent with prior findings (Behera & Dash, 2017; Naseem et al., 2020). Fossil fuel use has a long-term negative significant effect on pollutant emissions, contrary to the short-run result. Generally, due to increasing prices of fuels and natural gas, demand for fossil fuels reduces especially for those who cannot afford such energy sources. Likewise, people tend to judiciously use fossil fuels and some resort to renewable energy sources in episodes of higher fuel prices. As a consequence, the level of CO_2 emission from fossil fuels reduces.

From the short-run estimation, the error correction term's (ECT) coefficient of -1.599 indicates that on annual basis, the disturbance in the model decreases by 159.9% towards the equilibrium.

15.4.5 Diagnostic Tests Results

The diagnostic tests are shown in Table 15.7. According to the findings, the error term is free from heteroskedasticity, serial correlation, and non-normality. The Ramsey test confirms that the model is well-specified. The CUSUM and CUSUM of Squares tests demonstrate that at 5% significance, the model parameters are stable and the test plots are within the critical bounds.

15.5 Conclusion and Recommendations

This research aims to look into the effects of urbanization and fossil fuel consumption on CO_2 emission in Ghana using the STIRPAT model. The study applies the ARDL technique to observe the long- and short-term influence of the variables of interest on CO_2 emission with time series data spanning 1971–2017. The findings indicate that urbanization exerts a significant negative influence on CO_2 emission regardless of the period. This confirms the compact city theory which highlights on the pollution abatement impact of urbanization. Affluence significantly reduces environmental pollution in the short-run. For both time periods, we found that technology intensifies environmental pollution in Ghana. In the long-term, the demand for fossil fuels reduces CO_2 emissions. However, the short-run results show that fossil fuel consumption increases pollution.

The results offer some policy implications. Given that urbanization reduces pollution, policymakers are urged to espouse policies that will assist in urban development. These policies, however, must be focused on pollution control agendas to mitigate the environmental threats as urban population surges. Also, adopting environmentally friendly technologies and increasing the dependence on renewable energy sources are imperative to enhancing environmental quality. This can be accomplished by subsidizing low-carbon technologies in order to increase its demand and raising the prices of fossil fuels. Besides, it is crucial to ensure that the general public is properly educated about the value of environmental conservation.

Table 15.7 Diagnostic tests results

Specification	F-stats.	Prob.
X^2SERIAL	0.833	0.460
X^2HETERO	0.834	0.635
X^2NORM	1.211	0.546
X^2RESET	0.189	0.672

References

Abbasi, M. A., Parveen, S., Khan, S., & Kamal, M. A. (2020). Urbanization and energy consumption effects on carbon dioxide emissions: Evidence from Asian-8 countries using panel data analysis. *Environmental Science and Pollution Research, 27*(15), 18029–18043.

Abokyi, E., Appiah-Konadu, P., Abokyi, F., & Oteng-Abayie, E. F. (2019). Industrial growth and emissions of CO_2 in Ghana: The role of financial development and fossil fuel consumption. *Energy Reports, 5*, 1339–1353.

Adams, S., Adom, P. K., & Klobodu, E. K. M. (2016). Urbanization, regime type and durability, and environmental degradation in Ghana. *Environmental Science and Pollution Research, 23* (23), 23825–23839.

Ahmed, Z., Wang, Z., & Ali, S. (2019). Investigating the non-linear relationship between urbanization and CO_2 emissions: An empirical analysis. *Air Quality, Atmosphere & Health, 12*(8), 945–953.

Akorede, Y. F., & Afroz, R. (2020). The relationship between urbanization, CO_2 emissions, economic growth and energy consumption in Nigeria. *International Journal of Energy Economics and Policy, 10*(6), 491–501.

Ali, R., Bakhsh, K., & Yasin, M. A. (2019). Impact of urbanization on CO_2 emissions in emerging economy: Evidence from Pakistan. *Sustainable Cities and Society, 48*, 101553.

Anwar, A., Younis, M., & Ullah, I. (2020). Impact of urbanization and economic growth on CO_2 emission: A case of far east Asian countries. *International Journal of Environmental Research and Public Health, 17*(7), 2531.

Apergis, N., & Ozturk, I. (2015). Testing environmental Kuznets curve hypothesis in Asian countries. *Ecological Indicators, 52*, 16–22.

Baloch, M. A., & Suad, S. (2018). Modeling the impact of transport energy consumption on CO_2 emission in Pakistan: Evidence from ARDL approach. *Environmental Science and Pollution Research, 25*(10), 9461–9473.

Behera, S. R., & Dash, D. P. (2017). The effect of urbanization, energy consumption, and foreign direct investment on the carbon dioxide emission in the SSEA (south and southeast Asian) region. *Renewable and Sustainable Energy Reviews, 70*, 96–106.

Belaïd, F., & Zrelli, M. H. (2016, August). Renewable and non-renewable electricity consumption, carbon emissions and GDP: Evidence from Mediterranean countries. *Economic Research Forum, Working Paper, 1037*:1–25).

Chen, S., Jin, H., & Lu, Y. (2019). Impact of urbanization on CO_2 emissions and energy consumption structure: A panel data analysis for Chinese prefecture-level cities. *Structural Change and Economic Dynamics, 49*, 107–119.

Dietz, T., & Rosa, E. A. (1994). Rethinking the environmental impacts of population, affluence and technology. *Human Ecology Review, 1*(2), 277–300.

Dogan, E., & Aslan, A. (2017). Exploring the relationship among CO_2 emissions, real GDP, energy consumption and tourism in the EU and candidate countries: Evidence from panel models robust to heterogeneity and cross-sectional dependence. *Renewable and Sustainable Energy Reviews, 77*, 239–245.

Franco, S., Mandla, V. R., & Rao, K. R. M. (2017). Urbanization, energy consumption and emissions in the Indian context a review. *Renewable and Sustainable Energy Reviews, 71*, 898–907.

Ibrahiem, D. M., & Hanafy, S. A. (2020). Dynamic linkages amongst ecological footprints, fossil fuel energy consumption and globalization: An empirical analysis. *Management of Environmental Quality: An International Journal, 31*(6), 1549–1568.

International Energy Agency. (2008). *World energy outlook 2008*.

Jones, D. W. (1989). Urbanization and energy use in economic development. *The Energy Journal, 10*(4), 29–44.

Khan, M. K., Khan, M. I., & Rehan, M. (2020). The relationship between energy consumption, economic growth and carbon dioxide emissions in Pakistan. *Financial Innovation, 6*(1), 1–13.

Khan, K., & Su, C. W. (2021). Urbanization and carbon emissions: A panel threshold analysis. *Environmental Science and Pollution Research*, 1–9.

Khoshnevis, Y. S., & Golestani, D. A. (2019). CO_2 emissions, urbanisation and economic growth: Evidence from Asian countries. *Economic Research-Ekonomska Istraživanja, 32*(1), 510–530.

Kwakwa, P. (2012). Disaggregated energy consumption and economic growth in Ghana. *International Journal of Energy Economics and Policy, 2*(1), 34–40.

Kwakwa, P. A., & Alhassan, H. (2018). The effect of energy and urbanisation on carbon dioxide emissions: Evidence from Ghana. *OPEC Energy Review, 42*(4), 301–330.

Lv, Z., & Xu, T. (2019). Trade openness, urbanization and CO_2 emissions: Dynamic panel data analysis of middle-income countries. *The Journal of International Trade & Economic Development, 28*(3), 317–330.

Marcotullio, P., & Lee, Y. S. (2003). Urban environmental transitions and urban transportation systems: A comparison of the north American and Asian experiences. *International Development Planning Review, 25*(4), 325–355.

McGee, J. A., & York, R. (2018). Asymmetric relationship of urbanization and CO_2 emissions in less developed countries. *PLoS One, 13*(12), e0208388.

Mirzaei, M., & Bekri, M. (2017). Energy consumption and CO_2 emissions in Iran, 2025. *Environmental Research, 154*, 345–351.

Musah, M., Kong, Y., Mensah, I. A., Antwi, S. K., & Donkor, M. (2020). The connection between urbanization and carbon emissions: A panel evidence from West Africa. *Environment, Development and Sustainability*, 1–28.

Naseem, S., Guang Ji, T., & Kashif, U. (2020). Asymmetrical ARDL correlation between fossil fuel energy, food security, and carbon emission: Providing fresh information from Pakistan. *Environmental Science and Pollution Research*, 1–14.

Osobajo, O. A., Otitoju, A., Otitoju, M. A., & Oke, A. (2020). The impact of energy consumption and economic growth on carbon dioxide emissions. *Sustainability, 12*(19), 7965.

Pata, U. K. (2018). Renewable energy consumption, urbanization, financial development, income and CO_2 emissions in Turkey: Testing EKC hypothesis with structural breaks. *Journal of Cleaner Production, 187*, 770–779.

Pesaran, M. H., Shin, Y., & Smith, R. J. (2001). Bounds testing approaches to the analysis of level relationships. *Journal of Applied Econometrics, 16*(3), 289–326.

Rafindadi, A. A., Yusof, Z., Zaman, K., Kyophilavong, P., & Akhmat, G. (2014). The relationship between air pollution, fossil fuel energy consumption, and water resources in the panel of selected Asia-Pacific countries. *Environmental Science and Pollution Research, 21*(19), 11395–11400.

Raheem, I. D., & Ogebe, J. O. (2017). CO_2 emissions, urbanization and industrialization: Evidence from a direct and indirect heterogeneous panel analysis. *Management of Environmental Quality: An International Journal, 28*(6), 851–867.

Sadorsky, P. (2014). The effect of urbanization on CO_2 emissions in emerging economies. *Energy Economics, 41*, 147–153.

Salim, R., Rafiq, S., Shafiei, S., & Yao, Y. (2019). Does urbanization increase pollutant emission and energy intensity? Evidence from some Asian developing economies. *Applied Economics, 51*(36), 4008–4024.

Sarkodie, S. A., & Strezov, V. (2019). Effect of foreign direct investments, economic development and energy consumption on greenhouse gas emissions in developing countries. *Science of the Total Environment, 646*, 862–871.

Sarwar, S., & Alsaggaf, M. I. (2019). Role of urbanization and urban income in carbon emissions: Regional analysis of China. *Applied Ecology and Environmental Research, 17*(5), 10303–10311.

Shahbaz, M., Chaudhary, A. R., & Ozturk, I. (2017). Does urbanization cause increasing energy demand in Pakistan? Empirical evidence from STIRPAT model. *Energy, 122*, 83–93.

Sheng, P., & Guo, X. (2016). The long-run and short-run impacts of urbanization on carbon dioxide emissions. *Economic Modelling, 53*, 208–215.

Shuai, C., Chen, X., Wu, Y., Tan, Y., Zhang, Y., & Shen, L. (2018). Identifying the key impact factors of carbon emission in China: Results from a largely expanded pool of potential impact factors. *Journal of Cleaner Production, 175*, 612–623.

Usman, O., Olanipekun, I. O., Iorember, P. T., & Abu-Goodman, M. (2020). Modelling environmental degradation in South Africa: The effects of energy consumption, democracy, and globalization using innovation accounting tests. *Environmental Science and Pollution Research, 27*(8), 8334–8349.

Wang, S., Li, Q., Fang, C., & Zhou, C. (2016). The relationship between economic growth, energy consumption, and CO_2 emissions: Empirical evidence from China. *Science of the Total Environment, 542*, 360–371.

Xia, C., & Wang, Z. (2020). The effect of fossil fuel and hydropower on carbon dioxide emissions: EKC validation with structural breaks. *Journal of Environmental Engineering and Landscape Management, 28*(1), 36–47.

Yakubu, I. N. (2020). Institutional quality and foreign direct investment in Ghana: A bounds-testing cointegration approach. *Review of International Business and Strategy, 30*(1), 109–122.

Yi, Y., Ma, S., Guan, W., & Li, K. (2017). An empirical study on the relationship between urban spatial form and CO_2 in Chinese cities. *Sustainability, 9*(4), 672.

Zaidi, S. A. H., Zafar, M. W., Shahbaz, M., & Hou, F. (2019). Dynamic linkages between globalization, financial development and carbon emissions: Evidence from Asia Pacific economic cooperation countries. *Journal of Cleaner Production, 228*, 533–543.

Zaman, K., & Abd-el Moemen, M. (2017). Energy consumption, carbon dioxide emissions and economic development: Evaluating alternative and plausible environmental hypothesis for sustainable growth. *Renewable and Sustainable Energy Reviews, 74*, 1119–1130.

Zhang, Y. J., Yi, W. C., & Li, B. W. (2015). The impact of urbanization on carbon emission: Empirical evidence in Beijing. *Energy Procedia, 75*, 2963–2968.

Zhang, N., Yu, K., & Chen, Z. (2017). How does urbanization affect carbon dioxide emissions? A cross-country panel data analysis. *Energy Policy, 107*, 678–687.

Zhou, C., Wang, S., & Wang, J. (2019). Examining the influences of urbanization on carbon dioxide emissions in the Yangtze River Delta, China: Kuznets curve relationship. *Science of the Total Environment, 675*, 472–482.

Chapter 16
RPA in Energy and Utilities

Özge Doğuç

Abstract Robotic Process Automation (RPA) is an effective technology that uses software robots that typically mimic an employee to automate everyday tasks. These tasks are often related to the business processes of the E&U companies such as billing, payments, check-in and check-out and other office tasks. While RPA is gaining increasing popularity in different industries, the E&U sector lags behind others in terms of maturity. Energy and utilities (E&U) is a customer-centered industry where each individual is dependent on the provided services for their daily needs. The likelihood of human error in the sector is high, given the large number of transactions happening every day, so RPA becomes essential to help companies manage transactions efficiently and improve the customer experience which is often missed by the companies. This chapter discusses how RPA can positively impact both internal processes of E&U companies and their customer-facing operations.

16.1 Introduction

Energy and utilities (E&U) companies have been enjoying use of smart devices in supply-demand optimization and infrastructure design for the last two decades. Internet of things (IoT) allowed the E&U companies move from traditional processes to technology-driven environments. Also, Artificial Intelligence (AI) has been utilized by these companies to analyze the data collected by smart devices to assist with generating competitive rates for consumers, performing predictive maintenance for the equipment and pipelines; and following the energy stock market with high degree of accuracy.

AI has specifically played an important role in assisting employees through the companies' data and business models. These AI-assisted models have caused

Ö. Doğuç (✉)
The School of Business, İstanbul Medipol University, İstanbul, Turkey
e-mail: odoguc@medipol.edu.tr

S. Yüksel, H. Dinçer (eds.), *Strategic Approaches to Energy Management*, Contributions to Management Science,
https://doi.org/10.1007/978-3-030-76783-9_16

significant changes in E&U companies' operations while providing insights about customer needs and enhancing business capabilities. However, generating a large amount of data, smart devices increase the risk of human errors significantly. Robotic Process Automation (RPA) is an effective technology that uses software robots that typically mimic an employee to automate everyday tasks. These tasks are often related to the business processes of the E&U companies such as billing, payments, check-in and check-out and other office tasks.

While RPA robots do not necessarily replace human employees directly, it is expected to offload work from them; and the total amount of work that is offloaded from the employees is formulated in terms of full-time employee (FTE) savings. It is used by the companies for estimating the return on investment (ROI) for their RPA projects. The process that utilizes (occupies) the RPA robots the most, doesn't necessarily be the one that also provides the most FTE savings; and the reverse is also true. In a research from Capgemini (Capgemini Research Institute, 2019) the low-complexity and high-benefit use cases from E&U industry are listed as follows:

- Forecasting: E&U companies use advanced data science and machine learning techniques to develop local weather forecast systems. Weather models apply to all solar and wind farms, 6% and 9% for solar average forecast error for wind energy can be reduced. For every 1% reduction, customers are better allocated resources and saved $1 million per year.
- Network Behavior Interface: In order to understand the seasonal consumer behavior; optimize power usage to reduce peak loads in the electrical grid, balance loads and balances the distribution, E&U companies use intelligent algorithms to measure and predict parameters such as grid load, electricity generation and consumption and electricity prices. Effective energy network management includes correct planning of load demand, adequate maintenance schedule for generation, transmission and distribution lines, and efficient load distribution through supply lines. Therefore, accurate load estimation has an important place in the power generation industry to maximize the efficiency of the planning process. As a way of improving the accuracy of the electricity demand estimation, E&U companies apply various computational and statistical techniques to improve their estimation models.
- Yield Optimization: E&U companies use machine learning on cloud-based platforms to simulate real-world energy production environments. Its models run real-world wind and solar models and calculate electrical output to optimize production in separate turbines. In this way, up to 20% increase in energy production can be achieved. They can use the output of the models to optimize production at the turbine level. Also, after training the model on output parameters, the algorithm will begin to suggest improvements to newly developed wind farms. The algorithm can be varied in up to 20 different turbine configurations and data is collected from dozens of sensors located in the wind turbine and related equipment. These sensors send data back to the model in real time for updated recommendations on optimum wind farm configuration. (IoT) (Capgemini Research Institute, 2019)

- Energy Trading: Automation allows E&U companies to combine data in the trading space using RPA to simulate repetitive processes. This changes the role of analysts and frees more and more of their time to focus on higher-value tasks. So instead of extracting and collecting data, the analysts can spend more time on analyzing data and generating useful insights.

16.2 Robotic Process Automation (RPA)

Robotic process automation (RPA) is based on software robots (bots) and on artificial intelligence (AI). In traditional automation tools, a software developer designs an automation for an existing task which interacts with the application system using their internal application programming interfaces (APIs) or dedicated scripting languages. In contrast, RPA systems develop the list of activities by recording while the user performs the task in the application's graphical user interface (GUI), and then perform the automation by repeating those tasks directly in the GUI (Doguc, Robotic Process Automation (RPA) Applications in COVID-19, 2021). This not only reduces the time and effort to develop the automations, but it also produces resilient results by mimicking the user precisely (Robotic Process Automation (RPA) Case Study: Utilities Industry, 2019). RPA is designed to assist office-style functions that often require the ability to perform various types of tasks in a specific sequence. It creates and distributes a software robot capable of launching and running other software. Compared to the robots that have been used in manufacturing plants, office work often requires the similar kind of repetitive effort, but a physical robot is not necessary as data is manipulated across platforms and applications (Frankenfield, 2020).

As with all automation, RPA is used to remove processes that were previously done by employees from them, but this time, robotic software is configured by interacting between different systems such as spreadsheets, Customer Relationship Management (CRM) systems or Enterprise Resource Planning (ERP) (Willcocks, 2016). In short, RPA provides tools to automate rule-based, logical processes that contain well-defined and structured data with a deterministic set of output values. RPA processes are often repetitive and highly manual (Willcocks, 2016). Such tasks can be called "swivel chairs", referring to moving inputs from one side to another without the need to take initiative. However, given a suitable process and a well-defined working logic, the robot is expected to perform much higher than humans in terms of quality, time and cost (Willcocks et al., 2015).

The purpose of RPA is not just to help people automate their processes; In addition, it aims to completely change the way people and processes operate (Willcocks et al., 2015). While applications like Excel help people with calculations, they still need human intervention. In RPA, the calculations are mostly done by the robot behind the scenes.

The term Robotic Process Automation was first used in 2012 (Hindle, 2018). It started to gain popularity in 2014 and 2015, as companies began to announce the significant savings achieved through automation. With RPA, the back-office automation market became more important in early 2016 but was still relatively small scale during this time (Willcocks, 2016). Based on Horses for Sources (HfS) research (Fersht, 2017) and Everest Group research (Robotic Process Automation (RPA): Technology Vendor State of the Market Report, 2017), the global RPA market has increased by approximately 64 times between 2015 and 2017. (Robotic Process Automation (RPA): Market Technology Provider Report, 2017; Fersht, 2017). According to HfS Research, an increase of 42% from 2017 to 2018, while approximately 94% from 2018 to 2021 a market share increase is expected (Fersht, 2017).

The most interesting of the benefits of RPA is that RPA does not require a change in the company's IT infrastructure (Willcocks, 2016). Every action taken by RPA robots can be easily recorded and audited; and therefore, the risk of non-compliance is minimal (Willcocks et al., 2015). In this way, it differs from most conventional business process automation that can process data directly in a database. Numerous RPA applications have been reported in the last 5 years in business processes such as accounts payable, accounts receivable, travel expenses, fixed asset accounting, master data management, invoicing, keeping employee records (Willcocks, 2016; Asatiani, 2016; Willcocks et al., 2015). Most of these processes are back office or support processes for services that often contain repetitive and mundane tasks. With a good understanding of the key benefits of RPA, it is natural to ask the question of where and when RPA is best applied. As with all automation applications, RPA robots need clear rules to follow, effectively eliminating the non-rule-based processes that apply to RPA (or any other automation). Lacity et al. say that for automation to have high efficiency, high standardization, well-defined implicit logic and high maturity, optimum target processes must be defined (Willcocks et al., 2015). While high volume, high uptime savings, standardization and proper logic ensure seamless development and robot configuration, maturity means that the process will likely not be eliminated and therefore invalidate the resources used to develop the RPA solution (Devarajan, 2019).

To demonstrate suitability for automation, reviewing the implementation of RPA at Telefónica O_2 in the UK, Lacity et al. introduced a concept called automated tape (Lacity et al., 2015a). It visualizes the relationship between transaction volume and transaction length; puts the processes into four categories: The basic logic is that a process must reach a certain time savings in order to be considered as a candidate for automation. High-volume and repetitive processes fall into this category (Doguc, 2019).

16.3 Benefits of Using RPA in E&U Industry

While RPA is gaining increasing popularity in different industries, the E&U sector lags behind others in terms of maturity. In a survey by Protiviti, Technology and Financial Services leads the industry in terms of RPA maturity (Taking RPA to the

Next Level, 2019). Companies in these industries have been using RPA solutions for a while, and by 2021, 30% of companies are in advanced stages with RPA. On the other hand, only 9% of companies in the E&U sector have reached the same level of maturity.

Compared to the technology and telecommunications companies, the E&U sector employs a higher number of blue-collared employees, as generation and movement of energy resources require highly skilled labor to be done by the field teams. In addition, the E&U companies utilize operational staff for serve their customers at their physical addresses to maintain their meters and fix issues related with their services. These tasks require human effort that cannot be replaced by the RPA bots.

RPA's leading 'application autonomy' feature is particularly important for most E&U companies that rely on legacy software that cannot be configured to coordinate well with other applications. As mentioned earlier, RPA bots can use the GUI to interact with software; and therefore, can integrate legacy systems that cannot communicate otherwise. Often in processes where such legacy systems are used, employees have to manually pass information from one application to another. This is time consuming, prone to errors and therefore costly. RPA can automate such data transfers quickly and without errors. As stated by the RPA and AI Institute, "Companies that use large-scale workforce for general computational work where people perform high-volume, high-level operational process functions will increase its capabilities and save money and time with RPA software" (What is Robotic Process Automation? 2019).

The benefits of RPA for E&U companies can be summarized as follows:

- Cost savings (FTE) by replacing mundane tasks done by human employees such as billing and accounting
- Minimizes errors and speeds up turnaround time in solving customer problems such as billing inquires and new account setups
- Helps to respond rapidly during disasters and crises
- Quickly adapts to the quickly changing regulations and market needs in the industry

16.3.1 Value Proposition of RPA in E&U Industry

To better understand the true value of RPA, we must consider the emerging role of technology in E&U companies through a number of use cases in the industry. As an example, consider a large oil and gas company in the United States with around 5000 employees and $10 billion in yearly revenue. Oil and gas companies that have poor management with partners and customers, complex operational processes and limited digital optimization competencies use RPA's automation capabilities to:

- Analyzing data models: RPA robots, which record and learn users' actions, collect large amounts of data on customers' purchasing preferences. RPA robots can reveal the time required to process the order, the number of pending transactions, or even processes with a high percentage of exceptions and requiring human intervention. With the analytical capabilities of RPA, companies can reduce process bottlenecks to increase operational efficiency, increase employee happiness, and achieve higher back-office productivity (Anagnoste, 2013).

- Support accounting processes: RPA robots can work effectively and efficiently with processes that are repetitive, time-consuming and doesn't change over time. Most accounting processes exhibit these characteristics and therefore considered as primary targets for RPA. Oil and gas companies, particularly those facing these challenges when closing books monthly, use RPA to automate their most basic financial accounting processes and provide analytical insights: tracking revenue and expense accounts, posting journal entries, reconciliation balance sheet accounts and pre-preparation of financial statements. In addition, by creating automation alerts, the company's finance analysts only have to intervene when exceptions are created and are able to focus instead on more value-adding responsibilities. Also, executive level is given better end-to-end visibility into the financing of the company (Doguc, Robot Process Automation (RPA) and It's Future, 2019).
- Facilitate joint ventures: Oil and gas companies often create joint ventures to reduce risk, meet regulatory requirements and optimize their supply chains, in addition to sharing their capital and resources. Initial phases of joint ventures often become obstacles for IT, as integrating platforms from multiple companies is time consuming, or completely impossible. This is especially true if the systems and platforms used by joining companies are not compatible. Moreover, maintaining different applications can be costly and time consuming. Thanks to its presentation layer interactions, RPA can integrate otherwise incompatible systems to enable collaboration between inventory systems, accounting processes, and customer databases—all without reconfiguring the existing installation (Deckard, 2017).

For E&U companies, RPA robots offer many other benefits as they can implement repetitive processes such as customer records management, resolution of complaints, and measurement and invoicing not only faster but also at lower costs. Also, employing such a technology can overcome challenges posed by the highly demanding and dynamic market conditions. In addition, RPA can be used to improve accounting workflows, increase efficiency for processes along the value chain, and minimize variability in operational output. Many of the E&U industry products use predictive algorithms that use historical data and data analytics techniques to determine future results. RPA robots can work with very large datasets, and generate analytics based on given unsupervised algorithms, without requiring any human intervention.

RPA helps automate processes that require fast response times, have high-volume transactions, and preferably repetitive process dependent on the workflow. Some of these processes include payroll, human resources, and IT backup processes. For example, in a survey done by E&Y in 2019, around 60% of the respondents think that companies have to retrain their employees to catch up with the new advancements in technology; and this will create new opportunities for employees to learn new abilities. But the forward-thinking of digital transformation has not been universally embraced by public companies (How much digital labor can drive oil and gas productivity in 2019). Some of the E&U companies still see technological development, including RPA, as a burden. But what is missing among opponents is the recognition that the RPA is definitely an innovation worth investing in, especially for utility companies looking to develop their core business.

Similar to the companies in the other industries, for the electricity, water and gas companies the key question when implementing RPA is to determine the best method for measuring return on investment (ROI). Businesses in industries typically looked at ROI as a financial measure and calculated it as the realized return on investment as a proportion of the costs incurred. A Capgemini study (Capgemini Research Institute, 2019) shows that the E&U sector is gaining significant value from RPA and other smart automation systems compared to other industries. The study shows that operations quality and work accuracy are the areas that benefit significantly from RPA in the E&U sector.

Businesses need to take into account 'return on effort' (ROE) or 'impact on experience' (IoE) when measuring the results of RPA initiatives in the energy and utilities sector. One way to measure ROE is to scale the total incremental return generated by the effort spent on an RPA project. It should include quantitative aspects such as reduced implementation costs and financial savings from employees' paid benefits, as well as qualitative aspects such as increased benefits overall, increased employee morale and improved end-user experience. The effort expended can be calculated in terms of the total project execution cost. This includes not only the usual implementation overheads, but also the additional costs of allocating internal resources. However, RPA is often seen by internal workers as a double-edged sword. On the one hand, RPA offers employees a free time saving by eliminating their tedious tasks, on the other hand, RPA can be seen as a threat to some employees who are satisfied with their current activities. The implementation of RPA can also change the profile of the most desired recruits for data processing and management roles (Use of robotic process automation in energy trading, 2021).

In addition to effort and returns, two other aspects that play an important role in calculating return on investment are time and customer experience. Calculation of time from these points is simple: Cycle time can be measured approximately as 'mean solution time'/'average run time'. However, unlike the time and effort spent on processes, quantifying process experience follows a different process. To measure this, E&U services organizations need to delve into the various issues faced by their internal stakeholders, namely customers who will use robotic software.

This chapter discusses how RPA can positively impact both internal processes of E&U companies and their customer-facing operations.

16.4 RPA Applications in Energy and Public Services

Energy and utilities (E&U) is a customer-centered industry where each individual is dependent on the provided services for their daily needs. The likelihood of human error in the sector is high, given the large number of transactions happening every day, so RPA becomes essential to help companies manage transactions efficiently and improve the customer experience which is often missed by the companies. Because utilities generate large amounts of data with smart devices, it significantly increases the likelihood of human error. RPA is an effective technology that uses software robots that typically mimic an employee to automate everyday tasks. These tasks are often related to the business processes of utilities such as billing, payments, check-in and check-out, and other office tasks. E&U companies are experiencing the first wave of automation with the advent of RPA bots. RPA creates a personalized customer experience for energy consumers by enabling the integration of data flows and strengthens companies' existing analysis platforms with it. At this point, the main advantage of RPA implementations is the scaling that results in a significant reduction in audit costs.

Service delivery is the worst performing sector in customer service in most developing countries. Here, RPA can play a more important role in improving customer service and lowering operating cost. In the case of electrical networks that require constant monitoring, human intervention is required to deenergize problem areas of the grid. Any mistake can result in large fines and potential risk of life and property. RPA can check this by logging into relevant systems (SCADA, GIS, and other applications) to identify problems that will further aid engineers in decision making. This can save a co-player from a human error, timeless scheduling of outages and accidents during maintenance work (Dixit, 2018). Although it has not yet been widely implemented, E&U firms recognize the significant opportunities that RPA offers in energy trade to reduce costs and increase organizational efficiency. Given the high business volume, particularly in energy trading operations, RPA shows its potential to significantly reduce errors and costs across different processes and systems.

Some of the common application areas of RPA in E&U companies can be summarized as follows:

16.4.1 Back Office Processes

To demonstrate the value creation potential of RPA, London School of Economics and Political Science has prepared a case study demonstrating the adoption of RPA in a major European institution (Lacity, Willcocks, & Craig, Robotic Process Automation: Mature Capabilities in the Energy Industry, 2015b). In this case study, an E&U company automates approximately 25% of their back-office process with around 300 RPA robots. These back-office processes included customer billing, exception and complaint handling, and data consolidation. The robots were managed by two of the employees to perform around one million transactions every month—a task that has been done by about 600 employees. Against implementation costs, the RPA initiative achieved payback on the initial investment within 12 months and generates approximately 200% ROI annually.

16.4.2 Energy Trade

Basically, the energy trading operation is the execution of a large number of transactions, both buying and selling, in the energy market. For each of these transactions, various data elements (product specifications, price, volume, delivery time and delivery location) should be stored, which will be processed in different ways later. For example:

- Payments—Buyers must be billed, vendors must be paid, and the accounting team must be made aware of cash flow in and out of the company.
- Coordination—For every transaction with the counterparty, coordination must be maintained with organizations responsible for physical delivery of energy, nomination and logistics.
- Risk management—After each transaction, the company's trading net position changes, and the values at which various units within the company are exposed to risk against company limits should be evaluated periodically.
- Analysis—As the portfolio changes in each transaction, analytical teams regularly evaluate the economic and financial results of the transactions that can be made.

Energy trading operations employ a fairly large number of employees to manage the versatile and multidimensional data flow across the organization. The operations that these employees focus on are often repetitive, low-value and simple, as well as very time-consuming. As shown in a case study by Greysoft, RPA can reduce transaction times by nearly 50% while saving millions of dollars in operating expenses (RPA for Asset Management Officer—Case Study, 2018). Many companies involved in the energy business—indeed most companies in the E&U industry—have yet to reap many of the potential benefits RPA can provide.

To date, no proven way to implement enterprise-wide RPA in an energy trading firm has been released. Therefore, while there was an opportunity to create significant value provided by RPA in energy trade, E&U companies wishing to seize this opportunity had to make a special effort with regard to energy trading. Ideally, a company trying to adopt RPA will first use it in high-value applications and then consider gradually rolling out RPA to other applications. Inevitably, not every source of inefficiency caused by manual effort will be problematic enough to comply with RPA or achieve the predicted gain.

Considering the risks in the energy trading operation, it will be important that use of RPA can adapt to the internal processes of the company and have an organizational structure that allows change accordingly. However, similar to standard AI implementations, the unpredictable consequences of RPA robots require an in-house RPA team to be set up to ensure bots don't cause transactions to fail. E&U companies that transfer their processes to RPA should plan the development of their in-house capabilities and act in coordination with human resources in order to ensure continuous optimization of RPA usage, as they will choose the way to address the problematic areas inside.

16.4.3 Resolving Infeasible Customer Meter Readings

E&U companies read meters before billing customers for consumption. Occasionally human errors lead to incorrect bills, resulting in excessive or undercharging for the consumer. Such mistakes have serious consequences for the company's reputation. Therefore, verifying meter reading prior to invoicing is a very sensitive issue for EU companies. EU companies have seen the consequences of false meter reading risk over the years, and so over time they have started to integrate RPA into the process. In this case, RPA verifies the consumer's meter reading and then processes it as input to the billing process. In case of an error, it reports the problem to the relevant department. This significantly reduces the burden of the meter reading verification team and the number of complaints from incorrect billing (CRMT, 2020).

Millions of residential customers need their meters to be read at least four times a year for billing. A customer's meter reading can be self-reported or done by a rental meter reader, but in both cases thousands of readings arrive every day. Before RPA, E&U companies determine whether a meter reading was infeasible or not through their mainframe systems that applied definite rules to the readings that arrived every day. There can be many reasons to question meter readings. For example, if the meter reading for that quarter was lower than the meter reading of the previous quarter, it would indicate the unlikely situation that the customer added to the grid instead of consuming electricity. Infeasible meter readings are often given to employees to resolve them manually. Depending on the situation, people applied rules or judgments to correct mistakes. The rule-based decisions on infeasible meter readings are suitable for automation. Employees would still continue to handle exceptions that require human judgment. This business scenario improves resolution quality, consistency, and speed by reducing the number of FTEs that are required for this task on average by about 60%.

16.4.4 New Account Setup

Every E&U organization follows a strict process for creating new customer accounts. Sometimes the team processing new account requests overlooks certain factors in the process when allowing the consumer. This can sometimes lead to permissions, even for those who do not meet all the requirements set by the organization. RPA assists the team in reducing such errors by checking the information provided by the consumer against the established rules and only processes the installation request if all fields obey the rules. In the event of a problem, the software program will flag it to the appropriate department, which will fix the problem later (RPA for Energy and Utilities Industry, 2020).

An example for setting up new accounts and disconnecting existing ones can be given from Turkey. An energy distribution company with more than 3 million subscribers have been receiving around 800 requests for new accounts or disconnecting existing services from homes and companies. Each request had to go through a number of checks before an action could be taken. For example, new account opening requests required checks for the service availability in the requested area, financial background of the applicant, any previous balances on the same address, proofs of residency, etc. Any missing information or failed check result had to communicated back to the applicant; therefore, both successful and unsuccessful applications required considerable amount of manual effort to be done by a team of 16 employees. On average, an application required around 10 min of employees' time for the checks; which accumulate to 8000 min (133 h or roughly 17 man/days) every day.

The company employed 5 RPA robots that can simultaneously accept new applications and requests for disconnecting services. These robots worked around the clock to go through every check that needs to be done for each application and request. Robots sends emails and SMS information to the applicants if more information is needed. Otherwise, they approve the requests and create internal tickets for the other department that are responsible to opening and disconnecting service.

16.4.5 Intraday Updates

Requests that may include E&U companies' daily activities such as addition, replacement and shutdown for gas meters are pretty common; in fact, the number of such requests is usually in the thousands and is handled manually by a team of 20–30 people. They schedule appointments and create internal tickets for field teams that take the necessary actions. These are sometimes done through asset management applications that have the ability to optimize the field team's schedule according to the tasks created for them. Recently, RPA bots have been used to receive requests through the application and create tickets for field teams. They can optimize tasks even outside working hours. Companies, tasks of manual effort to create and optimize saves up to 85%, and regulatory compliance slam the higher rate 30% faster to meet the demand for jobs. RPA robots also provide 100% data entry accuracy (Robotic Process Automation (RPA) Case Study: Utilities Industry, 2019).

For example, a utilities company in Turkey used to employ a team of 6 people to generate daily schedules of the field teams, who were responsible for physically replacing and removal of the meters. The team worked after hours to manually go through the service requests that arrived in the last 24-hour period. Team required to find the optimal daily schedules for the field teams and had to complete their task by 6 am to allow the field teams enough time to prepare for their daily tasks. The utilities company operated one RPA robot to create daily schedules and revise them as new requests arrive during the day. This way, the schedules became always up-to-date and available to the field team—and 6 FTE savings was achieved.

16.5 Conclusion

The global energy industry is undergoing tremendous change. Traditional, centralized energy supply from sources such as coal, gas and nuclear faces an economic, political and social backlash as a result of climate change, health and safety and security concerns. At the same time, new uses such as electric vehicles must be met, as well as growing energy demand from developing countries. All these trends mean that distributed and weather-dependent renewable energy is an increasingly critical and growing part of the energy portfolio for the E&U companies.

Adopting advanced analytical solutions offered by RPA in meeting the growing demand for reliable energy has become essential for the E&U companies in the last decade. Promoting RPA within the company plays an important role in achieving greater efficiency and customer satisfaction. It will also ensure that aging business models are redesigned, and their productivity increased. However, research shows that many E&U organizations have difficulty in succeeding in smart automation, or automation is not becoming widespread within the company, and there is falling behind in creating added value. The main reason for this can be shown as the companies not seeing the critical usage gains in their basic processes and missing the potential benefit to be provided with RPA. The benefits of smart automation are constantly underestimated and misjudged, and the smart automation capability cannot penetrate the company's capillaries.

E&U companies can take a number of measures to eliminate the problems mentioned above. First, companies are evaluating uses that are most mature, most accessible to the majority and generate the most value, such as billing and data consolidations. By placing RPA robots in these areas of use, they reinforce the gains there. In addition, E&U companies position RPA robots in high value-added processes such as renewable energy and automatic meter reading. Indeed, it is easier to increase the value of these advanced technology-requiring processes than to create new processes. Second, E&U companies need to take a pragmatic approach when choosing areas to focus on automation and AI, and make sure they maximize ROI by redesigning processes before implementing them. Finally, they need to recognize intelligent automation as a component of RPA technology as a necessary component to complement the human workforce and identify the required change management.

Traditional business models that include stable and long-term contracts such as high-volume and capital-backed energy projects are gradually being replaced by smaller players and more flexible projects: shale oil, solar panels, microgrids are just a few examples. If the E&U sector is to grow alongside the challenges of this age, it requires adoption of automation and AI. This type of talent and culture organizations develop, more aggressively with new business models that will make innovation and change at the expense of failing to organizations that can grow in a lively manner (Capgemini Research Institute, 2019).

In summary, changing market needs, developments in electric vehicles, investments in solar and wind technology forces the E&U companies to implement changes in their processes. AI-assisted RPA bots offer significant savings in billings, complaint management, meter readings, usage forecasting by replacing mundane and repetitive tasks that are otherwise done by employees. While RPA adaptation rate in the E&U sector is relatively low, companies have been realizing the significant FTE savings and high ROI offered by RPA.

References

Anagnoste, S. (2013). The road to intelligent automation in the energy sector. *Management Dynamics in the Knowledge Economy, 6*(3), 489–502.

Asatiani, A. P. (2016). Turning robotic process automation into commercial success – Case OpusCapita. *Journal of Information Technology Teaching Cases, 6*(2), 67–74.

Capgemini Research Institute. (2019). *Intelligent automation in energy and utilities.*

CRMT. (2020, May 28). *Robotic process automation in energy & utilities industry*. Retrieved from RPA for Energy and Utility: https://www.crmt.com/en/resources/blog/robotic-process-automation-in-energy-utilities-industry/

Deckard, M. (2017). *RPA's impact on the utilities industry*. UiPath.

Devarajan, Y. (2019). A review on intelligent process automation. *International Journal of Computer Applications, 182*(36), 40–44.

Dixit, P. (2018). *AI & RPA in energy & utilities*. HCL Technologies Limited.

Doguc, O. (2019). Robot process automation (RPA) and It's future. In *Handbook of research on strategic fit and design in business ecosystems* (pp. 469–492). Igi-Global.

Doguc, O. (2021). Robotic process automation (RPA) applications in COVID-19. In *Financial strategies in competitive markets*. Springer.

Fersht, P. S. (2017, January 21). *The Robotic process automation market will reach $443 million this year.* Retrieved from https://www.horsesforsources.com/RPA-marketsize-HfS_061017

Frankenfield, J. (2020). *Robotic process automation—RPA*. Investopedia.

Hindle, J. L. (2018). *Robotic process automation: Benchmarking the client experience*. Knowledge Capital Partners.

Lacity, M., Willcocks, L., & Craig, A. (2015a). *Robotic process automation at Telefónica O2*. The Outsourcing Unit Working Research.

Lacity, M., Willcocks, L., & Craig, A. (2015b). *Robotic process automation: Mature capabilities in the energy sector*. The Outsourcing Unit.

Robotic Process Automation (RPA) case study: Utilities industry. (2019, November 28). Retrieved from CiGen: https://www.cigen.com.au/cigenblog/robotic-process-automation-rpa-case-study-utilities-industry

Robotic Process Automation (RPA): Technology vendor state of the market report. (2017). Everest Global.

RPA for asset management institution – Case study. (2018, January 25). Retrieved from Greysoft: http://greysoft.co/blog/index.php/case-studies/rpa-asset-management-institutions-case-study/

RPA for energy and utilities industry. (2020). Retrieved from Comtec: https://www.comtecinfo.com/rpa/rpa-in-energy-and-utilities-industry/

Taking RPA to the next level. (2019). Protiviti.

What is Robotic Process Automation? (2019). Retrieved from Institute of Robotic Process Automation and Artificial Intelligence: https://irpaai.com/what-is-robotic-process-automation/.

Willcocks, P. L. (2016). *Service automation – robots and the future of work* (1st ed.). Steve Brookes Publishing.

Willcocks, L., Lacity, M., & Craig, A. (2015). *The IT function and Robotic process automation*.

Chapter 17
Energy Oriented Management Approach as a Market Activity Tool in Achieving Competitive Advantage

Fulya Almaz

Abstract Today, all for-profit or non-profit businesses carry out many activities in order to provide a competitive advantage and to make it sustainable. These businesses that want to achieve success in competition and being sustainable have to manage energy, which is an important input, correctly. With energy management, which has been on the agenda in all world markets since the 1970s, they integrate practices that include both sustainable competitive advantage and environmental issues into the strategies of their businesses. In this study, in order to contribute to the relevant literature, energy-oriented management approach as a market efficiency tool in gaining competitive advantage is conceptually examined. In this context, a deep literature review has been made. As a result, it has been seen that a fully integrated energy management process from top management to the lowest level should be developed in order for a successful energy management approach to take root in enterprises. It has been determined that the energy management, which will be applied successfully, has an effect on the competitive advantage of the enterprises.

17.1 Introduction

Energy has a critical role in terms of both being effective in the economic growth of countries and being an important input for businesses. Movements such as rapid growth, industrialization, population growth and urbanization in the world economy have led to an increase in energy demands. In particular, the energy crisis in 1973 created a significant tension in all countries of the world (Öztürk & Saygın, 2017). Increase in oil prices, decrease in traditional energy resources, increase in fossil fuel consumption and air pollution brought some problems around the world. In addition to resource reduction, the limitations imposed on greenhouse gas emissions within the scope of the United Nations Framework Convention on Climate Change and the

F. Almaz (✉)
Mediterranean University, Famagusta, Cyprus
e-mail: fulyaalmaz@akdeniz.edu.tr

S. Yüksel, H. Dinçer (eds.), *Strategic Approaches to Energy Management*, Contributions to Management Science,
https://doi.org/10.1007/978-3-030-76783-9_17

Kyoto Protocol have led the world states to renewable energy resources and made it mandatory to formulate effective policies in energy use. In this context, measures have been taken for the production and use of clean energy as well as the supply of safe, quality and cheap energy (Flavin & Lenssen, 1994).

The legal obligations and technical, economic, global and social problems that have been encountered have affected the countries as well as the enterprises and pushed the enterprises to an energy-oriented management approach. In this context, activities such as efficient use of energy in enterprises, providing savings, reducing energy density, and setting some standards in energy management have come to the fore. Thus, in a world where industrial competition is intense, the struggle of businesses for survival has also changed direction. Therefore, businesses have had to find and use strategic tools to help them develop their strategic competitive advantage in order to win the competitive battle.

Sustainable competitive advantage is defined as a strategy that cannot be easily imitated by the competitors of the resources owned and used by a company and thus creates value (Barney, 1991; Williams, 1992). When the relevant literature is examined, it is seen that there are two different sustainable competitive advantage approaches: Position Approach emphasizes the structure of the industry and the position of the enterprise in this industry in providing competitive advantage (Porter, 1985), while the Resource-Based Approach emphasizes the competencies that the enterprise has developed within itself (Hamel, 1994; Barney, 1991; Wernerfelt, 1984). Although both approaches maintain their validity, it is clear that business performance is at a level that cannot be neglected. At this point, the following question comes to mind: can energy and its effective management, which is an important input for almost all industries, serve businesses to gain sustainable competitive advantage?

In this study, it is discussed whether energy management, which affects the whole world, is effective in providing competitive advantage for businesses. In this context, energy-oriented management approach is analyzed conceptually as a market efficiency tool in providing competitive advantage. First, the process that leads businesses to an energy-oriented management approach and the subject of competitive advantage were presented with definitions, and then the energy-oriented management approach as a market efficiency tool in providing competitive advantage was discussed.

17.2 Literature Review

17.2.1 From Energy Concept to Energy Focused Management Approach

The two major world wars in the twentieth century and the ongoing global conflicts are seen as access to energy resources (especially oil) and sharing of these resources.

It is thought that the feature that defines the development of countries, makes a country different from other countries and makes it superior is the energy abundance of that country. Therefore, energy has begun to play a critical role in gaining competitive advantage and creating employment.

When the relevant literature is examined, at the end of the 1800s, England turned to oil resources with the idea of using oil as an alternative to coal in naval ships (Sevim, 2018). During this period, Germany realized the importance of oil and tried to access oil reserves. By the 1900s, the coal age turned into an oil age and the struggle to share oil resources between Britain and Germany started the First World War. In the 1940s, the world was the scene of the Second World War, especially when America wanted a share of the cake. The oil crisis came to the fore with the Arab and Israel War in 1973, and a paradigmatic transformation was experienced in energy. America embarks on new pursuits to find oil and rationale oil, Europe conducts research to obtain energy from coal and nuclear reactors, the Netherlands takes savings measures by penalizing excess electricity use, automobile manufacturers produce cars that save fuel, Japan invests in the electricity industry (Öztürk & Saygın, 2017). This paradigmatic transformation brought the concepts of energy diversification (Büyükmıhcı, 2003) to the agenda and led countries to alternatives in energy supply.

While the development process continues, environmental problems that alarming have pushed the world to seek solutions. In the face of this quest, a Sustainable Development Model, which is defined as meeting the needs of today without risking the needs of future generations, was established (Burtland Commission, 1987). This model was first discussed at the Stockholm Conference (1972). This conference was convened to raise awareness on environmental protection by drawing attention to the interaction between human and environment. Recommendations on the production and use of energy to protect the environment were also included in the conference. In this context, decisions were taken such as rationalizing energy management, holding scientific meetings and sharing the available information in order to understand the existing problems and produce solutions (United Nations Documents, 1972).

In 1992, the Rio Conference was held and it was mentioned in this conference that environmental threats can be alleviated by minimizing the amount of energy used in production, and that efficiency and competition can be increased. At this point, it was emphasized that energy resources can be used both economically and environmentally with the measures to be taken by the governments (United Nations Sustainable Development, 1992). After these conferences, the Johannesburg Conference was held in 2002, and within the scope of this conference, it was discussed that many applications such as energy management, production of technologies using efficient and renewable clean energy, and regulation of energy programs are the responsibility of governments (United Nations Millenium Project, 2002).

Energy, which has a critical role in terms of both being effective in the economic growth of countries and being an important input for businesses, is defined as a commodity that can be produced, purchased, stored, subjected to a specific process, recovered and used in a process (Uzun & Değirmen, 2018). Energy is classified as non-renewable energy resources or renewable energy resources according to various parameters. Traditional fossil fuels such as oil, coal and gas, which are expected to be exhausted over time, are non-renewable energy sources. Due to the use of these energy resources, the whole world has faced the following problems:

1. Increasing greenhouse gas emissions trigger climate changes,
2. Fluctuation in oil prices puts industries in a bottleneck,
3. Geopolitical and energy security issues,
4. Increase in food prices.[1]

Studies show that global fossil fuel consumption will increase by 1.6% annually between 2000 and 2030 (Pamir, 2007). It is necessary to access renewable and clean energy resources and to ensure energy efficiency in order to combat energy and environmental problems affecting the whole world (Hamelinck et al., 2005). Renewable energy is defined as alternative energy that can be self-renewed, can be obtained from limited resources, reduce dependence on fossil fuels, alleviate the environmental impact of fossil fuels, and increase energy efficiency (Saavedra et al., 2018). Solar energy, wind energy, biomass energy, geothermal energy and hydroelectric energy are widely used renewable energy sources around the world.

In its most general form, solar energy is defined as the use of solar rays and heat by converting them into energy (Kannan & Vakeesan, 2016; Wee et al., 2012). It is a preferred renewable energy source due to its unlimited resource, not being exhausted and low installation cost. Wind energy is a renewable energy source that converts the kinetic energy of air (wind) into mechanical energy and then into electrical energy (Wee et al., 2012). Biomass energy, which is defined as organic matter obtained from living organisms and organic wastes and can be converted into energy, is converted into solid, liquid and gaseous fuels (Mafakheri & Nasiri, 2014; Wee et al., 2012; Khan, 2009). Geothermal energy, which is defined as the thermal energy collected at a certain depth to the earth's surface, is obtained by drilling wells in certain regions and is used to meet the needs such as heating and electricity (Jassim, 2013; Wee et al., 2012). The most widely used renewable energy source worldwide is hydroelectric energy and it is defined as the conversion of water energy to electricity (Bahadori et al., 2013).

The mentioned renewable energy sources should be included and put into practice within the scope of sustainable energy management programs and policies at national and international level because they are clean, contribute to energy production safety, reduce gas emissions, are environmentally sensitive and have economic value (Bayraç, 2010). In this context, besides renewable energy sources, activities such as efficient use of energy, providing savings and reducing energy density are also on the agenda.

While energy efficiency aims to protect and/or increase product quality with optimum energy use, energy saving aims to reduce energy consumption regardless of product quality and production processes (Van Den Wymelenberg et al., 2013;

[1]The bioethanol system used to ensure the continuity of oil based technology causes an increase in food prices.

Çalıkoğlu, 2004; Sandberg & Soderstrom, 2003). Energy efficiency includes efficiency-enhancing measures such as prevention of possible energy losses in steam, gas, air, heat and electricity generation, recovery of waste, reduction of energy demand, finding more efficient energy resources and energy recovery (Bayraç, 2010). These two concepts are not alternatives to each other and the existence of one does not require the existence of the other (Munguia et al., 2018). Energy intensity, on the other hand, is a concept that indicates the amount of energy consumption needed to produce unit output and is inversely proportional to energy efficiency (Bayraç, 2010).

Efficient energy management that reduces energy use at the same production level or provides more economic benefits from the same energy input is directly related to sustainable development goals (Smith & Kelly, 2015; Thollander & Palm, 2013). At this point, the World Energy Council framed the concept of energy efficiency as an energy trilemma (Paliekhova & Simon, 2016), which includes energy equality, energy security and environmental sustainability. With this framing, the World Energy Council has referred to all three goals of sustainable development.

The 1973 energy crisis, which created an important tension in all countries of the world, led countries to both seek new production technologies and renewable energy resources. Therefore, countries have taken steps to increase energy diversity and find economic solutions. Increasing global energy consumption has brought some environmental problems such as greenhouse gas emission, air pollution and fossil fuel consumption, and thus the energy-oriented management approach has been brought to the agenda. Therefore, a number of legal obligations and technical, economic, global and social problems that have been shaped during the process are considered as factors that push countries and businesses to an energy-oriented management approach.

In its most general form, energy management is defined as the planning necessary for an organization to use the energy it needs in all organizational and technical processes at the optimum level to carry out its activities (Fiedler & Mircea, 2012). Lee et al. (2011) argue that many disciplines such as management, engineering, architecture, and finance should be utilized in order to achieve success in planning. This situation points out that energy management should be handled with an interdisciplinary approach.

When the relevant literature is examined, it is seen that energy management is a cyclical process that involves the top management and requires effective effort (Kannan & Boie, 2003). This cyclical process starts with the decision mechanism taking the energy management issue into the agenda. At this point, first of all, the current energy status of the enterprise is determined in detail. Then, the ways of saving energy are determined and the technical and economical feasibility of the saving application is determined. Finally, the application is started, the expected results are compared with the results obtained, and if there are any defects, they are corrected and progress is made on a cyclical basis.

In studies conducted, it has been found that energy losses, especially in production enterprises, manifest themselves as the use of old technology, during the transfer from one process to another in production processes, or as heat losses in poorly insulated facilities (Glaser, 1992). Some standards have been established to prevent these wastages (Paliekhova & Simon, 2016). When the relevant area is examined, it is seen that the International Energy Agency (IEA) and the International Standardization Organization (ISO) lead the creation of these standards.

It is seen that the ISO 50001 standard, MSE 2000 standard and KSA 4000 standards are applied worldwide. In the member states of the European Union, the ISO 50001 standard and the EN 16001 standard are widely used. These standards, whose technical features are similar, ensure that energy management practices are in high compliance with ISO 9001 (quality management), ISO 14001 (environmental management) and OHSAS 18001 (occupational health and safety management) standards.

These standards can be designed for all industries or for specific industries (Paliekhova & Simon, 2016). At the same time, these standards can be in mandatory (required) or voluntary (recommended) formats. When the relevant literature is examined, it is seen that countries' approaches to these standards also differ. For example, Japan, one of the best practice examples, has linked energy management to the Energy Conservation Law. The Netherlands deals with energy management under the Voluntary Agreement Program, not under a standard. Thailand, on the other hand, considered energy management as an obligation and made the ISO 50001 standard mandatory, especially for large companies.

These standards contribute to regulating and standardizing the energy management systems of enterprises. Especially the ISO 50001[2] standard is important in eliminating the inconsistencies in national standards and unified energy management in different countries (Paliekhova & Simon, 2016; Anisimova, 2013).

17.2.2 Sustainable Competitive Advantage

In its most general form, competitive advantage is defined as the resources that a business has and the position that the business obtains against competing businesses as a result of the effective use of these resources (Hofer & Schendel, 1980). The resources that an enterprise has and uses to gain competitive advantage are considered as organizational resources (such as management, control mechanisms), physical resources (such as building, facility, equipment) and human resources (such as human, education, relations) (Barney, 1991). Perceiving these resources differently in the eyes of the stakeholders of the business adds value to that business.

Sustainable competitive advantage is defined as a strategy that cannot be easily imitated by the competitors of the resources owned and used by a company and thus creates value (Barney, 1991; Williams, 1992). In order for businesses to have a

---

[2]At this point, it is beneficial to mention the ISO 50001 Energy Management System. ISO 50001 Energy Management System is a standard that aims to reduce costs and environmental damages by providing energy efficiency and energy savings in facilities and buildings.

sustainable competitive advantage, the resources they have must have the characteristics of being valuable, scarce, imitative and non-substitutable (Barney, 1991). Businesses holding resources with these characteristics use cost leadership, product differentiation and focusing strategies to provide sustainable competitive advantage (Porter, 1985).

When the relevant literature is examined, it is seen that there are two different sustainable competitive advantage approaches: Position School/Approach (Porter, 1985) and Resource Based Approach (Hamel, 1994; Barney, 1991; Wernerfelt, 1984). The Position School/Approach argues that the strategies that will provide competitive advantage of the enterprises operating in a particular industry are determined by the competitive forces in the industry by linking the differences between the enterprises to external factors (Porter, 1985). On the other hand, Resource Based Approach argues that the strategies that will provide a competitive advantage to the enterprise depend on the resources and capabilities of that enterprise (Hamel, 1994; Barney, 1991; Wernerfelt, 1984).

The Position Approach emphasizes the structure of the industry and the position of the business in this industry in providing competitive advantage, while the Resources Based Approach emphasizes the competencies that the business has developed within itself. According to the results of the studies conducted in America, external environmental factors are effective 20% and internal environmental factors are around 46% in providing competitive advantage. Although these results defend that both approaches are valid, they indicate that the performance of the company is not negligible. At this point, the following question comes to mind: can energy and its effective management, which is an important input for almost all industries, serve businesses to gain sustainable competitive advantage?

17.3 Discussion

After the global energy crisis, nations focused on the efficient use of energy and energy management has been accepted as one of the main business processes in almost all industries. In the industrial competitive environment, energy-oriented management approach has turned into an important resource in obtaining a sustainable competitive advantage (Grant, 2011). By tending towards an energy-oriented management approach, businesses can not only achieve cost advantage, but also build a positive image by reflecting themselves as responsible businesses in the eyes of all stakeholders (Di Somma et al., 2015; Cartes, 2011; Buitelaar & Perez, 2000). Thus, businesses improve their competitive positions by minimizing costs with efficient and rational management of energy, which is an important input.

At this point, it is necessary not to ignore the arguments of the two strategic approaches in order to provide competitive advantage and make it sustainable. As stated before, Position Approach emphasizes the structure of the industry and the position of the business in this industry in providing competitive advantage (Porter,

1985), while the Resources Based Approach emphasizes the competencies that the business has developed within itself (Hamel, 1994; Barney, 1991; Wernerfelt, 1984). According to the results of the studies conducted in the USA, external environmental factors (structure of the industry) are 20% effective and internal environmental factors (business structure) are around 46% in providing competitive advantage. Although these results defend that both approaches are valid, they indicate that the performance of the company is not negligible. Therefore, businesses that want to improve and strengthen their market position in the market in which they operate have to develop their own unique talents in order to win the competitive battle. In this context, alignment of decision-makers' strategies with energy measures will support them to gain an advantage over their rivals and to make this superiority sustainable. It is clear that the energy-oriented management approach contributes to businesses in this way. Neagle stated that the reasons for businesses to adopt an energy-oriented management approach are to reduce costs, increase profitability and improve public image, and emphasized that energy is a great opportunity for businesses to create value by providing competitive advantage over their competitors. In his study, energy efficiency will provide sustainable competitive advantage and supply security by minimizing high energy costs and will also reduce environmental concerns (Munguia et al., 2018).

A proven energy management system enables businesses to improve their product quality, save in many areas, increase their profitability and reduce the additional investment expenditures needed for development (Paliekhova & Simon, 2016). The purpose of using energy efficiently is not only to increase performance but also to ensure economic growth, to ensure compliance of businesses with inclusive and efficient industry value chains, to provide significant competitive advantages for access to certain markets, and ultimately to ensure sustainability.

Kaman (2002) has succeeded in effectively integrating the energy management process into business activities and tried to identify the characteristics of the best energy management practices by comparing the energy management programs of the best enterprises in the field. According to the results of this study, the best energy management practices have the following characteristics:

1. Strong corporate leadership in energy efficiency.
2. Resource allocations striving to continuously improve energy efficiency.
3. A fully integrated energy management process from top management to the lowest level.
4. Continuous measurement of production areas and budget responsibility to units.
5. Energy efficient designs of production processes and facilities.
6. The use of natural workgroups of the facility to continually fine-tune the working principles of its equipment, improve the energy efficiency of the equipment, and maintain the improvement of energy efficiency.

The researcher reached the conclusion that businesses that are positioned as the best in their field have strengthened their competitive positions by adopting an energy oriented management approach, making efforts to improve energy efficiency, and sustaining these improvement efforts consistently. This study of Kaman (2002) is inspiring for other businesses to build an efficient and effective energy management system as well as showing that energy management is effective in gaining competitive advantage.

Munguia et al. (2018), in their study examining the opportunities of Mexican maquiladoras to gain competitive advantage through energy management, stated that energy management should become a part of the corporate visions of enterprises. According to the authors, the energy oriented management approach supports the optimum use of energy resources, the promotion of sustainable practices among employees, the inclusion of stakeholders' needs in decision-making processes, the adoption of energy measures, reduction of CO_2 emissions, implementation of permanent energy policies and sustainable economic growth. The authors also stated that besides these effects of the energy oriented management approach, it contributes to the businesses in building a positive corporate image, creating a solid stakeholder network and gaining a real competitive advantage, which is not visible at first sight but becomes visible in the long term. In addition, the researchers emphasized that if businesses adopt an energy oriented management approach and adapt their economic, technical and intellectual capital to the same goal, they will achieve a sustainable, long-lasting and permanent competitive advantage.

Paliekhova and Simon (2016) pointed out in their studies that energy management and sustainable development goals are closely linked and that the voluntary certification of energy management systems is a market efficiency tool. In this context, the researchers emphasized that by applying the ISO 50001 standard, companies can provide a real competitive advantage in addition to energy savings.

17.4 Conclusion

In this study, whether energy management, which affects the whole world, is effective in providing competitive advantage for businesses is opened to discussion. In this context, energy-oriented management approach has been conceptually examined as a market efficiency tool in providing competitive advantage. It has been seen that the energy-oriented management approach in the industrial competitive environment has turned into an important resource in obtaining a sustainable competitive advantage (Grant, 2011). By tending towards an energy-oriented management approach, businesses have not only achieved a cost advantage, but also have been able to build a positive image by reflecting themselves as responsible businesses in the eyes of all stakeholders (Di Somma et al., 2015; Cartes, 2011; Buitelaar & Perez, 2000). Thus, businesses have been able to develop their competitive positions by minimizing costs with efficient and rational management of energy, which is an important input (Capehart et al., 1997).

Within the scope of the study, it was concluded that a fully integrated energy management process from top management to the lowest level should be developed in order to establish a successful energy management approach in enterprises. In addition, it has been determined that the energy management, which will be applied successfully, has an effect on the competitive advantage of the enterprises.

References

Anisimova, T. Y. (2013). Analysis of standards in energy management. *Middle-East Journal of Scientific Research, 13*(5), 654–657.

Bahadori, A., Zahedi, G., & Zendehboudi, S. (2013). An overview of Australia's hydropower energy: Status and future prospects. *Renewable and Sustainable Energy Reviews, 20*, 565–569.

Barney, J. (1991). Firm resources and sustained competitive advantage. *Journal of Management, 17* (1), 99–120.

Bayraç, H. (2010). Enerji kullanımının küresel ısınmaya etkisi ve önleyici politikalar. *Eskişehir Osmangazi Üniversitesi Sosyal Bilimler Dergisi, 11*(2), 229–259.

Buitelaar, R. M., & Perez, R. P. (2000). Maquila, economic reform and corporate strategies. *World Development, 28*(9), 1627–1642.

Burtland Commission. (1987). *Our common future*.

Büyükmıhcı, M. K. (2003). Yenilenebilir enerji kaynakları avrupa birliği ülkelerindeki uygulamalar ve enerji ve tabii kaynaklar bakanlığı tarafından hazırlanmakta olan kanun tasarısı taslağı çerçevesinde planlanan önlemler. *Yeni ve Yenilenebilir Enerji Kaynakları Sempozyumu*, TMMOB, 3–4 Ekim Kayseri, ss.15–22.

Çalıkoğlu, E. (2004). Enerji verimliliği ve EİEİ tarafından yürütülen çalışmalar. 23. Ulusal Enerji Verimliliği Kongresi, EİE İşleri Genel Müdürlüğü, Enerji Tasarrufu Koordinasyon Kurulu Yayını, Ankara.

Capehart, B. L., Turner, W. C., & Kennedy, W. J. (1997). *Guide to energy management*. Fairmont Press.

Cartes, D. (2011). Energy systems, economics, and sustainability. In R. M. Isaac & D. A. Norton (Eds.), *Experiments on energy, the environment, and sustainability, research in experimental economics* (Vol. 14, pp. 213–222). Emerald Group Publishing Limited.

Di Somma, M., Yan, B., Bianco, N., Graditi, G., Luh, P. B., Mongibello, L., & Naso, V. (2015). Operation optimization of a distributed energy system considering energy costs and exergy efficiency. *Energy Conversion and Management, 103*, 739–751.

Fiedler, T., & Mircea, P. M. (2012). *Energy management systems according to the ıso 50001 standard challenges and benefits*. Paper presented at the international conference on applied and theoretical electricity (ICATE), Craiova, Romania.

Flavin, C., & Lenssen, N. (1994). Enerjide arayışlar, Çeviren: Yaman Köseoğlu Tema Vakfı, Yayın No: 12.

Glaser, C. (1992). *Industrial energy auditing: An opportunity for ımproving energy efficiency and ındustrial competitiveness*. Department of Energy.

Grant, N. (2011). *Being energy efficient is key to being competitive*. Guard. http://www.Theguardian.Com/Sustainable-Business/British-Gas-Energy-Efficiencycompetitive-Advantage. Accessed 10.03.2021.

Hamel, G. (1994). In G. Hamel & A. Heene (Eds.), *The concept of core competence* (pp. 11–33). John Willey and Sons.

Hamelinck, C., Suurs, R. A. A., & Faaij, P. C. (2005). International bioenergy transport costs and energy balance. *Biomass and Bioenergy, 29*, 114–134.

Hofer, C. W., & Schendel, D. (1980). *Strategy formulation: Analytical concepts*. West Publication.

Jassim, H. (2013). Geothermal energy and possibility use it in ıraq. *The Islamic College University Journal, 8*(25), 193–213.

Kaman, J. A. (2002). Best in class energy management practices provide competitive advantage. *Energy Engineering, 99*(4), 42–56.

Kannan, R., & Boie, W. (2003). Energy management practices in sme. Case study of a bakery in Germany. *Energy Conversion and Management, 44*(6), 945–959.

Kannan, N., & Vakeesan, D. (2016). Solar energy for future world: A review. *Renewable and Sustainable Energy Reviews, 62*, 1092–1105.

Khan, A. A. (2009). Potential to use biomass for bio-energy in Ontario. *Guelph Engineering Journal, 2*, 39–44.

Lee, S. K., Teng, M. C., Fan, K. S., Yang, K. H., & Horng, R. S. (2011). Application of an energy management system in combination with fmcs to high energy consuming ıt industries of Taiwan. *Energy Conversion and Management, 52*(8–9), 3060–3070.

Mafakheri, F., & Nasiri, F. (2014). Modeling of biomass-to-energy supply chain operations: Applications, challenges and research directions. *Energy Policy, 67*, 116–126.

Munguia, N., Vargas-Betancourt, N., Esquer, J., Giannetti, B. F., Liu, G., & Velazquez, L. E. (2018). Driving competitive advantage through energy efficiency in mexican maquiladoras. *Journal of Cleaner Production, 172*, 3379–3386.

Öztürk, S., & Saygın, S. (2017). 1973 Petrol krizinin ekonomiye etkileri ve stagflasyon olgusu. *Balkan Sosyal Bilimler Dergisi, 6*(12), 3.

Paliekhova, L. L., & Simon, S. (2016). Competitive advantages through the implementation of international energy management standards. *Вісник Придніпровської Державної Академії Будівництва Та Архітектури, 3*(216), 42–52.

Pamir, N. (2007). Enerji arz güvenliği ve türkiye. *Stratejik Analiz Dergisi.*

Porter, M. E. (1985). *Competitive advantage: Creating and sustaining superior performance.* The Free Press.

Saavedra, M. R., Fontes, C. H., & Freires, G. M. (2018). Sustainable and renewable energy supply chain: A system dynamics overview. *Renewable and Sustainable Energy Reviews, 82*, 247–259.

Sandberg, P., & Soderstrom, M. (2003). Industrial energy eficiency: The need for ınvestment decision support from a manager perspective. *Energy Policy, 31*(15), 1623–1634.

Sevim, C. (2018). Geçmişten günümüze enerji güvenliği ve paradigma değişimleri. *Stratejik Araştırmalar Dergisi, 13*, 93–105.

Smith, C. B., & Kelly, E. P. (2015). *Energy management principles: Applications, benefits. savings* (2nd ed.). Elsevier.

Thollander, P., & Palm, J. (2013). *Improving energy efficiency in industrial energy systems: An ınterdisciplinary perspective on barriers, energy audits, energy management, policies and programs.* Springer.

United Nations Documents. (1972). *Report of the united nations conference on the human environment.*

United Nations Millenium Project. (2002). *Report of the world summit on sustainable development.*

United Nations Sustainable Development. (1992). *Agenda 21.* United Nations Conference on Environment & Development.

Uzun, A., & Değirmen, M. (2018). Energy efficiency and energy management in industrial. *Internatıonal Journal of Economıc Studies, 4*(2), 84–97.

Van Den Wymelenberg, K., Brown, G. Z., Burpee, H., Djunaedy, E., Gladics, G., Kline, J., Loveland, J., Meek, C., & Thimmanna, H. (2013). Evaluating direct energy savings and market transformation effects: A decade of technical design assistance in the northwestern USA. *Energy Policy, 52*, 342–353.

Wee, H. M., Yang, W. H., Chou, C. W., & Padilan, M. V. (2012). Renewable energy supply chains, performance, application barriers, and strategies for further development. *Renewable and Sustainable Energy Reviews, 16*(8), 5451–5465.

Wernerfelt, B. (1984). A resource based view of the firm. *Strategic Management Journal, 5*(2), 171–180.

Williams, J. R. (1992). How sustainable is your competitive advantage. *California Management Review, 34*, 29–51.

Chapter 18
The Core of Business: Is It Energy Management or Management Energy?

Ercan Karakeçe

Abstract This study focuses on the significance of energy by considering it as a strategic element in competitive environments. Contrary to what thought, when discussing energy, it will try to highlight not only physical description but also other issues to be addressed in various other contexts. For this purpose, while emphasizing energy as both a necessity and a power in production, it is interpreted from a business perspective at the same time. This study expresses how energy notion may be evaluated from the perspective of managerial and entrepreneurial actions. Thus, this inquiry evaluates the concept of energy from a different perspective by bringing together engineering and management inspirations.

18.1 Introduction

Since the existence of human beings, its relationship with the concept of energy has been changing. It can be argued that there is essentially no difference between the person who needed fire for feeding and heating, which was necessary for the survival of his life in the early days, and those who needed electricity due to their needs in today's modern world technology. As a reflection of changing requirements and forms, there are variations in this direction in the perception of energy. Considering starting from Göbeklitepe, the oldest structure discovered (Agan, 2016), energy always appears as a rumination of the effort put forward in every situation that man cannot cope with alone. Both the constructions in Göbeklitepe (Karacalı & Urfalıoğlu, 2019) and the pyramids (Heldal et al., 2016; Charvátová et al., 2011) found in various parts of the world, the elevation of those massive stones in harmony reminds one of how it had made, namely energy. One of the things that a person dreaming of that day needs is the energy he requires for his actions.

E. Karakeçe (✉)
Department of Foreign Trade, Vocational School of Social Sciences, Istanbul Medipol University, İstanbul, Turkey
e-mail: ekarakece@medipol.edu.tr

S. Yüksel, H. Dinçer (eds.), *Strategic Approaches to Energy Management*, Contributions to Management Science,
https://doi.org/10.1007/978-3-030-76783-9_18

With the influence of discoveries in the human adventure that progresses from individual to society, many new phenomena occur in human life. While every innovation flags the way for the next, humanity's need for energy is increasing day by day. The searches for the energy required for humankind who established civilizations diversified and even conflicts encountered. Especially in human history, the industrial revolution can seem like a breaking point. With this step, not only production and consumption have changed, but also the change in the expectations of humanity and society has been bounced. Thus, the relationship of humans with energy has deepened more and more (Chow et al., 2003).

It is essential to recognize the energy we discuss today from various aspects to glance at the concept by keeping the historical flux aside. For this purpose, our study commences with the principal features related to the notion of energy, and we focus on the content of the concept. Then, intersection points in researches related to the concept will be defined. Subsequently, the energy thought will be viewed from a managerial philosophical perspective. And as well as physical characteristics associated with the idea of energy, other abstract features will be reviewed. Then, the parts required for sustainable production will be examined and the future of energy will be focused on. In the next division, energy, which is handled with management thought, will be analyzed from an entrepreneurial viewpoint. Lastly, the research will be concluded with recommendations and observations that will require a holistic view of the energy issue.

18.2 Understanding of the Nature of Energy Concept

When we investigate energy in reputable databases, it is noteworthy that we come across pages of studies mostly quoted in the fields of physics, chemistry, astronomy, and biology. It is quite ordinary that energy studies revealed in mathematics and engineering constitute the first part of the definition of energy that recurs to mind. In this respect, it should be stated that the studies deal with the physical properties of energy. However, on the other hand, it is seen that the concept of energy concerns economic, political, geopolitical, and even military fields. For this reason, this part of the book will try to express the aspect that supports another extent related to energy. So, when the connotation of the concept is so rich, and it is under the influence of many areas, what should this concept remind us, how should we understand and interpret the concept of energy? Of course, since this book is not a work of fundamental sciences, it will not be included in the section related to that fields. In this study, the main point of which is strategic energy management, we will endeavor to evaluate the concept from the perspective of a manager and entrepreneur.

When we examine the concept in dictionaries, we come across the following expressions:

> "*the power and ability to be physically and mentally active; the power from something such as electricity or oil that can do work, such as providing light and heat*" (Cambridge Dictionary); "*a fundamental entity of nature that is transferred between parts of a system in the production of physical change within the system and usually regarded as the capacity for doing work, the capacity of acting or being active, usable power*" (Merriam-Webster).

In the definitions we encounter, it is essential to focus on the concept other than basic sciences. From this point on, the necessity of a more holistic approach to the notion emerges. It can be seen in the dictionary expressions, as it comes to mind first, the term is related to both nature and physical elements. This field has already been reinforced in the references made on the Web of Science. However, as it is tried to be stated, the aspect of the issue facing social sciences falls far behind the natural ones. Nevertheless, it should be couched in that the topic is similarly vital for social sciences as reflected in its dictionary meanings. If we consider energy as a solid science element, issues such as production, transfer, storage, and usage attract attention. It is regarded as the area of interest of basic sciences, as stated above. However, this issue concerns the field of management as well as engineering.

When revealing energy, meaning energy production, it is essential to focus on the topic of energy types. The energy needed can be obtained from various sources. For instance, numerous alternatives such as hydroelectric power plants, coal power plants, nuclear power plants, wind panels, natural gas power plants, solar panels are utilized to generate electrical energy (Twidell & Weir, 2015). Of course, it would not be correct to interpret the emerging power only as electrical energy. The power obtained takes its place in our lives by being transformed into many different formats. Thus, it is seen that various effects will affect many points in our lives, from eating, to warming, from production to transportation, and many aspects of our lives. Thus, the necessity of energy in our lives is apprehended (Demirel, 2012).

Considering the meaning of energy we use in daily life, we see that the concept is sometimes used synonymously with the concept of power (Liu & Zhong, 2016). This concept, which expresses the state of having the things we need to be able to do a job, can also be seen as an indicator of necessity or incompetence when not possessed. Whether at the micro or macro level, energy and power terms can be used to describe all the details necessary to tackle a job. In another aspect, macrosystems can also be considered as a reflection of the accumulation of microsystems. Or we can expect the power of macrosystems to be reflected in microsystems.

Although not only with the concept of energy, when it is associated with one of power, many different thoughts can draw our attention (Saghir, 2005). For example, while evaluating energy as an input in the sense we have mentioned, we may neglect to consider the other elements that will enable us to reach the output as if it is a form of energy. However, each power can be interpreted as energy to achieve the targeted goal. When considering for an individual, physical strength, endurance, agility, verbal and numerical aptitude, social abilities, business networks, experiences, education etc. can be interpreted as power factors. On the other hand, for an organization, many physical and non-physical elements such as production facilities, research and development abilities, sales and marketing activities, storage, distribution, customer relations management, loyal suppliers, strong business partners, knowledge, experience, business relations, brand value, technological capabilities, qualified employees, visionary management, innovativeness, and a solid organizational culture can be registered in the power zone of the institution (Fuchs, 2013).

Likewise, when evaluated from the perspective of a country, various values such as underground and aboveground wealth, geo-political and geo-strategic location, historical background, sociological values, education system, demographic structure, infrastructure and superstructure investments, political and economic stability, military power, financial statement, technology, and intellectual accumulation etc. can be interpreted as power/energy (Singhania & Saini, 2018; Makino et al., 2004).

Of course, it should be expressed that a price must be paid for all these mentioned elements. Efforts, or investments, are required for an establishment or progress. It can be interpreted as the transformation of the available or borrowed energy into another form. While one side is made/born into a new form, there is an extinction/transformation for the other side. In other words, when considering both concrete and abstract terms, progress should be taken into consideration while talking about exhaustion on the one hand. The important thing is whether the value that emerges against the spent value is worth it. In other words, the opportunity costs of these targets should be considered as well as the targets.

As mentioned, if we degrade the production of value to a simple equation such as combining raw materials, capital, and labor, then we would stick within the tactical field. However, to create value, severe feasibilities are required starting from the determination of the customer/target audience. As well as customers, human resources should be treated as a power, just as the "source" expression mentioned in the term. And with the effort to put forth, this energy will accumulate as intellectual capital. The effects of education cost for progress will not be easily measured as physical factors. However, it will make itself felt both in the functioning of the works and in the long-term performance with the trained workforce and knowledge. Therefore, investments in knowledge and management should be thought of for strategic purposes. With the transformation of targets from labor-intensive to knowledge-intensive ones, there will be an increment in performance indicators. Of course, the role of developing information technology in this field will be helpful for progress (Kamat, 2007).

When we review the academic studies, a very detailed picture emerges. When examined within the scope of our study, there are various studies focused on energy types and energy investments (DeCanio, 1993; Boyle, 2004; Dincer et al., 2019). Many studies associated with the concept of clean energy have been published, and it has been discussed from different aspects (Dincer & Acar, 2015; Brown et al., 2001). Yildiz and Kazimi (2006) discussed the usage of nuclear energy in hydrogen generation systems. And some proposed it as an alternative to solid waste fuels. Miyamoto (1997) offers biological systems to generate environmentally-kind methods. Makogon et al. (2007) explained the details about natural gas in generating power in their studies. Blanco (2009) expressed wind power as a source of future energy. Antonio (2010) offered wave absorption for clean energy production. And it was compared with wind power by Falnes (2007). Shenoy et al. (2012), stated photovoltaic structures to utilize from solar radiation. In some, sunlight is used for generating electricity (Buresch, 1983; Kamat, 2007). Some researchers investigated biofuels to generate biomass energy alternatives for a better environment (Klass, 1998; McKendry, 2002; Srirangan et al., 2012).

18.3 Energy as a Power or Problem

Energy is a vital requirement for human beings. They need power to both survive and enrich their life. Whether at the individual or organizational level, they must reach the energy in order to act. When we look at it from reconception, energy seems like the symbol of technology and civilization. In this sense, energy appears as both a symptom and a necessity of life (Saghir, 2005).

When evaluated from a managerial point of view, first of all, energy should be considered as an input. In the equation of supply as handled in economics, energy must be counted as a required element as necessary as raw materials, capital, and labor for today's producers. Whatever input is used for generating energy, a factory will likely experience a halt of production if there is an energy shortage. For this reason, energy support systems are seen as a requirement in some production facilities for short-term power outages. It is one of the indicators of how difficult it will be to produce without energy (Farret & Simoes, 2006).

Another issue is that costs must be incurred to power requirements. If an individual or organization does not generate its energy, it tries to cater to power requirements via outsourcing like raw materials. It can result in severe consequences for the firm in terms of both controlling the resource by others and pricing it. When we look at the issue in a macro dimension, we may face more critical consequences. Accessing direct energy or the raw material needed to generate power through imports means both the foreign trade deficit and budget imbalances for most countries (Murat et al., 2014; Yalta & Yalta, 2017).

While generating energy, of course, some legal obligations arise. First of all, we may need to put energy production in a renewable framework. If traditional methods are preferred in energy production, it is desirable to control and neutralize the wastes. Besides, we may be subject to control for energy production by representatives of both the central administration and the local authority. In terms of legal obligations, energy consumption is taken into account as much as energy production. Considering the imbalances in terms of foreign trade deficit, foreign dependency, environmental pollution, and the obligations brought by international agreements, energy production may cause various restrictions (Leal-Arcas et al., 2014). In this regard, the state authority can impose obligations that support environmentally friendly energy and thrifty consumption. Prohibiting devices' usage above a defined wattage or promoting types of equipment leveled high energy class can be shown as examples of these decisions (Sorrell, 2015).

When we look at the energy issue strategically, the requirement of addressing the matter on both micro and macro basis emerges. We first examine energy on a micro basis that keeps the individual or organization alive. The problem of power must be solved to survive his life and continue activities. It requires detailed etudes in this area. Also, the important thing is not only to make the preliminary preparation but also to perform properly. There are other things to be done in long-term planning besides these issues. The current market should be monitored perpetually, and the determination and agility to make revisions should be ensured. Of course, knowing when and how to take a step is as vital priorly. If reviewed in this way, it becomes possible to see every element contributes to value as a form of energy.

On the other hand, when we contemplate energy on a larger/macro scale, it can be predicted that a more complex composition will emerge. Alliances can come into play, as some exchanges are done to obtain energy. These initiatives can unite or separate political and military interests. Because although it is not mentioned, states are in a struggle just like corporations. Every macrostructure or group accounts for a better position on the world platform. And in today's world, it is seen that this contest is sometimes conducted over direct energy sources and sometimes over energy transmission channels. As a result of these international conflicts of interest, once in a while, annoying consequences may arise. Unresolved disputes can bring states to the brink of war and lead to severe disasters (Farret & Simoes, 2006). In this world where the weak try to survive, the strong strive to add more to their strength. Thus, energy is a determinant that affects not only production and finance but also geopolitical and geostrategic events. For this reason, countries that having raw materials for energy, generating power, and transmitting energy, demand that ones in the requirement of power remain constantly dependent on them. In this way, they both obtain financial returns by controlling energy and desire to establish authority over their interlocutors with the power they have. For this reason, every state takes action to change the rules of the game in its favor and to increase its existing authority/power. It shows that the struggle between energy supplier countries and energy demanders longs continually.

For all these reasons, it would be more accurate to consider energy as a process rather than accepting an output. The energy will continue to preserve the existing situation or take it further. It requires sustainable energy studies instead of taking one-off steps. Thus, it becomes obligatory in terms of environmental concerns, cost elements, and evaluating the vision. To be able to contend on both micro and macro bases, it is required to manage all elements seen as energy sustainably. Otherwise, power and reputation will be failed by not being able to protect the existing position.

From this point of view, it can be argued that the resource-based approach, which fills a significant gap in the literature, might be insufficient for evaluating this field. According to this view, which adopts having the relevant resources for the execution of the works and performance outputs, the whole issue is degraded to the relationship to be established with the resource. However, countries such as Venezuela from South America, Iraq, and Syria from the Middle East, have rich natural resources, shows us that only a resource-based view cannot be the only reason for success. Of course, it is vital to have energy and energy resources. However, being able to manage it is as crucial as having energy.

Here, perhaps, we may oblige to resort to another view. We may demand to get guidance from the strategy tripod approach developed by Peng et al. (2009). In this way, it may be possible to make more realistic evaluations about the energy issue. This view has been put forward by adding the institutional view to the resource-based and industry-based approaches, which have a substantial duty in the literature. We just mentioned that resource-based thought is about controlling resources. The industry-based view can be summarized as success creates a synergistic environment for organizations doing similar work in the country/market. In other words, if success can occur somewhere, it can be inspired as they can create support for another structure and advance each other. Based on this view, the fact that success is teachable makes the market more competitive, even if it carries a natural selectionist view. However, although Peng et al. supports these two views, they describe them as incomplete. According to them, other factors should be considered for success, they are not sufficient to reach the target. This leads them towards the environment and values. From this point on, it is necessary to address the energy issue in a long-term and more comprehensive way, not with a short-term perspective. Associating the gain/success with the environment and conditions can save actors from blind imitation. As stated above, an investor may explore his energy elements and construct his development by looking from a broader perspective (Panwar et al., 2011).

18.4 Past, Present and Future of Energy

As human being progressed, he produced more, and the more he produced, the more energy he required. He worked harder for more energy and attacked nature more and consumed it. While this is the summary of the adventure up to now, anymore he has started to realize the results of his actions. In fact, he was consuming his residence while consuming nature. It was also polluting the air, water, and soil but he was not aware of it until the threshold was crossed. Maybe he is still not fully aware (Smil, 2019).

But today, unlike in the past, the cost is not the only task he has to think about. Whether he wants it or not, he now must take the environment into account. Because nature can no longer cope with human beings and is running out. As it is consumed, it cannot renew itself, the species it hosts disappear one by one. In short, the ecosystem is slowly deteriorating. As the works of human civilization rise and produce, the people consume their world. Maybe they think saving themself with short-term tactical solutions. They produce and continue to consume in places where there are no or few restrictions. But nothing is eternal, and the world is not limitless, it is depleting and deteriorating (Poizot & Dolhem, 2011).

It forces people to choose between cost and ethical values. He must think about the future of himself and the world while he cares about energy. Because its system is not sustainable. Maybe he salvages the day, but now he is worried about tomorrow. Because now there is global warming, we have a scary hole in the top of us in the atmosphere, the glaciers are melting, and we are facing many chained disasters (Dresselhaus & Thomas, 2001). For these reasons, administrators now must think about this issue even if they find it restrictive, producers have to take precautions even if they bear the costs. And academics are trying to contribute to this field by studying in this field. Energy production requires investment. Considering its installation, execution, and results, it requires detailed engineering and management knowledge (Bhattacharyya, 2011).

It is known that each form of energy generation has its advantages and disadvantages (Ellabban et al., 2014). For example, investments made through dams might become questionable due to the deteriorating global climate balance. Although coal plants are operated inexpensively, they fuel environmental concerns (Alrikabi, 2014). Even though natural gas power plants are environmentally friendly, they can trigger external dependency and current account deficit problems. While nuclear power plants are an effective form of energy for some, investment costs and the possibility of accidents can be alarming for others. For establishing geothermal power plants, it is necessary to be close to the source (Barbier, 2002). While solar panels and wind farms are company-friendly in terms of maintenance costs, they are shaped according to climatic requirements. Photovoltaic structures are accepted as firm-friendly systems because of low maintenance costs (Buresch, 1983). Due to these differences, energy investments can also be spread over a wide range. We come across a wide variety of studies examining this area in the literature (DeCanio, 1993; Dincer et al., 2019; Dincer & Yuksel, 2019; Wang et al., 2019; Qiu et al., 2020).

18.5 A Ray of Hope: Eco-Entrepreneurship

As mentioned above, human beings keep consuming for production. However, there are now different alternatives to traditional production (Dresselhaus & Thomas, 2001). Environment-friendly variations can be seen in each final product, raw material, and energy preferences. For some, this is considered just a marketing illusion (Zehner, 2012), for others a compensation approach (Panwar et al., 2011). Whatever it is evaluated as we see that the environment is now considered.

While thinking about what we can offer as an answer to the question of how to make a difference in this process, the solution may be a new type of entrepreneur: eco(logic)-entrepreneurs. Unlike others, these entrepreneurs take action by taking into account nature, rather than just focusing on their profits. It is possible to say that they are respectful to the environment due to their emergence (Wagner, 2009). According to others, they care about environmental disasters and do not leave their waste to nature randomly. They even try to make it useful if possible, at least harmless, or less harmful. They plan their investments and actions according to nature-friendly factors. While it can sometimes be a production facility, sometimes it can appear as a reclamation facility. There are some who see the concepts of recycling and upcycling as part of their work (Mathiesen et al., 2011).

When we consider the sum of all the details necessary for the production as an energy cost, it is impossible not to give entrepreneurs the right. For this reason, we can say that they have a different mentality than others (Mars & Lounsbury, 2009). Perhaps what we need is to instill this awareness in all manufacturers. Because within the definition of entrepreneurship, only being profit-oriented should not be stated. The entrepreneur has to act in a strategic manner while planning business. It is possible to associate the concept of sustainability with many entrepreneurship elements.

Since the entrepreneur is the one who discovers, plans, organizes, and executes the opportunities, then it will affect nature while shaping these preferences (Eckhardt & Shane, 2003). If a value chain is created by the army of entrepreneurs at all stages from production to consumption (Valliere & Peterson, 2009), then entrepreneurs themselves take care of energy. As we have stated, the entrepreneur arranges the requirements in both concrete and abstract terms, the problem of the system arises from entrepreneurship. And of course, the solution will also emerge from entrepreneurship.

If we need to mention Schumpeter's creative destruction when it comes to entrepreneurship, the destruction we will express does not mean the demolition caused by humanity today. However, the entrepreneurship system that has overcome the emerging environmental and sustainability problems and has set a new goal may have the effect of destroying the order of others (Abernathy & Clark, 1985). As awareness and popularity increase in these issues, it will not be surprising that competition will shift to these ecological areas. For this reason, environmentally sensitive entrepreneurship activities will continue to be talked about in the coming years (Lund & Mathiesen, 2009).

At this point, it can be said that there is a struggle between the traditional systems and environmental systems (York, 2012). Productions and organizations that do not harm people and nature will find more space in the search for new systems. For the entrepreneur, whom we can consider as the initiating spirit (Karakeçe & Çemberci, 2020), the color of the competition can turn green (Poizot & Dolhem, 2011). Developed countries can see this problem as postponed by continuing their production in developing countries (Jorgenson, 2007). However, with the current environmental conditions causing difficulties, problems regarding sustainability arise, which leads them to seek other solutions (Goldemberg, 2006).

Even when we do not include our abstract concepts, we see that energy production changes with the concept of renewability. We see that parties with different available resources embrace appropriate environmentally friendly energies (Twidell & Weir, 2015). For example, we can claim that states give incentives to those who invest in this field. We can say that waste-free energy investments such as wind energy and solar panels have made progress with these incentives (Simsek & Simsek, 2013). On the other hand, entrepreneurs are involved in the competition in storage and economic consumption as well as obtaining power.

18.6 Conclusion

This chapter, written to examine the energy needed for the continuation of life from a managerial perspective, recommends this chapter, written to examine the energy required for the continuation of life from a managerial perspective, offers thinking about various aspects of energy. For example, when we use energy, it should be thought that it can be related to all the elements that need to be organized to reach the targeted output. It can be interpreted in the same way, whether they are individuals or organizations at the micro-level or countries at the macro one. For example, it can be thought that the companies are energy accumulation points, and ones closed down are not able to transfer their energy to the next generations. Of course, it is unimaginable for all companies established to live forever. However, it should make us think that some companies live through almost 1000 years, and others vanished (Napolitano et al., 2015).

On the other hand, it should be stated that energy is not only be handled in the field of natural sciences and engineering but is also an important research field within social sciences. From the point of view of economics, energy is both an input and an output. It should be considered as one of the fundamental elements of development and production (Bhattacharyya, 2011). Considering the constraints theory, while making curative actions on the areas that will cause problems in achieving the goal, both abstract and concrete factors should be regarded (Goldratt, 1990). In this respect, energy is a significant input, and an energy cut will mean the termination of the activity. Also, it should be seen that the power of each element is transformed into another form, which can be associated with energy at every stage from the beginning to the end (Farret & Simoes, 2006).

Energy management is a grounded concept in life that requires serious effort as well as energy production. While it is essential to determine the most optimum solution among complex choices in the generation, it should be used efficiently during the usage phase (Hamedi & Gandomkar, 2012). It can be viewed as a competitive element for energy, which is of interest to both natural and social sciences, from energy production to distribution, from storage to usage. This concept, which is used as a synonym for power in many places, needs long-range and strategic examination. Besides, energy is the scene of conflicts at the level of states like in individual and organizational ones. With the help of the strategy tripod approach, it can be thought that different situations can strengthen possible alternatives in energy. In this case, the investments to be made should turn into an original identity rather than imitation. Of course, while environmentalist approaches accepted as universal would not be denied, the diversity of the existing wealth will also make itself felt in the energy field (Boyle, 2004).

In this direction, entrepreneurs have a great responsibility. We expect them to pay attention to the issues we talked about in the area of energy while fulfilling their great duties in meeting the requirement of society. First of all, the desired entrepreneurs are expected to be sensitive not only to financial outputs but also to social and environmental issues. This reveals the essential for entrepreneurs to act sustainably and strategically. If we acknowledge the entrepreneurs as the essence of the system, we can put the notion of energy in the same place for them. Of course, both material and spiritual wealth are mentioned here, and we must admit that the entrepreneurs bring together numerous determinants. We need to embrace that they are different from other people in this sense and will still be a part of the solution for the puzzle. Whether it is the production of energy-saving devices or renewable energy investments, the steps taken in this area can be described as hopeful (Panwar et al., 2011; Lee & Cheng, 2016).

We do not inherit the Earth from our ancestors, we borrow it from our children.
A Native American Proverb.

References

Abernathy, W. J., & Clark, K. B. (1985). Innovation: Mapping the winds of creative destruction. *Research Policy, 14*(1), 3–22.

Agan, C. (2016). A preliminary study on the conservation and polishing performance of Sanliurfa limestones as a traditional building material. *Bulletin of Engineering Geology and the Environment, 75*(1), 13–25.

Alrikabi, N. K. M. A. (2014). Renewable energy types. *Journal of Clean Energy Technologies, 2* (1), 61–64.

Antonio, F. D. O. (2010). Wave energy utilization: A review of the technologies. *Renewable and Sustainable Energy Reviews, 14*(3), 899–918.

Barbier, E. (2002). Geothermal energy technology and current status: An overview. *Renewable and Sustainable Energy Reviews, 6*(1–2), 3–65.

Bhattacharyya, S. C. (2011). *Energy economics: Concepts, issues, markets and governance.* Springer Science & Business Media.

Blanco, M. I. (2009). The economics of wind energy. *Renewable and Sustainable Energy Reviews, 13*(6–7), 1372–1382.

Boyle, G. (2004). *Renewable energy* (p. 456). Oxford University Press.

Brown, M. A., Levine, M. D., Short, W., & Koomey, J. G. (2001). Scenarios for a clean energy future. *Energy Policy, 29*(14), 1179–1196.

Buresch, M. (1983). *Photovoltaic energy systems: Design and installation.* McGraw-Hill.

Cambridge. (n.d.). Energy. In *dictionary.cambridge.org dictionary.* Retrieved March 1, 2021, from https://dictionary.cambridge.org/dictionary/english/energy

Charvátová, I., Klokočník, J., Kolmaš, J., & Kostelecký, J. (2011). Chinese tombs oriented by a compass: Evidence from paleomagnetic changes versus the age of tombs. *Studia Geophysica et Geodaetica, 55*(1), 159–174.

Chow, J., Kopp, R. J., & Portney, P. R. (2003). Energy resources and global development. *Science, 302*(5650), 1528–1531.

DeCanio, S. J. (1993). Barriers within firms to energy-efficient investments. *Energy Policy, 21*(9), 906–914.

Demirel, Y. (2012). Energy and energy types. In *Energy* (pp. 27–70). Springer.

Dincer, I., & Acar, C. (2015). A review on clean energy solutions for better sustainability. *International Journal of Energy Research, 39*(5), 585–606.

Dincer, H., & Yuksel, S. (2019). Balanced scorecard-based analysis of investment decisions for the renewable energy alternatives: A comparative analysis based on the hybrid fuzzy decision-making approach. *Energy, 175*, 1259–1270.

Dincer, H., Yüksel, S., & Martinez, L. (2019). Balanced scorecard-based analysis about European energy investment policies: A hybrid hesitant fuzzy decision-making approach with quality function deployment. *Expert Systems with Applications, 115*, 152–171.

Dresselhaus, M. S., & Thomas, I. L. (2001). Alternative energy technologies. *Nature, 414*(6861), 332–337.

Eckhardt, J. T., & Shane, S. A. (2003). Opportunities and entrepreneurship. *Journal of Management, 29*(3), 333–349.

Ellabban, O., Abu-Rub, H., & Blaabjerg, F. (2014). Renewable energy resources: Current status, future prospects and their enabling technology. *Renewable and Sustainable Energy Reviews, 39*, 748–764.

Falnes, J. (2007). A review of wave-energy extraction. *Marine Structures, 20*(4), 185–201.

Farret, F. A., & Simoes, M. G. (2006). *Integration of alternative sources of energy*. John Wiley & Sons.

Fuchs, D. (2013). Theorizing the power of global companies. In J. Mikler (Ed.), *Handbook of global companies* (pp. 77–95). Wiley-Blackwell.

Goldemberg, J. (2006). The promise of clean energy. *Energy Policy, 34*(15), 2185–2190.

Goldratt, E. M. (1990). *Theory of constraints* (pp. 1–159). North River.

Hamedi, H., & Gandomkar, M. (2012). A straightforward approach to minimizing unsupplied energy and power loss through DG placement and evaluating power quality in relation to load variations over time. *International Journal of Electrical Power & Energy Systems, 35*(1), 93–96.

Heldal, T., Storemyr, P., Bloxam, E., & Shaw, I. (2016). Heritage stone 6. Gneiss for the Pharaoh: Geology of the Third Millennium BCE Chephren's Quarries in Southern Egypt. *Geoscience Canada, 43*(1), 63–78.

Jorgenson, A. K. (2007). The effects of primary sector foreign investment on carbon dioxide emissions from agriculture production in less-developed countries, 1980–1999. *International Journal of Comparative Sociology, 48*(1), 29–42.

Kamat, P. V. (2007). Meeting the clean energy demand: Nanostructure architectures for solar energy conversion. *The Journal of Physical Chemistry C, 111*(7), 2834–2860.

Karacalı, A. O., & Urfalıoğlu, N. (2019). An evaluation of restitutions prepared for the architecture of the Neolithic site of Göbeklitepe and a proposition//Göbeklitepe Neolitik Alanı Mimarisi İçin Hazırlanmış Restitüsyonlar Üzerine Bir Değerlendirme ve Bir Öneri. *Megaron, 14*(1), 63.

Karakeçe, E., & Çemberci, M. (2020). The Spirit of business life: Entrepreneurship. In *Strategic priorities in competitive environments* (pp. 125–139). Springer.

Klass, D. L. (1998). *Biomass for renewable energy, fuels, and chemicals*. Elsevier.

Leal-Arcas, R., Filis, A., & Gosh, E. S. A. (2014). *International energy governance: Selected legal issues*. Edward Elgar Publishing.

Lee, D., & Cheng, C. C. (2016). Energy savings by energy management systems: A review. *Renewable and Sustainable Energy Reviews, 56*, 760–777.

Liu, D., & Zhong, S. (2016). L2 vs. L1 use of synonymy: An empirical study of synonym use/acquisition. *Applied Linguistics, 37*(2), 239–261.

Lund, H., & Mathiesen, B. V. (2009). Energy system analysis of 100% renewable energy systems—The case of Denmark in years 2030 and 2050. *Energy, 34*(5), 524–531.

Makino, S., Beamish, P. W., & Zhao, N. B. (2004). The characteristics and performance of Japanese FDI in less developed and developed countries. *Journal of World Business, 39*(4), 377–392.

Makogon, Y. F., Holditch, S. A., & Makogon, T. Y. (2007). Natural gas-hydrates—A potential energy source for the 21st century. *Journal of Petroleum Science and Engineering, 56*(1–3), 14–31.

Mars, M. M., & Lounsbury, M. (2009). Raging against or with the private marketplace? Logic hybridity and eco-entrepreneurship. *Journal of Management Inquiry, 18*(1), 4–13.

Mathiesen, B. V., Lund, H., & Karlsson, K. (2011). 100% renewable energy systems, climate mitigation and economic growth. *Applied Energy, 88*(2), 488–501.

McKendry, P. (2002). Energy production from biomass (part 2): Conversion technologies. *Bioresource Technology, 83*(1), 47–54.
Merriam-Webster. (n.d.). Energy. In *Merriam-Webster.com dictionary*. Retrieved March 1, 2021, from https://www.merriam-webster.com/dictionary/energy.
Miyamoto, K. (Ed.). (1997). *Renewable biological systems for alternative sustainable energy production* (no. 128). Food & Agriculture Org.
Murat, S., Hobikoğlu, E. H., & Dalyancı, L. (2014). Structure and sustainability of current account deficit in Turkish economy. *Procedia-Social and Behavioral Sciences, 150*, 977–984.
Napolitano, M. R., Marino, V., & Ojala, J. (2015). In search of an integrated framework of business longevity. *Business History, 57*(7), 955–969.
Panwar, N. L., Kaushik, S. C., & Kothari, S. (2011). Role of renewable energy sources in environmental protection: A review. *Renewable and Sustainable Energy Reviews, 15*(3), 1513–1524.
Peng, M. W., Sun, S. L., Pinkham, B., & Chen, H. (2009). The institution-based view as a third leg for a strategy tripod. *Academy of Management Perspectives, 23*(3), 63–81.
Poizot, P., & Dolhem, F. (2011). Clean energy new deal for a sustainable world: From non-CO_2 generating energy sources to greener electrochemical storage devices. *Energy & Environmental Science, 4*(6), 2003–2019.
Qiu, D., Dinçer, H., Yüksel, S., & Ubay, G. G. (2020). Multi-faceted analysis of systematic risk-based wind energy investment decisions in E7 economies using modified hybrid modeling with IT2 fuzzy sets. *Energies, 13*(6), 1423.
Saghir, J. (2005). *Energy and poverty: Myths, links, and policy issues*. The World Bank.
Shenoy, P. S., Kim, K. A., Johnson, B. B., & Krein, P. T. (2012). Differential power processing for increased energy production and reliability of photovoltaic systems. *IEEE Transactions on Power Electronics, 28*(6), 2968–2979.
Simsek, H. A., & Simsek, N. (2013). Recent incentives for renewable energy in Turkey. *Energy Policy, 63*, 521–530.
Singhania, M., & Saini, N. (2018). Determinants of FPI in developed and developing countries. *Global Business Review, 19*(1), 187–213.
Smil, V. (2019). *Energy in world history*. Routledge.
Sorrell, S. (2015). Reducing energy demand: A review of issues, challenges and approaches. *Renewable and Sustainable Energy Reviews, 47*, 74–82.
Srirangan, K., Akawi, L., Moo-Young, M., & Chou, C. P. (2012). Towards sustainable production of clean energy carriers from biomass resources. *Applied Energy, 100*, 172–186.
Twidell, J., & Weir, T. (2015). *Renewable energy resources*. Routledge.
Valliere, D., & Peterson, R. (2009). Entrepreneurship and economic growth: Evidence from emerging and developed countries. *Entrepreneurship & Regional Development, 21*(5–6), 459–480.
Wagner, M. (2009). Eco-entrepreneurship: An empirical perspective based on survey data. In *Frontiers in eco-entrepreneurship research*. Emerald Group Publishing Limited.
Wang, S., Liu, Q., Yuksel, S., & Dincer, H. (2019). Hesitant linguistic term sets-based hybrid analysis for renewable energy investments. *IEEE Access, 7*, 114223–114235.
Yalta, A. Y., & Yalta, A. T. (2017). Dependency on imported oil and its effects on current account. *Energy Sources, Part B: Economics, Planning, and Policy, 12*(10), 859–867.
Yildiz, B., & Kazimi, M. S. (2006). Efficiency of hydrogen production systems using alternative nuclear energy technologies. *International Journal of Hydrogen Energy, 31*(1), 77–92.
York, R. (2012). Do alternative energy sources displace fossil fuels? *Nature Climate Change, 2*(6), 441–443.
Zehner, O. (2012). *Green illusions: The dirty secrets of clean energy and the future of environmentalism*. U of Nebraska Press.

Chapter 19
Renewable Energy and Environment in the Context of Sustainability

Halil İbrahim Uzun

Abstract The demand for energy in the world is increasing every year. For this reason, future energy demands should be planned within the framework of efficiency and trust principles. In addition to the environment and natural resources, sustainability of development and justice in economic prosperity are also necessary for all societies. The priority in sustainable development is undoubtedly the issue of energy production and consumption. It is not possible to meet the increasing energy demand with economic growth and population growth with traditional energy production systems that have destroyed the environment so far. For this reason, renewable energy systems, which are the most important alternative, should be considered from the perspective of sustainable development. In this study, the development processes of the concepts of sustainability, sustainable development and the basic foundations of today were put forward and the current situation in energy production was analyzed. Finally, the relationship and interaction between renewable energy sources and sustainable development were discussed.

19.1 Introduction

Industrialization, which gained speed in the eighteenth century, brought many new concepts to the literature. Increasing production and consumption caused the excessive consumption of resources that form the basis of production. The negative effects of this century, when the concept of development gained importance for all countries, began to be seen strongly in the middle of the twentieth century. The immoderate usage of resources created societies that exploit and exploited financially, while at the same time destroying the environment that is considered the common heritage of all humanity. In the political environment of the 1970s, a need for establishing a balance between development and environment arose almost all over the world.

H. İ. Uzun (✉)
The Faculty of Engineering and Architecture, Muş Alparslan University, Muş, Turkey
e-mail: ha.uzun@alparslan.edu.tr

S. Yüksel, H. Dinçer (eds.), *Strategic Approaches to Energy Management*, Contributions to Management Science,
https://doi.org/10.1007/978-3-030-76783-9_19

After 1970, development started to be mentioned with the concept of sustainability by many countries, especially the United Nations (Caradonna, 2014). The concept of sustainability emerged and used at the beginning of the nineteenth century in areas that were regarded as primary resources under the conditions of the period such as agriculture, forests, and fishery. The concept used in different fields is generally related to the future of humanity and is focused on the protection of resources. It is seen as an approach that brings together disciplines such as economics, sociology, environment, business, politics and law (Tıras, 2011). According to (Gilman, 1992) sustainability is the functioning of the system, including the society and ecosystem, without consuming basic resources until an uncertain end. (Ruckelshaus, 1989) defines sustainability as a doctrine in which economic growth and development will interact with ecology and all parties will be protected (Ozmehmet, 2008). Sustainability presents a concept that requires various changes in the way of thinking without lowering the life standards. At the heart of these changes there are the abandonment of the consumer society understanding, environmental management where global solidarity is ensured, social responsibilities and economic solutions (Seydiogullari, 2013). (Hart & Milstein, 1999) accepts the economy as a subset of society by adopting the approach that considers economy, society, and environment as 3 basic components of sustainability. The society that includes the economy is a subset of the environment.

In the Brundtland Report, which was prepared by the UN in 1987, the concept of sustainability was defined for the first time, which is used today, and was used identically to the concept of "sustainable development" in the first time. According to this definition, sustainability is "development that meets today's requirements without compromising the ability to meet the needs of future generations". In the Dictionary of Urban Science Terms, it is expressed as "environmentalist world view aiming at ensuring economic development without sacrificing the principle of using environmental values and natural resources with rational methods in a way that does not cause wastefulness, taking into account the rights and benefits of present and future generations" (Keleş, 1998).

There are many turning points that lead the concept of sustainability to the definition included in the Brundtland Report. The first of these is the report "The Limits of Growth" prepared by the Club of Rome in 1972. In the report, it was emphasized that there is a dependency relationship between the economy and the environment, and development creates a great burden on the environment (Meadows et al., 1972). The United Nations Conference on the Human Environment, held in Stockholm in 1972, was the first global organization to address the concept of sustainable development in the context of economy and environment. At the conference, discussions were held on the environmental damage caused by industrialization and the effects of this destruction on development and solutions to the problems. Although the focus of these discussions was development and the environment, the concept of sustainability has been limited to the environment in the future.

The concept of Sustainable Development was used as a term for the first time in the study named "World Conservation Strategy" prepared by the International Union for the Conservation of Nature and Natural Resources (IUNC). In 1983, the World Commission for Environment and Development was established with the mission of securing the interests of future generations. Especially developed countries realized that the negative effects of environmental destruction are not seen at the place and time of occurrence but affect the whole world. While the phenomenon of development was perceived as the economic activities carried out by the poor countries in order to sustain their existence before the developments, it is seen that discourses were produced that the environment was the common value of all humanity. The report named 'Our Common Future', also known as the Brundtland Report, seems to be an important example of this approach (Brundtland et al., 1987). At the UN Conference on Environment and Development held in Rio de Janeiro, Brazil in 1992, an action plan to be implemented on a voluntary basis was announced to member countries. Additional conventions such as the UN Convention on Biological Diversity, the UN Framework Convention on Climate Change are shown as moves that expand the dimension of the issue. The Rio conference aimed to realize the action plans of the UN Stockholm Declaration on the Human Environment accepted in 1972. Cooperation of country administrations, sectors and non-governmental organizations was envisaged to establish a global partnership in carrying out the actions. In Rio, the importance of international collaborations for the efficient usage of resources in the world was especially emphasized. As a result of the conference, in the declaration approved by the participating countries, it was also stated that states should reduce and gradually eliminate the unsustainable production and consumption patterns in order to ensure a continuous and balanced development and to create a livable environment for people (Ozmehmet, 2008).

Conferences such as the UN International Conference on Population and Development held in Cairo in 1995 and the UN Conference on Human Settlements—Habitat II held in Istanbul in 1996 were organizations that included the issue of population movements together with the environment in the concept of sustainability (Tıras, 2011). In 1976, in the Habitat II Conference held after the UN Conference on Human Settlements, which is shortly Habitat, held in Vancouver, the negotiations were made on the provision of sustainable human settlements in a world where adequate housing and urbanization took place for everyone, referring to the UN Convention. The Rio +5 Summit was held in New York in 1997 to evaluate what kind of actions took place in the last 5 years regarding the decisions taken in the 1992 Rio Conference and the applicability of the decisions. At the end of this meeting organized as a UN Special Session, it was seen that the expectations in the Rio Conference were not realized and it was emphasized that more realistic and applicable goals should be set. In 2002, the Sustainable Development Summit was organized by the UN in Johannesburg.

At the summit, where detailed action plans on combating poverty and protecting the environment were discussed, governments made commitments in five priority areas. An agreement for 2 billion people who cannot access modern energy facilities reached on providing energy opportunities, but there is no agreement on the transition to renewable energy. Participating countries made a commitment to adopt green energy and not support energy types for sustainable development (Seydiogullari,

2013). Considering the process that the concept of sustainable development went through in gaining its current meaning, it can be said that it covers the economy, environment, and social dimensions. Basically, it is emphasized that economy, environment, and sociology are in a relationship with one another and the balance between them should be considered.

The most obvious binding factor in the relationship between development and the environment is undoubtedly the need for energy. The ever-increasing need for energy, which is the main input of social and economic development, the limited and constant decrease of widely used energy resources created a trend towards the reassessment of energy policies by countries and the effective and efficient usage of energy resources. Especially until the oil crisis that took place in 1973, fossil fuels were widely used due to the advanced production technologies and the period when fossil fuels were relatively inexpensive to use. However, the oil crisis created a serious insecurity. Despite the dramatic decrease in oil and natural gas prices in the 1980s, developments directed societies to different energy resources. The key concept in energy policies became "energy security". In ensuring energy security, increasing energy diversity was accepted as a strategy and renewable energy resources started to come to the fore more frequently and widely. The environmental awareness that emerged in the 1970s increased with the environmental issues that became sanctions in international agreements in the 1990s. Environmental awareness also triggered an awareness about issues such as traditional energy generation models and inefficient usage of energy have direct negative effects on natural resources and the environment at micro and macro levels. Renewable energy types were accepted as clean energies that do not create a pollution load in the atmosphere (Qiu et al., 2020; Li et al., 2020; Seydiogullari, 2013).

While sustainability was mentioned with many different issues, especially natural resources, the concept of sustainable development started to differ from the concept of sustainability previously used in the same definition. Sustainable development is defined in the literature as ensuring the efficient management of natural resources in a way that allows the continuity of economic development by preserving human health and natural balance, and leaving a natural, physical, and social environment to future generations. It is defined as leaving the world to future generations by meeting the needs of people without making natural resources irreplaceable and without destroying the environment irreparably. Thus, it was accepted that the economic system is included in the ecological system in meeting human needs (Tıras, 2011).

Sustainable development requires a sustainable supply of energy in the long term, which can be obtained easily and sustainably at a reasonable cost and can be used for all necessary needs without causing negative sociological effects. Energy sources such as fossil fuels (coal, oil and natural gas) and uranium are generally considered to be based on limited resources. Other energy sources such as sunlight, wind and hydropower are considered renewable and therefore sustainable in the relatively long term (Zhong et al., 2020; Li et al., 2021). Waste and biomass are also mentioned among sustainable energy sources (IEA, 2018a). Environmental concerns are an important factor in sustainable development. The negative effects of energy production and consumption activities on the environment cause various health and ecological problems over time. Ideally, it is necessary to use energy resources that have no environmental impact or are kept to a minimum in energy production and consumption, which is the most important instrument of sustainable development.

It is noted that the transition to renewable energy in the context of sustainable development will of course be a faster and more environmentally friendly approach than tightening the controls on fossil fuel-based traditional energy production systems (Zhou et al., 2021; Yüksel et al., 2020). Although the differences in the regional production resource diversity and quantities of renewable energy resources are seen as a disadvantage, it is also emphasized that it can create awareness in the society as it offers local solutions. The small scale of the equipment shortens the installation times while it provides quick intervention to changes in demand.

19.2 Environmental Degradation and Energy

Today, energy is shown as the most important input of all human activities, especially the economy. It is known that with economic growth, the population grows at an annual average rate of 2% for almost every country. Increasing living standards, increasing dependence on technology and population growth increase the energy demand (UN, 2019). Although a final problem such as global climate change is shown at the beginning of the problems related to obtaining and using energy, there are many issues such as air pollution, acid rain, dilution in the ozone layer, forest destruction and increased radioactive material emission and radioactive pollution. In recent years, it was reported that while the environmental sensitivity in energy industries and societies increased in the production phase and the costs are moving upward, the predictions that the world population will double in 2050, the primary energy demand will increase 1,5–3 times (WEC, 2016). Acid rain and acid precipitation, the depletion of the ozone layer in the stratosphere and the increase in sensitivity to energy-related issues such as global climate change, and on the other hand, cause a trend towards renewable energy sources and technologies as a solution to the approaching energy shortage. Based on all these data, it can be said that energy is one of the main factors to be considered in sustainable development discussions. There are many factors that contribute to achieving sustainability. One of the most important ones is the need for a fully sustainable energy supply. The security of energy resources is necessary in terms of sustainability, but it will not be sufficient for development. It is also important to use energy, which is safe in terms of resources, effectively and efficiently without causing negative social effects (Anastassios, 2012).

In recent years, the effects of environmental damage have become more pronounced. Environmental effects of human activities, population growth and its irregular distribution, increase in production and consumption are also seen as the main causes of environmental destruction. In addition to traditional pollutants such as SO_2, NO_x, particulate matter and CO, micro-pollutants, or hazardous air pollutants, which are often toxic chemicals and harmful in small doses, as well as pollutants such as CO_2 are also increasing in the environment. With the change of equipment and technology used in industrial activities, especially transportation activities have a great impact on the distribution of pollutant elements. NO_x pollution became more pronounced because of the widespread usage of road transport.

It is important to classify the main topics of environmental damage in general before the destruction caused directly by energy supply and demand in the environment.

- Major accidents causing environmental damage
- Natural disasters
- Water pollution
- Sea pollution
- Wrong land usage
- Radiation and radioactivity
- Solid waste disposal
- Dangerous air pollutants
- Acid rain
- Ozone depletion in the stratosphere and
- Global climate change

It is noted that there is an international awareness on energy supply and demand, and there are three main issues: precipitation of pollutants caused by acid rain, dilution of the ozone layer in the stratosphere and global climate change.

19.2.1 Acid Rain

Pollutants that occur in mobile and fixed sources using fossil fuels such as smelting plants, industrial boilers, transportation where metal content in ores other than iron is reduced at high temperature are carried and accumulated in the atmosphere over long distances (Du et al., 2020). Especially suspended SO_2 and NO_x emissions react with water and oxygen in the atmosphere to form acid types such as sulfuric and nitric acid and fall to the earth as rain. The source of the acid rain problem, SO_2 and NO_x control is undoubtedly required for its solution. Although acid rain was seen as local problems for a long time, the presence of complex elements such as volatile organic compounds (VOCs), chloride, ozone and trace metals that cause air pollution and acid rain also increased the extent of the damage (IEA, 2018a). Some damages caused by precipitation due to acid rain are listed below.

- Acidification in large water bodies such as lakes, streams, and groundwater
- Plant poisoning (acidity)
- Poisoning in aquatic life forms and deterioration in the food chain
- Damage to forests and agricultural products
- Deterioration and corrosion in materials such as buildings, metal structures and fabrics

– Negative effect of sulphate aerosols on physical and optical features of clouds for living things

Electricity generation, heating, industrial energy usage, coal usage for different reasons constitutes 80% of SO_2 emissions. Most of the activities related to energy are also the main source of acid rain today. In OECD countries, 48% of the total emission of NO_x, another source of acid rain, is caused by road transport [9]. Most other NO_x emissions are due to fossil fuels used in fixed sources for various purposes. The countries with the greatest contribution to acid rain and precipitation are naturally the USA, the Post-Soviet Union Countries and China, where energy-related activities are common (WEC, 2016).

As mentioned before, acid rain shows detrimental effects on the ecology of water systems, forests, and building materials in historical and cultural monuments. It is a great misconception that the effects of acid rain are generally seen in the countries that cause it. The complex nature of the atmosphere, winds and different climatic effects show that acid rain is directed towards very different areas. This situation makes the "polluter pays" principle meaningless, which is the basic principle of environmental damage. Therefore, the basic principle of environmental protection should be to prevent damage at its source. For this reason, researches on "clean coal technologies" are widely carried out today. Measures are taken such as cleaning coal before burning, using techniques such as high-efficiency combustion and flue gas filtering technologies. Measures taken for road transport, which is another important source of acid rain, are not sufficient. Because especially three-way catalytic converters increase the fuel consumption of some dangerous pollutants and thus other emissions such as CO_2. Practices such as electric vehicles, popularizing public transport and rail transport, and increasing the accessibility of cities are important solution items in reducing emissions from road transport.

19.2.2 Ozone Depletion

The ozone layer is a structure that holds ultraviolet (UV), infrared radiation (240 $\pm$ 320 nm), located in the stratosphere, approximately 12–25 km above the earth's surface [8]. CFC, chlorofluorocarbons (CFC), halons (chlorinated and brominated organic compounds) and NO_x emissions react with ozone, causing ozone depletion, degradation, and regional depletion. The depletion of the ozone layer causes an increase in ultraviolet and infrared radiation levels reaching the ground, causing an increase in skin cancer rates, eye damage and various biological deteriorations. The most important factors in ozone degradation are coolers used in air conditioning and cooling equipment, CFCs in foams used for isolation, NO_x emissions from fossil fuel and biomass combustion processes, natural denitrification, nitrogenous fertilizers and emissions from airplanes. The Montreal Protocol, signed in 1987, was a turning point as a result of many years of debates on the causes of the destruction of the ozone layer. As a result of the presentation of scientific evidence that stratospheric ozone is destroyed by CFCs and halons, countries undertook commitments at the London Conference in 1990 to take measures to reduce these emissions [10]. Accordingly, a consensus was reached on the production of CFC-free backup equipment, development of technologies that do not use CFCs, gradual reduction of CFCs, and a complete ban, especially for technologies using CFCs (WWI, 2010).

19.2.3 Global Climate Change and Greenhouse Effect

The greenhouse effect is a natural phenomenon that warms the earth with the help of water vapor and clouds. However, today, the increase in the emissions of various gases, especially CO_2, which increase the greenhouse effect, caused the world to heat more and the balance to deteriorate. Human-induced CO_2 meets approximately 50% of the greenhouse effect experienced today. Some other gases such as CH_4, CFC, halons, N_2O, which are released as a result of human and industrial activities, and ozone and peroxyacetylnitrate produced for different purposes contribute to the greenhouse effect and cause the temperature to rise.

The greenhouse effect associated with global climate change is shown as the most important environmental damage caused by energy usage. The increase in the concentration of greenhouse gases in the atmosphere prevents the reflection of the rays from the sun from the earth and causes the temperature to increase. The surface temperature of the Earth increased by approximately 0.68 °C in the last century, and it is estimated that the sea level increased by an average of 20 cm as a result of the melting of glaciers. Unless conditions change and current trends in emissions continue, it is estimated that the Earth temperature will increase by 2–4 degrees at the end of the twenty-first century. In this case, the sea level will rise by 30–60 (IEA, 2018b). These physical changes may cause the destruction of coastal areas, displacement of agricultural lands, shifting of fertile regions to upper latitudes and salinization of freshwater resources. Discussions are being held on various measures such as increasing afforestation, actions to reduce existing CO_2 emissions, and reduction of carbon emissions in order to reduce the greenhouse effect. However, from the perspective of developing countries, cost-benefit analyses and policies for rapid economic growth should also be considered. For this reason, there is a need for international policy implementations such as increasing efficiency in energy usage and compliance with fuel orientation. However, the existence of many different views on climate change, which makes international consensus difficult, is also seen as an important handicap. The most important argument to be made against the views that temperature increase is a natural process and that activities are not the cause of climate change is that the environment is destroyed under all conditions (IEA, 2017).

There are steps to be taken by focusing on environmental damage and its consequences adversely affect all humanity along with future generations. It is necessary to measure and estimate the required data accurately and effectively, with short- and long-term main goals, to create new, continuous, and reliable data, and to interpret the results objectively.

In general, there are some solution suggestions for the prevention of environmental damage and disposal of the consequences of the destruction.

- Continuous development of renewable energy technologies
- Energy efficiency
- Development of battery technologies
- Producing clean alternatives for transportation
- Promoting public transport
- Improving fossil fuels and combustion systems
- Recycling
- Protection and enhancement of forest assets
- Carbon or fuel taxes
- Increasing environmental awareness
- Environmental protection-oriented lifestyle

19.2.4 Renewable Energy Sources and Technologies

Renewable energy refers to the energy source that can exist in the natural cycle of nature later. With the oil crises in the 1970s, there was a breakthrough in finding alternative energy sources and developing renewable energy technologies. The limitation and high cost of fossil energy sources, including petroleum, made it attractive to use and expand renewable energy systems. As a result, the perception of the consequences of environmental damage such as ozone layer deterioration, acid rain and global climate change by all societies led to a strong social acceptance regarding the transition to renewable energy systems (SUSCHEM, 2018).

Renewable energy is the use of natural processes to transform them into energy. Direct solar energy, which can be obtained without the need for any production process such as mining activities in the environment, or indirectly from solar, with very low environmental damage, continuously renewable, ready to use at any time, all of the energy sources such as hydraulic, photon, wind, biomass, biogas, geothermal, wave, current, tide and hydrogen are defined as renewable energy (Seydiogullari, 2013). However, there are problems, such as not being permanent and being present in different regions, in access to these energy resources. Despite these difficulties, investments are increasing every year with research activities on renewable energy resources and technologies (Athukoralaa et al., 2015). Strategically about renewable energy, issues such as harvesting, improving conversion efficiencies, lowering installation and maintenance costs, and security are important.

Renewable energy technologies gain importance due to the increase in service costs and subsequent environmental concerns. As a result of renewable energy technologies, which are low operating costs, the increase in advanced technology areas, and the developments in modelling systems, the initial investment cost is also an area that decreases. For this reason, developed countries around the world support innovative ideas in the field of renewable energy every year. Renewable power plants have an advantage due to their modularity, low operating costs, and the relative unlimitedness of basic generation resources compared to traditional energy generation methods. The global proliferation of renewable energy systems will certainly reduce pollution levels. As a result of the great developments in solar energy technologies in recent years, large-scale solar power plants are also becoming widespread (Salvarli & Salvarli, 2020).

Renewable energy sources are widely used especially in developed countries within the scope of transition policies to a low carbon economy. Among the IEA countries, the share of renewable energy sources increased by 5.7% per year between 1970 and 2002, and this share is expected to increase by another 60% by 2030, according to IEA estimates (Batmaz et al., 2019). Today, constituting 24,5% of the world's electricity production, renewable energy is shown as the second biggest contributor to global electricity generation (Salvarli & Salvarli, 2020). During 2018–2023, renewable energy types such as solar PV, wind, hydroelectric and bioenergy are expected to meet about 70% of global electricity generation growth. Until 2023, it is predicted that global electricity demand will be met by hydroelectric (16%), wind (6%), solar PV (4%) and bioenergy (3%). However, it is stated that biofuels in road transport were 3.4% in 2017 and will have the lowest share of renewable energy with 3.8% in 2023. Fossil fuels take the first place with a share of 79% in the world's final energy production. This is followed by renewable energy sources with a share of 18% and nuclear energy with a share of approximately 3%. While traditional biomass constitutes most of the renewable energy share, it is followed by hydroelectricity, hot water/heating, power generation and biofuels, respectively. Renewable energy sources are used in many areas such as electricity generation, water heating, heating of greenhouses, drying, lighting, heating, chemical processes (Salvarli & Salvarli, 2020).

19.3 Conclusion

The increasing need for energy every year due to factors such as economic growth, technological developments and population growth, the limitation of fossil fuels such as oil, coal and natural gas used in traditional energy production systems and the depletion scenarios make it necessary to increase efficiency and find alternative energy sources. Fossil fuels still have the largest share of energy consumption and are on the rise worldwide. In order to realize a sustainability-related energy supply, some criteria such as diversity in energy source and generation systems, efficiency, renewable energy resource management, social cost-based pricing, service management and fairness must be met. These criteria can be classified in environmental, economic, and social dimensions. Therefore, fossil fuelled resources can be substituted with different resources depending on the increase in the need for energy but ensuring sustainability without harming the environment is vital. Every humanitarian action naturally brings a new burden to the environment. Efforts to protect and substitute natural resources stand as a necessity when future generations are also considered. In backward countries exploited by developed countries and in developing countries, development moves put more pressure on an already destroyed environment. For the environment that has already exceeded its self-renewal capacity, additional loads also cause problems that cannot be compensated. The large population growth rates of developing countries compared to developed countries make them more dependent on their resources. Therefore, two different solution proposals are shown in response to the increasing energy consumption in the world. The first of these is making improvements in traditional energy production systems and resources (especially coal) where countries use available or easily accessible resources, increasing efficiency, keeping pollutant emissions in the production phase. The second is the gradual transition to renewable energy systems. The potentials of countries should be considered in the context of sustainable development. However, studies show that investing in renewable energy systems is more appropriate in terms of cost benefit rather than system improvement. Renewable energy systems are effective in many areas such as low cost, job creation, developing future industries, and meeting energy and environmental goals. The development and usage of renewable energy will contribute to the creation of new business areas with subjects such as energy security, environmental protection, economic development, mechanical production, construction, transportation, and industrial improvement. Solar, wind and biomass energies, which can be installed on a local scale, will provide on-site solutions to increasing energy demands and enable regional actions to be taken in the protection of the environment. The current situation and future scenarios regarding the increase in energy demand make a great development possible for the renewable energy market. It is estimated that the share of renewable energy in meeting the global energy demand will reach 12.4% in 2023. It is also clear that if investments in renewable technologies continue in the long term, they will make a significant contribution to the renewable energy need. The share of fossil fuels in the total primary energy supply is expected to constitute approximately 81% of the total in 2023. By 2050, renewable energy is expected to reach approximately 30% of the world's energy share. In the light of these predictions, countries should increase the share of domestic and renewable energy resources in their production systems to the maximum extent in order to ensure a balanced resource diversification. The most practical way for many countries to achieve their goals in their current strategic planning will be to support, develop and encourage environmentally friendly and clean practices in production and services. Most of the renewable energy technologies developed are used in leading developed countries such as the USA, Japan, and Europe. Using less and cleaner energy in power plants, buildings, industrial facilities, and transportation systems can reduce costs by up to 80%. These measures, where 30% less energy can be consumed, will also prevent environmental damage. Thus, countries can maintain cost-effectiveness and make sustainable progress. There is a very close relationship in terms of environmental degradation and sustainable development in energy usage. In order to ensure the sustainability of resources, development, human and living generation, it is necessary to increase the usage of renewable energy resources and systems, to prevent environmental damage or to dispose it at the source, and efficiency in both energy production and consumption. Minimizing the pollutants that occur in the production and consumption stages of energy and destroy the environment, economical usage of resources and safety issues should be taken into consideration in the planning stage. The main source of global climate change and the greenhouse gases that cause it is energy production and consumption. Approximately 80% of greenhouse gases originate from energy production and consumption, and this situation reveals the necessity of renewable energy for sustainable development. The current energy problem in the world, which is predicted to be exacerbated in the coming years, can be overcome with renewable energy resources. Renewable energy sources are the biggest alternative for both sustainable development and protection of natural resources and the environment.

References

Anastassios, P. (2012). *Technology trends: A review of technologies and policies study on technology trends DTI*. Retrieved from www.dti.gov.za

Athukoralaa, A., Jayasuriyaa, W. J. A., Ragulageethana, S., Pereraa, A. T. D., Sirimannaa, M., & Attalagea, R. A. (2015). A techno-economic analysis for an integrated solar PV/T system with thermal and electrical storage – case study. In *Moratuwa engineering research conference* (pp. 182–187).

Batmaz, T., Bayraç, H. N., & Güllü, M. (2019). Türkiye'de Yenilenebilir Enerji Kaynaklarının Büyüme ve karbon Emisyonu İlişkisi. *Osyal ve Ekonomi Araştırmaları Dergisi, 6*(3), 645–658.

Brundtland, G. H., Khalid, M., Agnelli, S., Al-Athel, S., & Chidzero, B. J. N. Y. (1987). *Our common future* (p. 8). New York.

Caradonna, J. L. (2014). *Sustainability: A history*. Oxford University Press.

Du, L., Dinçer, H., Ersin, İ., & Yüksel, S. (2020). IT2 fuzzy-based multidimensional evaluation of coal energy for sustainable economic development. *Energies, 13*(10), 2453.

Gilman, R. (1992). *Sustainability*. Retrieved from https://www.context.org/about/definitions/#sustainability

Hart, S. L., & Milstein, M. B. (1999). Global sustainability and the creative destruction of industries. *MIT Sloan Management Review, 41*(1), 23.

IEA. (2017). *Energy technology perspectives*.

IEA. (2018a). *Renewables 2018 analysis and forecasts to 2023*.

IEA. (2018b). *World energy outlook*.

Keleş, R. (1998). *Kentbilimleri Sözlüğü*. (İ. Kitapevi, Ed.). Ankara.

Li, Y. X., Wu, Z. X., Dinçer, H., Kalkavan, H., & Yüksel, S. (2021). Analyzing TRIZ-based strategic priorities of customer expectations for renewable energy investments with interval type-2 fuzzy modeling. *Energy Reports, 7*, 95–108.

Li, X., Zhu, S., Yüksel, S., Dinçer, H., & Ubay, G. G. (2020). Kano-based mapping of innovation strategies for renewable energy alternatives using hybrid interval type-2 fuzzy decision-making approach. *Energy, 211*, 118679.

Meadows, D. H., Meadows, D. L., Randers, J., & Behrens, W. W. (1972). *The limits of growth*. Universe Books.

Ozmehmet, E. (2008). Dünyada ve Tükiye Sürdürülebilir Kalkinma Yaklaşimlari. *Journal of Yaşar University, 3*(12), 1853–1876.

Qiu, D., Dinçer, H., Yüksel, S., & Ubay, G. G. (2020). Multi-faceted analysis of systematic risk-based wind energy investment decisions in E7 economies using modified hybrid modeling with IT2 fuzzy sets. *Energies, 13*(6), 1423.
Ruckelshaus, W. D. (1989). Toward a sustainable world. *Scientific American, 261*(3), 166–175.
Salvarli, M. S., & Salvarli, H. (2020). For sustainable development: Future trends in renewable energy and enabling technologies. *Resources, Challenges and Applications, 3*.
Seydiogullari, H. (2013). Renewable energy for sustainable development. *Planlama, 23*(1), 19–25.
SUSCHEM. (2018). *Key enabling technologies in Horizon Europe paper*. Retrieved from www.suschem.org
Tıras, H. (2011). *Sustainable development and the environment: A theoretical review*. Erciyes University.
UN. (2019). *World population prospects 2019 highlights*.
WEC. (2016). *World energy resources 2016 summary*.
WWI. (2010). *Renewable energy and energy efficiency in China*. Retrieved from www.worldwatch.org
Yüksel, S., Dinçer, H., & Uluer, G. S. (2020). The role of technological development on renewable energy usage: An econometric analysis for G7 countries. In *Handbook of research on sustainable supply chain management for the global economy* (pp. 136–153). IGI Global.
Zhong, J., Hu, X., Yüksel, S., Dinçer, H., & Ubay, G. G. (2020). Analyzing the investments strategies for renewable energies based on multi-criteria decision model. *IEEE Access, 8*, 118818–118840.
Zhou, P., Luo, J., Cheng, F., Yüksel, S., & Dinçer, H. (2021). Analysis of risk priorities for renewable energy investment projects using a hybrid IT2 hesitant fuzzy decision-making approach with alpha cuts. *Energy, 224*, 120184.

Chapter 20
Determining Optimal Islamic Financing Methods for Small-Scale Sustainable Energy Investments Regarding Socio-Economic Welfare

Hakan Kalkavan and Serkan Eti

Abstract The aim of this study is to determine appropriate Islamic financing methods (IFMs) for sustainable energy investments to reach a wider community. Since energy investments are costly projects, their financing should be carried out in the most efficient way. Likewise, the provision of environmentally friendly energy resources is essential for a sustainable human-society-nature balance. As a human duty to prevent climate change, everyone should be aware of personal consumption according to ecologically friendly. In this context, the use of Islamic financing instruments can help spread sustainable energy investments, particularly across Muslim countries. In terms of small-scale investments, it would be significant for the sustainable energy investments to reach people with religious sensitivity, as Islamic financial institutions have a closer business relationship with both SMEs and end-users. To put it briefly, it is vital to make the most appropriate decision regarding sustainable energy investments, considering the economic limitations of countries and people. According to the analysis result of the MCDM based AHP method; Murabaha financing comes to the fore in small-scale energy investments. Accordingly, Murabaha (cost-plus) financing can be a good alternative to interest-bearing credits with a variable interest rate. Respectively, the second-best weighted IFM is found as Mudarabah which is a special type of partnership where one party is a financier, and the other is an entrepreneur. In reverse, the least suitable method has been found as Ijarah. Thus, the Islamic banking system should develop various business models for energy investments through these two most appropriate IFMs. So much so that, thanks to small-scale sustainable energy investments, both environmentally friendly carbon emissions will be reduced, and the country's external energy imports will be reduced via the production of extra energy. As a result, shifting from anthropocentric to ecocentric lifestyle that is self-sufficient in terms of energy and compatible with the environment can be generated for the sake of sustainability.

H. Kalkavan (✉) · S. Eti
The School of Business, İstanbul Medipol University, İstanbul, Turkey
e-mail: hkalkavan@medipol.edu.tr; seti@medipol.edu.tr

S. Yüksel, H. Dinçer (eds.), *Strategic Approaches to Energy Management*, Contributions to Management Science,
https://doi.org/10.1007/978-3-030-76783-9_20

20.1 Introduction

Energy can be an important tool for countries to ensure their economic development. So much so that energy is an important asset that is passed down both as a wealth that can be transferred between generations and as a budget deficit caused by lack of energy (Yüksel et al., 2019). Due to the growing world population and the rise in the consumption of people, the need for energy also increases. Conventional energy sources will be insufficient in this respect at a time in the future. Therefore, energy resources need to be increased. Otherwise, serious economic bottlenecks await countries that cannot diversify their energy production and produce their own energy (Bhattacharya et al., 2016; Wang et al., 2020).

Moreover, climate change associated with carbon emissions due to greenhouse gas emissions is one of the most vital problems of humanity since the twentieth century (Du et al., 2020). As a human duty to prevent climate change, everyone should be aware of global social responsibility; on the one hand, it is a matter that it can support more or less by making its personal consumption environmentally friendly, on the other hand, by enhancing public awareness and influencing policymakers. Inasmuch as, this is an issue that needs to be handled altogether from individual to society—from states to international organizations (Sandler, 2004; Bauwens & Eyre, 2017).

However, since energy investments are costly projects, their financing should be carried out in the most efficient way. Likewise, the provision of environmentally friendly energy resources is essential for a sustainable human-society-nature balance (Li et al., 2021a, 2021b). Therefore, there is a serious need for countries to make sustainable energy investments (Xie et al., 2021). The efficient realization of these investments is possible by using the most suitable financing tools. In this context, the use of Islamic financing instruments can help spread sustainable energy investments, particularly across Muslim countries (Kalkavan, 2020; Yang et al., 2021). In terms of small-scale investments, it would be significant for the sustainable energy investments to reach people with religious sensitivity, as Islamic financial institutions have a closer business relationship with both SMEs and end-users. In this respect, Islamic financing methods such as "murabaha, mudarabah, musharakah, ijarah, sukuk, and waqf crowdfunding", which are determined to be appropriate for small-scale energy investments, have been included in the analysis.

The aim of this study is to determine appropriate Islamic financing methods for sustainable energy investments to reach a wider community. First, the importance of energy for countries and the meaning of environmentally friendly energy consumption are discussed in the introduction. Following, the significance and types of sustainable energy investments were discussed. In the third part, descriptions of Islamic finance instruments convenient for sustainable energy investments are specified. In the fourth part of the study, information on the "multi-criteria decision-making" (MCDM) based Analytic Hierarchy Process (AHP) methodology and analysis results of determining optimal Islamic finance instruments for sustainable energy investments are given. The conclusion part of the study covers discussing the results of the analysis and the strategies developed.

20.2 Sustainable Energy Investments

Renewable energy can be defined as the energy source obtained from natural resources such as sun, wind, waterpower et cetera, and particularly does not cause carbon emissions. In this regard, it should also be emphasized that renewable energy sources can renew themselves and do not run out in nature (Haiyun et al., 2021). The amount of energy needed and consumed in the world is increasing day by day. The main reasons for this are population growth and the widespread use of technological products depending on consumption habits. For all these reasons, it is necessary to increase the energy production capacity. However, another important issue is the necessity to do this without harming the environment while generating energy and thus ensuring sustainable energy investment (Yüksel & Ubay, 2021). Because the carbon emission spreading to the environment as a result of fossil fuel consumption increases the greenhouse gas emission by damaging the environment. Coal, petroleum products and nuclear power plants, which are types of energy produced by traditional methods, cause radioactive effects, chemical waste and carbon emissions that harm the environment, damaging the nature and the sustainability of natural life (Cheng et al., 2021; Li et al., 2021a, 2021b).

In order to provide sustainable energy, there are various alternatives such as "solar, wind, geothermal, biomass and hydropower", and there are large-scale energy investments and small-scale energy investments in terms of size (Cheng et al., 2021). Large-scale energy investments are made on the basis of states with the support of large capital, and the energy obtained is produced for use both domestically and abroad. The purpose of small-scale energy investments is to provide an environmentally friendly house that produces self-sufficient energy from its own house and land (Burton & Hubacek, 2007).

Studies on sustainable energy investments vary widely in terms of scale, energy source, methodology, and strategies. Evaluating sustainable energy investments based on quality-house, Tang and Dinçer (2019) have determined that the most important problem in terms of technical requirements of energy investments is the capacity problem. From another aspect, in an interesting study examining the literature of quantitative and semi-quantitative methods used to model risks and uncertainties in sustainable energy planning studies, Ioannou et al. (2017) have determined that the probability aspect of the technical and economic parameters of risks are taken into account in quantitative methods, while non-statistical parameters such as socio-economic factors are considered in methods such as scenario analysis and multi-criteria decision analysis. Moreover, in the research conducted by Steg et al. (2018), which investigates the factors that drive energy consumers, various factors should be considered -apart from economic prices- such as social status, social norms, identity difference, environmental sensitivity according to consumer preferences at a global level for a successful transition to a sustainable energy system.

Furthermore, some studies using "multi-criteria decision-making" (MCDM) methods are as follows. In a study using the fuzzy DEMATEL technique, Zhou et al. aimed to identify the issues that financial institutions should pay attention to in their decision to finance large-scale energy projects. According to the results of the research, the most important criterion that financial institutions should pay attention to in providing loans to the company that will invest in energy is "technological and financial competence". Correspondingly, in an akin research aimed at determining innovative strategies for different renewable energy alternatives based on the Kano model, Li et al. (2020) found that the most important strategic implementation criterion is "availability of equipment and technological infrastructure" for technical competence of alternative energy investments. In another study conducted by a similar method, Qiu et al. (2020) analyzed the systematic risks of wind energy investments within the scope of E7 countries and determined that "volatility in exchange rates and interest rates" are the most critical hazards. Lastly, in a study using an agent-based simulation technique that proposes a "Philanthropic-Crowdfunding-Partnership" (PCP) model for renewable energy investments, Ari and Koc (2021a) projected that an alternative financing model of the PCP reduces wealth inequality.

To put it briefly, it is vital to make the most appropriate decision regarding sustainable energy investments, considering the economic limitations of countries and people. In this context, the method of determining and evaluating appropriate criteria through MCDM has been widely used in the literature. Similar to the literature reviewed above, the MCDM based AHP method is conducted in this study. As its original aspect of the research, it is proposed to apply Islamic financing methods to sustainable energy investments.

20.3 Islamic Financing Methods

20.3.1 Murabaha

Murabaha is the most used financial instrument in Islamic financing. Islamic banks make use of the funds they collect mostly through the murabaha method. Murabaha is simply defined as adding a certain amount of profit to a good and selling it. In murabaha transactions, the Islamic bank enters into commercial activity as the purchaser of raw materials, accommodation units, land, machinery, vehicles, and such goods. The product is then sold to the customer by adding the agreed profit amount to the purchased price. In order for a murabaha contract to be valid, both parties must know obviously type, quality, and price of the product before the contract is made (Warde, 2000; El-Gamal, 2006).

In the murabaha transaction, a contract is made between the customer and the Islamic bank for the goods requested by the customer. The Islamic bank undertakes to purchase the requested good and sells it at a fixed price, with a profit added to it. The customer also makes a commitment to buy the goods at the agreed price from the Islamic bank (Tunç, 2010). However, according to Islamic law, the customer is not obliged to purchase the goods purchased by the bank. This is considered as the risk of the trade and the bank must have this risk in the murabaha transaction. Various price predictions can be determined according to the payment method and term. The customer can make cash or deferred payment according to his/her own preference. But, once the price is determined, it can neither be increased nor decreased according to early payment or late payment (Usmani, 2002).

20.3.2 Mudarabah

Mudarabah, as a financial transaction, has been used not only in Muslim countries but also in European countries, primarily Italy, and was named as commenda. Commenda was applied as a common method in commercial transactions in medieval Europe (Çizakça, 1996). Mudarabah is a special partnership type that one party is the financier (rabbul-mal) who provides the money, and the other party conducts commercial activities with this money as a mudarib (working partner). The mudarib is responsible for the administration and operation of the business. The financier (rabb-ul-mal), as the owner of capital, can stipulate the mudarib to do a certain job, and such a job is called a restricted mudarabah. If the mudarib is free to do the economic activity he wants, provided that it complies with Islamic rules, it is called unrestricted mudarabah (Usmani, 2002).

Due to the difficulties in practical application, the mudarabah method takes a very small place in the financing applications of Islamic banks while providing loans. In fact, the most known and widely applied form of partnership in commercial activities in Muslim societies has been mudarabah. In mudarabah, the profit rate is distributed according to the request of the partners and the dividend rate is specified in the contract made at the beginning (Tunç, 2010). There are two types of mudarabah contracts in Islamic banking applications. In the first, the fund owner (rabb-ul-mal) deposits his/her money to the Islamic bank. The Islamic bank (mudarib) conducts economic activity by operating this money and shares the profit or loss with the fund owner at the end of the business. In the second, the Islamic bank is the financier (rabb-ul-mal) and the bank gives the entrepreneur (mudarib) the capital money to operate. Profit/loss resulting from economic activity is shared between the bank and the entrepreneur. Mudarabah is mostly used in production or investment projects. The aim of mudarabah financing instrument is to establish a partnership between an Islamic bank that provides project financing and the entrepreneur who does not have sufficient economic capital but has the necessary knowledge and skills regarding the business to be done (El-Gamal, 2006). After the project is over, the entrepreneur and the financier share the profit of the venture made according to the terms of the contract. If the enterprise makes a loss, the capital owner bears the responsibility for the damage, but the liability is limited to the money invested in the business. If there is bad management or negligence of the business in the undertaking, then the entrepreneur is obliged to pay back the money to the capital owner. But if there is no negligence, the entrepreneur does not have to pay any compensation, as he/she has already lost time and effort (Ayub, 2007).

20.3.3 *Musharakah*

Unlike mudarabah, the most distinctive feature of musharakah is that both parties also can put either different amounts of money into a project. Even if the parties put an equal amount of money, the dividend rate may differ. Both parties have the right to take part in the administration of the business but do not have to take an active role. Briefly, the musharakah financing instrument applied by Islamic banking is to finance a commercial activity or invest in a project based on a profit and loss partnership between the expert entrepreneur and the Islamic bank (Ayub, 2007; Tunç, 2010). As in murabaha, after the project is over, the profit is distributed according to the profit share rate in the contract. But if the project has made a loss, the loss is apportioned exactly according to the proportion of capital invested. Musharakah contract can be made on a short- or long-term basis. There are two types of musharakah: constant and diminishing. If the Islamic bank does not change its initial money until the end of the contract, it is called the constant musharakah. On the other hand, the diminishing musharakah is defined as the transfer of the money invested by the Islamic bank to its partner (customer) over time (El-Gamal, 2006; Usmani, 2002).

20.3.4 *Ijarah*

Ijarah conceptually refers to the lease of an asset to take advantage of the usufruct and includes also wage labor, i.e., labor lease. Ijarah, also known as Islamic leasing, is a common type of financing used in Islamic banking. It is the bank's purchase of a commercial good such as vehicle, machine, land, etc., and allocating it for the use of the customer in return for a rent. The type of the commodity, the amount of rent and installments to be paid for that commodity, and the payment duration must be specified in the leasing contract. The price of the leased commodity is calculated from the benefit of the good and over the equivalent market value. There are a few differences between ijarah and conventional leasing. Firstly, since the ownership of the merchandise remains with the Islamic bank, the bank is mainly responsible for it. Secondly, the sale price of the rented commercial good is not determined in advance, but the sales price is determined at the end of the lease contract (Warde, 2000; Ayub, 2007).

20.3.5 Sukuk

The word sukuk was used as a paper/bill showing obligations arising from commercial activities in Muslim countries. Today, they are similar to "securities" as an Islamic finance instrument which is also called a lease certificate. Investment sukuk are certificates that are issued at equal value, representing the ownership and usufruct rights specific to a particular project. In accordance with Islamic law, the investment sukuk are distributed to the market, unilaterally to be operated by the mudarib, and a fixed yield is given in return for the property right. While stocks represent indefinite ownership in a company, certificates in sukuk represent ownership for a certain period of time. Additionally, it had been necessary to raise large funds for large projects, and for that reason to develop an interest-free finance model, the formation of the sukuk concept occurred. Moreover, sukuk issuance is similar to the method of securitization in the traditional economic system, provided that the economic activities are considered halal by Islamic law (Ayub, 2007; Cebeci, 2019).

20.3.6 Waqf Crowdfunding

In Islamic societies, institutions of zakat and waqf have been the practices of redistribution to eliminate economic inequality from past to present. Waqf (foundation) is a social responsibility mechanism where the wealth obtained from donations is used for the good of society (Çizakça, 2000; Kahf, 2014). In order for any asset to be a waqf, it must be an immovable asset, but later this requirement was loosed in order to include liquid assets known as cash waqf (Kuran, 2001). Waqfs fulfilled their functions by simultaneously providing financial resources through religious, political, social, and cultural connections. In this respect, the economic ties of a waqf with its related parties played an important role. So much so that, those economical relationships have ensured continuity of its existence in line with its establishment purpose and contributed both materially and spiritually to the society (Yediyıldız, 2012; Genç, 2014). Cash waqfs are the foundations in which all or a portion of their capital consists of cash. Fundamentally, cash waqf used to operate the capital available in the foundation institution according to Islamic financing rules. Hereby, it continued its services with the profit obtained from various financing applications, and in this context also constituted the basis of modern Islamic finance instruments (Bulut & Korkut, 2016). The crowdfunding method is a fintech application that provides banking-like financial intermediation between fund owners and fund seekers through online platforms. Crowdfunding can be applied either in philanthropic projects or on a debt financing basis by bringing the aforementioned three parties together (Al-Bashir, 2019). In this context, waqf-based Islamic crowdfunding is the structuring of fintech as a foundation crowdfunding platform following Islamic law. In line with this, funds could be raised for environment-friendly and socially responsible projects such as renewable energy and housing investments by utilizing crowdfunding financing blended with waqf institutions (Mohsin & Muneeza, 2019; Ari & Koc, 2021b).

20.4 An Application with the AHP Method

In this study, it is aimed to determine the most suitable Islamic financing methods (IFMs) for the sustainability of small-scale energy investments (Aguarón et al., 2021). For this purpose, six IFMs have been taken into consideration. In order to compare these methods with each other, multi-criteria decision-making techniques based on expert opinion were used. AHP and DEMATEL are some of these models (Düzgün, 2021). Analytic Hierarchy Process (AHP) is a multi-criteria decision-making technique for weighing the alternatives. This method was developed in the 1970s by Thomas Saaty (Saaty, 1990). Thanks to this method, different options are listed according to their importance weight (Dinçer & Yüksel, 2018). IFMs are "Murabaha, Mudarabah, Musharakah, Ijarah, Sukuk, and Waqf Crowdfunding". An application has been made on these methods. AHP method was used to enumerate the methods.

Expert opinions were collected as the first step of AHP. Opinions of three experienced academics working on alternative finance were consulted. Experts filled out the questionnaire with the standard 1–9 scale of AHP (Zhang et al., 2009). Initial matrix is the average of opinions matrices of three experts. Initial matrix is given in Table 20.1.

Column totals were taken for the normalization of the matrix obtained after the expert opinion was averaged. The sum of columns is given in Table 20.1. The normalized matrix is obtained by dividing each element of the matrix by its own column sum. The normalized matrix is shown in Table 20.2.

The row averages of this matrix were obtained. These averages show the weights of IFMs. The biggest weight is considered as the most suitable method. Obtained weights and their order are specified in Table 20.2. The CI/RI value is calculated for the validity of the weights obtained from the AHP method. This value must be less,

Table 20.1 Initial matrix

Islamic financing methods	Murabaha	Mudarabah	Musharakah	Ijarah	Sukuk	Waqf Crowdfunding
Murabaha	1	3.667	4.667	7.667	6.667	2.667
Mudarabah	0.273	1	3.000	6.333	5.000	1.278
Musharakah	0.214	0.333	1	4.667	2.333	0.844
Ijarah	0.130	0.158	0.214	1	0.800	0.412
Sukuk	0.150	0.200	0.429	1.250	1	0.889
Waqf crowdfunding	0.375	0.783	1.184	2.427	1.125	1
Sum	2.142	6.141	1.494	23.344	16.925	7.090

Table 20.2 Normalized matrix

	Murabaha	Mudarabah	Musharakh	Ijarah	Sukuk	Waqf Crowdfundig
Murabaha	0.467	0.597	0.445	0.328	0.394	0.376
Mudarabah	0.127	0.163	0.286	0.271	0.295	0.180
Musharakah	0.100	0.054	0.095	0.200	0.138	0.119
Ijarah	0.061	0.026	0.020	0.043	0.047	0.058
Sukuk	0.070	0.033	0.041	0.054	0.059	0.125
Waqf crowdfunding	0.175	0.127	0.113	0.104	0.066	0.141

Table 20.3 Weights and order of Islamic financing methods

Islamic financing methods	Weight	Rank
Murabaha	0.435	1
Mudarabah	0.220	2
Musharakah	0.118	4
Ijarah	0.043	6
Sukuk	0.064	5
Waqf crowdfunding	0.121	3

1. CI is the Consistency Index value. This value is calculated with Eqs. (20.1) and (20.2).

$$\lambda_{max} = \frac{1}{n} \sum_{i=1}^{n} \left(\frac{\sum_{j=1}^{n} a_{ij} w_j}{w_i} \right) \tag{20.1}$$

$$CI = \frac{\lambda_{max} - n}{n - 1} \tag{20.2}$$

Here n represents the number of criteria. w_j represents the weights. a_{ij} are elements of the expert opinion matrix. RI is the Random Index. For n = 6, its value is 1.24. (Unver & Ergenc, 2021; Al-Harbi, 2001). λ_{max} value is 6.308. Then, the CI value of the study was 0.062, and the CI/RI value was calculated as 0.05. This value is less than 0.1. Therefore, it can be said that the analysis is consistent (Aktaş et al., 2015). In other words, the weights obtained by the AHP method are valid.

When Table 20.3 is examined, the highest weight value is Murabaha with 0.435. In other words, Murabaha is the most suitable Islamic financing method for small-scale energy investments. After Murabaha, Mudarabah method has the biggest weight with 0.22. This method has been determined as the most suitable IFM after Murabaha. The third largest IFM is Waqf Crowdfunding. That method's weight is 0.121. Musharakah ranks fourth in weight. The weight of Musharakah is 0.118. Finally, the weight of Sukuk is 0.64 and the weight of Ijarah is 0.043, respectively. According to these weights, Sukuk ranks fifth and Ijarah is the last. In other words,

Ijarah has been determined as the least suitable IFM for small-scale energy investments. According to the analysis results, the best Islamic Financing Method for small-scale sustainable energy investment has been determined as Murabaha, and in reverse, the last option has been found as Ijarah.

20.5 Conclusion

Financing energy projects is as important as energy resources for countries and investors. Providing environmentally friendly energy resources is essential for a sustainable human-society-nature balance. Nowadays, energy financing is mostly provided through a banking system. However, especially investors or end-users, who are religiously sensitive and are abstain from using interest, have problems with engaging in the traditional financial systems. Therefore, the importance of Islamic banking methods comes to the fore. Here also, the determination of which method is more appropriate becomes necessary. According to the analysis result of the AHP method based on expert opinions, Murabaha financing method comes to the fore in small-scale energy investments. Accordingly, Murabaha financing can be a good alternative to interest-bearing loans. In fact, the principle of cost-plus financing by adding the fixed profit amount on the cost of the purchased product may be more advantageous than taking a loan with a variable interest rate. Moreover, it is ensured that the fund used is transformed into a real investment with the purchase-sale contract. Respectively, the second-best weighted Islamic Financing method is found as Mudarabah which is a special type of partnership where one party is a financier, and the other is an entrepreneur. At the end of the project, the entrepreneur and fund provider share the profit of the venture in accordance with the previously made contract. Establishing a partnership between an Islamic bank that finances the project and an entrepreneur who does not have sufficient capital but has the know-how signifies making efficient use of capital. Thus, it is possible to make energy investments with high added value for the benefit across the society.

Small-scale energy investments are the most popular type of energy for individual investors. Companies and individual investors considering investing in this field should use the cost-plus method or partnership-based methods to raise capital. Thanks to both financing methods, investments will be more effective, and returns will be more profitable. Thus, the Islamic banking system should develop various business models for energy investments through these two methods. Moreover, informing investors and the wider public about interest-free financing opportunities and creating various discount promotions will contribute to the spread of sustainable energy investments across the country. So much so that, thanks to small-scale sustainable energy investments, both environmentally friendly carbon emissions will be reduced, and the country's external energy imports will be reduced via the production of extra energy. As a result, shifting from anthropocentric to ecocentric lifestyle that is self-sufficient in terms of energy and compatible with the environment can be generated for the sake of sustainability.

References

Aguarón, J., Escobar, M. T., & Moreno-Jiménez, J. M. (2021). Reducing inconsistency measured by the geometric consistency index in the analytic hierarchy process. *European Journal of Operational Research, 288*(2), 576–583.

Aktaş, R., Doğanay, M. M., Gökmen, Y., Gazibey, Y., & Türen, U. (2015). *Sayısal Karar Verme Yöntemleri*. Beta Yayıncılık.

Al-Bashir, M. A. A. (2019). Crowdfunding in Islamic finance: Ensuring proper shariah oversight. In U. A. Oseni & S. N. Ali (Eds.), *Fintech in Islamic finance: Theory and practice*. Routledge.

Al-Harbi, K. M. A. S. (2001). Application of the AHP in project management. *International Journal of Project Management, 19*(1), 19–27.

Ari, I., & Koc, M. (2021a). Philanthropic-crowdfunding-partnership: A proof-of-concept study for sustainable financing in low-carbon energy transitions. *Energy, 222*, 119925. https://doi.org/10.1016/j.energy.2021.119925

Ari, I., & Koc, M. (2021b). Towards sustainable financing models: A proof-of-concept for waqf-based alternative financing model for renewable energy investments. *Borsa Istanbul Review*. https://doi.org/10.1016/j.bir.2021.03.007

Ayub, M. (2007). *Understanding Islamic finance*. John Wiley & Sons (Asia) PTE Ltd.

Bauwens, T., & Eyre, N. (2017). Exploring the links between community-based governance and sustainable energy use: Quantitative evidence from Flanders. *Ecological Economics, 137*, 163–172. https://doi.org/10.1016/j.ecolecon.2017.03.006

Bhattacharya, M., Paramati, S. R., Ozturk, I., & Bhattacharya, S. (2016). The effect of renewable energy consumption on economic growth: Evidence from top 38 countries. *Applied Energy, 162*, 733–741. https://doi.org/10.1016/j.apenergy.2015.10.104

Bulut, M., & Korkut, C. (2016). A comparison between ottoman cash Waqfs (CWs) and modern interest-free financial institutions. *Vakıflar Dergisi, 46*, 23–45.

Burton, J., & Hubacek, K. (2007). Is small beautiful? A multicriteria assessment of small-scale energy technology applications in local governments. *Energy Policy, 35*(12), 6402–6412. https://doi.org/10.1016/j.enpol.2007.08.002

Cebeci, İ. (2019). Sukuk. In *TDV İslâm Ansiklopedisi*. https://islamansiklopedisi.org.tr/sukuk. 28.03.2021.

Cheng, F., Lin, M., Yüksel, S., Dinçer, H., & Kalkavan, H. (2021). A hybrid hesitant 2-tuple IVSF decision making approach to analyze PERT-based critical paths of new service development process for renewable energy investment projects. *IEEE Access, 9*, 3947–3969. https://doi.org/10.1109/ACCESS.2020.3048016

Çizakça, M. (1996). *A comparative evolution of business partnerships: the Islamic world and Europe, with specific reference to the Ottoman archives* (Vol. 8). Brill.

Çizakça, M. (2000). *A history of philanthropic foundations: The Islamic world from the seventh century to the present*. Boğaziçi University Press.

Dinçer, H., & Yüksel, S. (2018). Çok Kriterli Karar Verme Yöntemlerinin Finans Sektöründeki Uygulamasına Yönelik Yapılmış Çalışmaların Analizi. *Ekonomi İşletme ve Maliye Araştırmaları Dergisi, 1*(1), 1–16.

Du, L., Dinçer, H., Ersin, İ., & Yüksel, S. (2020). IT2 fuzzy-based multidimensional evaluation of coal energy for sustainable economic development. *Energies, 13*(10), 2453.

Düzgün, M. (2021). Identification of the effective criteria for the selection of a warehouse site in the healthcare logistics industry and their placement in order of importance by the Dematel method. *Beykoz Akademi Dergisi, 8*(2), 363–375.

El-Gamal, M. A. (2006). *Islamic finance: Law, economics, and practice*. Cambridge University Press.

Genç, M. (2014). Klasik Osmanlı Sosyal-İktisadi Sistemi ve Vakıflar. *Vakıflar Dergisi, 42*, 9–18.

Haiyun, C., Zhixiong, H., Yüksel, S., & Dinçer, H. (2021). Analysis of the innovation strategies for green supply chain management in the energy industry using the QFD-based hybrid interval valued intuitionistic fuzzy decision approach. *Renewable and Sustainable Energy Reviews, 143*, 110844.

Ioannou, A., Angus, A., & Brennan, F. (2017). Risk-based methods for sustainable energy system planning: A review. *Renewable and Sustainable Energy Reviews, 74*, 602–615. https://doi.org/10.1016/j.rser.2017.02.082

Kahf, M. (2014). *Notes on Islamic economics: Theory and institutions*. CreateSpace Independent Publishing Platform.

Kalkavan, H. (2020). The relation between corporate social responsibility and financial performance: Reviewing empirical studies and discussing the ethical aspect. In H. Dincer & S. Yüksel (Eds.), *Strategic outlook for innovative work behaviors: Interdisciplinary and multidimensional perspectives* (pp. 179–191). Springer International Publishing. https://doi.org/10.1007/978-3-030-50131-0_10

Kuran, T. (2001). The provision of public goods under Islamic law: Origins, impact, and limitations of the waqf system. *Law and Society Review, 35*, 841–898.

Li, Y. X., Wu, Z. X., Dinçer, H., Kalkavan, H., & Yüksel, S. (2021a). Analyzing TRIZ-based strategic priorities of customer expectations for renewable energy investments with interval type-2 fuzzy modeling. *Energy Reports, 7*, 95–108.

Li, Y., Wu, Z., Dinçer, H., Kalkavan, H., & Yüksel, S. (2021b). Analyzing TRIZ-based strategic priorities of customer expectations for renewable energy investments with interval type-2 fuzzy modeling. *Energy Reports, 7*, 95–108. https://doi.org/10.1016/j.egyr.2020.11.167

Li, X., Zhu, S., Yüksel, S., Dinçer, H., & Ubay, G. G. (2020). Kano-based mapping of innovation strategies for renewable energy alternatives using hybrid interval type-2 fuzzy decision-making approach. *Energy, 211*, 118679. https://doi.org/10.1016/j.energy.2020.118679

Mohsin, M. I. A., & Muneeza, A. (2019). Integrating waqf crowdfunding into the blockchain: A modern approach for creating a waqf market. In U. A. Oseni & S. N. Ali (Eds.), *Fintech in Islamic finance: Theory and practice*. Routledge.

Qiu, D., Dinçer, H., Yüksel, S., & Ubay, G. G. (2020). Multi-faceted analysis of systematic risk-based wind energy investment decisions in E7 economies using modified hybrid modeling with IT2 fuzzy sets. *Energies, 13*(6), 1423. https://doi.org/10.3390/en13061423

Saaty, T. L. (1990). *Decision making for leaders: The analytic hierarchy process for decisions in a complex world*. RWS Publications.

Sandler, T. (2004). *Global collective action*. Cambridge University Press.

Steg, L., Shwom, R., & Dietz, T. (2018). What drives energy consumers? Engaging people in a sustainable energy transition. *IEEE Power and Energy Magazine, 16*(1), 20–28. https://doi.org/10.1109/MPE.2017.2762379

Tang, Z., & Dinçer, H. (2019). Selecting the house-of-quality-based energy investment policies for the sustainable emerging economies. *Sustainability, 11*(13), 3514. https://doi.org/10.3390/su11133514

Tunç, H. (2010). *Katılım bankacılığı felsefesi, teorisi ve Türkiye uygulaması*. Nesil Yayınları.

Unver, S., & Ergenc, I. (2021). Safety risk identification and prioritize of forest logging activities using analytic hierarchy process (AHP). *Alexandria Engineering Journal, 60*(1), 1591–1599.

Usmani, M. T. (2002). *An introduction to Islamic finance*. Kluwer Law International.

Wang, S., Ha, J., Kalkavan, H., Yüksel, S., & Dinçer, H. (2020). IT2-based hybrid approach for sustainable economic equality: A case of E7 economies. *SAGE Open, 10*(2), 2158244020924434. https://doi.org/10.1177/2158244020924434

Warde, I. (2000). *Islamic finance in the global economy*. Edinburgh University Press.

Xie, Y., Zhou, Y., Peng, Y., Dinçer, H., Yüksel, S., & an Xiang, P. (2021). An extended Pythagorean fuzzy approach to group decision-making with incomplete preferences for analyzing balanced scorecard-based renewable energy investments. *IEEE Access, 9*, 43020–43035.

Yang, F., Kalkavan, H., Dinçer, H., Yüksel, S., & Eti, S. (2021). Gaussian-based soft computing approach to alternative banking system for sustainable financial sector. *Complexity, 2021*, 4570936. https://doi.org/10.1155/2021/4570936

Yediyıldız, B. (2012). Vakıf. In *TDV İslâm Ansiklopedisi*. https://islamansiklopedisi.org.tr/vakif#2-tarih. 28.03.2021.

Yüksel, S., Dinçer, H., & Meral, Y. (2019). Financial analysis of international energy trade: A strategic outlook for EU-15. *Energies, 12*(3), 431. https://doi.org/10.3390/en12030431

Yüksel, S., & Ubay, G. G. (2021). Determination of optimal financial government incentives in wind energy investments. In *Strategic outlook in business and finance innovation: Multidimensional policies for emerging economies*. Emerald Publishing Limited.

Zhang, Z., Liu, X., & Yang, S. (2009, June). A note on the 1-9 scale and index scale in AHP. In *International conference on multiple criteria decision making* (pp. 630–634). Springer.

Chapter 21
The Impact of Energy Investments on Employment: A Study on the OECD

Halim Baş and İrfan Ersin

Abstract The energy sector is among the vital sectors in the economies of the country. The share of this sector in employment is also important. The aim of this study is to examine the impact of energy investments on employment in the OECD region, which has an important place in the world economy. In the study, in which VAR analysis was used as a method, the period 1990–2019 was analyzed annually. According to the analysis results, no concrete impact of energy investments on employment could be determined. Considering that the energy sector has an important share in the world economy, it is thought that energy investments should be planned in a way that increases employment.

21.1 Introduction

The primary objectives of energy policies are factors such as security of supply, environmental impacts and cost. In addition to these, direct employment and export opportunities created with the energy sector also stand out (Erdal, 2012). Dependence of countries to foreign countries in terms of energy puts a serious burden on the economies of the country in terms of both prices and costs. For this reason, it is very important for countries to determine their energy policies carefully (Dincer et al., 2019). Energy policies of countries are related to the balance of energy production and consumption. When there is a production deficit in a country, the needed energy will have to be imported. In summary, it is important for countries to plan their energy investments and to give investment incentives in terms of production (Wang et al., 2019).

One of the macroeconomic targets of the country's economies is employment. Employment is among the factors that affect economic growth. When countries cannot use their workforce adequately, they face the problem of unemployment. The

H. Baş (✉) · İ. Ersin
Vocational School of Social Sciences, İstanbul Medipol University, İstanbul, Turkey
e-mail: hbas@medipol.edu.tr; iersin@medipol.edu.tr

S. Yüksel, H. Dinçer (eds.), *Strategic Approaches to Energy Management*, Contributions to Management Science,
https://doi.org/10.1007/978-3-030-76783-9_21

unemployment problem causes both economic and social problems. Therefore, the country's economic administrations have to produce policies regarding two important issues such as energy and employment. Making investment planning in the energy sector can both increase energy production and reduce unemployment. At this point, country economies take employment into consideration based on energy resources and technologies.

Energy types are examined in two different types as renewable and non-renewable (Qiu et al., 2020). Countries that are weak in terms of non-renewable energy sources are highly dependent on foreign energy. Therefore, it is important for countries that are weak in non-renewable energy to turn to renewable energy sources and reduce their external dependency in this sense (Liu et al., 2021; Xie et al., 2021). This situation will both reduce foreign dependency and allow the creation of new employment areas. Investments made in order to benefit from renewable energy sources have important effects on the national economies (Zhao et al., 2021). The distribution of non-renewable energy sources is not the same everywhere. Therefore, many countries have to import energy (Dinçer & Karakuş, 2020a, 2020b; Yüksel et al., 2020).

OECD countries are recognized as the world's largest economies. This organization consists of 37 different countries. Hungary, Turkey and other member countries except Mexico is located in the developed countries category. Because of these features, countries play an active role in guiding international markets. Among the OECD countries, especially the USA draws attention in terms of energy production and consumption. Considering energy resources, it is among the top three countries in the world in the production, import and export of crude oil, natural gas, coal, nuclear energy, electric energy and petroleum products. Among other OECD countries, Canada, Norway, Australia, France and Sweden are the leading countries in the world in terms of energy production and consumption (Tezcan, 2015). Therefore, it is possible to say that the OECD region constitutes an important representation in revealing the relationship between energy and employment.

In our study, the impact of energy investments on employment is considered for the OECD region. In this context, the importance of energy investments for employment was mentioned in the second part of the study. In the third chapter, econometric applications and findings are given. In the last section, results and recommendations are presented.

21.2 The Importance of Energy Investments for Employment

It is mentioned that there is an increasing energy demand in the world depending on the population increase. Depending on population growth, the strongest evidence of climate change is greenhouse gases from human production and consumption. There are different parameters at the point of meeting the energy demand (Qi et al., 2020). Basically, these parameters are shaped around the economic conditions of the countries, the richness of raw materials and the availability of transportation facilities. On the other hand, countries produce different strategies and policies in order to meet their energy needs. In this context, direct and indirect effects of the policies and strategies produced emerge. It is possible to reduce the current account deficit arising from energy imports and to create job opportunities as a new investment area (Haiyun et al., 2021). In this framework, the effects of energy investments on job creation will be analyzed. First, the importance of renewable energy and employment of investment areas will be emphasized, and then the impacts of non-renewable energy on employment will be comparatively discussed.

21.2.1 Employment Impact of Renewable and Non-Renewable Energy Investments

Climate change, which has been on the world agenda for a long time, maintains its importance. Many scenarios regarding the destruction of climate change are mentioned. For example, warming of the atmosphere, melting of glaciers, rise in sea water levels. In this context, different effects of global warming occur in various parts of the world. While in some regions the warming weather and the decrease in the amount of snowfall are expected to cause drought and thirst (Cayan et al., 2008), in some places the damage of the projected sea level rise is on the agenda (Van den Hurk et al., 2007). The effects of these results mean new challenges that people will face in their lives in the long term (Climate Change Program US, 2014). These impacts have been and continue to be mostly human-induced. The acceleration of industrialization and the enormous increase in the use of fossil fuels, especially from the mid-1700s, is expressed as an important factor. In addition, the increasingly concentrated greenhouse gases used in some production branches are considered as an accelerating effect. However, climate change is not only affected by human-induced factors (Du et al., 2020). At the same time, the effects of active volcanic mountains with the eruptions and changes in the sun are in question (Karl et al., 2009).

On the other hand, two basic intervention types are known against the risks caused by climate change. The first of these is the climate change mitigation option. In this framework, it can be aimed to limit global climate change by reducing greenhouse gas emissions. The other option is to adapt to climate change. Adapting to climate change is expressed as mitigating the negative consequences of climate change with different actions (Füssel & Klein, 2006). It may be meaningful to center the goal of changing the environment to include both forms of intervention. At this point, people's behavior is also important. In general, there are behaviors that affect climate change directly and indirectly. Energy saving and climate change sensitive automobile purchasing, house insulation or preferring public transportation are within the scope of direct effects. In an indirect context, climate change policies and the attitudes developed against the regulations made by these policies are important. In addition, being psychologically motivated against climate change and acquiring an environmentally sensitive citizenship awareness are the factors. However, laws and agreements implemented to reduce the damages caused by production methods are included in the context of indirect effects (Zahran et al., 2006).

Climate change mitigation costs create inequality between countries' struggles. In this context, emission reduction related to the impacts of climate change can only be limited to cost-effective regions. There is little or no cooperation between countries regarding climate change and their level of compliance with policies play an active role. In the socio-economic context, the expected population, technological development, physical and human capital elements are at least as important as the consumption of energy and fossil resources (Taconet et al., 2020). On the other hand, the predicted results with climate change trigger many transformations. In this sense, the uncertainty brought about by physical climate change greatly strengthens the tendencies towards energy potential and resources. Accordingly, a radical transformation of energy systems appears as a strong possibility. It is stated that the supply and demand side changes may differ in the transformation of energy systems. Factors affecting supply such as raw material, price and availability are determinants. On the other hand, the demand changes depending on the transformation that takes place in the form of heating and cooling. Therefore, reliability, cost and local environmental factors are the structural determinants of physical climate change (Cronin et al., 2018).

On the other hand, increasing environmentalism movement due to the development of societies brings the neomalthus thought to the agenda. This thinking concentrates on some of the shortcomings that arise due to increased resource consumption. In this sense, issues such as food and water, where production will not be sufficient to keep up with consumption, are at the center of the focus. Therefore, the impact of climate change on scarce resources is interpreted in the context of prudence (Gleditsch, 2021). However, based on this understanding, an optimistic point of view is also put forward. Especially in the context of demographic transition, it is important to balance the world population in the long term, to produce using less resources and to discover new ways of recycling thanks to investment in science and technology (Gleditsch, 2021).

The supply of life-facilitating needs in the world is gaining momentum. People can make more economical choices in terms of time and cost while performing their routine work. In this context, it is important to take steps to facilitate people's lives. Access to renewable energy sources can be shown among the priorities of governments. Currently, there are markets formed by renewable energy sources. Consumers, government energy programs and incentive subsidies are known to be essential factors in the formation of these markets (Sørensen, 2004). However, market characteristics of different renewable energy types may vary depending on the nature of the resource concerned. Price, climatic conditions, use of space, the existence of businesses, competitive behavior and preferences can be determinants of variability in production (Sørensen, 2004).

On the other hand, it is not always possible to meet the increasing energy demand with countries' own resources. It is a desired result for countries to have development policies and sustainable economies. However, the insufficiency of internal dynamics may direct countries towards external financing resources. In this framework, achieving sustainable development goals requires countries to act within the framework of "responsible production and consumption" (Azam & Haseeb, 2021). Therefore, the standard economic perspective attaches importance to being renewable and sustainable with a "win-win" logic (Böhringer et al., 2013).

This understanding increases the interest in green energy investments in the world. Investments in green energy fields are not limited to alternative generation technologies. In addition, it means investment in traditional buildings and transport infrastructure (Hillebrand et al., 2006). However, the result of technological change is an increase in output rather than a noticeable increase in the employment level (Lehr et al., 2008). The outputs of the investment contribute significantly to both the social and economic development of countries and the protection of the world. The understanding of green investment envisages reducing greenhouse gas emissions and air pollution without significantly reducing production and consumption (Eyraud et al., 2013). On the other hand, the actors of green investment are public and private sectors. Economic development in a country remains the most important goal for the most part. Therefore, high GDP of countries directly affects green energy investments (Liao, 2018). One of the most important functions of the public actor is to prevent the effects of climate change by accelerating the implementation of green energy investments in line with the necessary legal regulations and to directly contribute to renewable energy production (Kečck et al., 2019; Gernaat et al., 2021).

In this framework, optimistic forecasts for the future predict the transformation of traditional production structures. The impact of a radical transformation in energy systems on the future of work and employment is important. In general, employment expansion caused by direct, indirect and compulsory works is mentioned. Accordingly, the works brought about by the installation of renewable energy systems are basic activities such as raw material production and marketing of the most important items of the supply chain, ensuring the sustainability of renewable energy systems. Depending on the expansion of the renewable energy field, it is possible to increase the economic activities of workers, shareholders, entrepreneurs, and governments (Fragkos & Paroussos, 2018).

On the other hand, there are estimates within the framework of different models regarding the employment outputs of investments to be made in renewable energy systems. Bulavskaya and Reynès (2018) predicts that by 2030, 50,000 full-time jobs will occur in the Netherlands due to investment in renewable energy systems. Garrett-Peltier (2017) reveals that an investment of $ one million in renewable energy will create up to three times more jobs than investment in fossil fuels. It also estimates that for every $ one million shifted from employment made up of brown jobs to green energy, there will be a net increase of 5 jobs. On the other hand, countries that attach importance to the production of energy equipment during the transition to renewable energy systems can prioritize exports. Thus, an additional employment area can be created due to the sector formed. In this context, in the scenarios produced for different regions of the world, additional employment opportunities are expected in some regions and job losses in some regions. Therefore, given the regional differences, it seems possible for countries that discover the employment effect of renewable energy and establish export infrastructure for it to expand employment (Ram et al., 2020). At the same time, action is taken by taking into account the additional employment effect within the internal dynamics of the countries and the monetary value of employment in rural areas rather than urban areas (Nagatomo et al., 2021). Indeed, based on the experiences of developed countries, the effects of wage subsidies are becoming increasingly important in developing countries where social inequalities are intensely experienced. Thus, it is considered as a powerful option for employment creation in rural areas and regions where poverty, inequality and impossibility prevail (Nasirov et al., 2021).

On the other hand, fossil fuels, which are considered as non-renewable energy sources, are still the dominant element of energy production. It is thought that this situation will continue for a certain period of time. In this framework, the change and transformation of an established attitude in production technique in a situation where alternatives are multiplied is considered as a time-consuming action (Ucal & Xydis, 2020). Therefore, it is still difficult to find out which of the renewable energy and non-renewable energy production systems will create more direct employment. At the same time, since renewable energy technology investments require more cost per unit than non-renewable energy fuels, it can also pose an obstacle to a rapid employment output (Bulavskaya & Reynès, 2018). However, there are comparative studies to estimate the employment effect of both systems. Barros et al. (2017) stated that renewable energy alternatives do not have a homogeneous contribution. In this context, some alternatives have a stronger direct employment effect, while others have a gradually decreasing effect in the near future. On the other hand, non-renewable energy alternatives are seen as highly competitive when certain conditions are met. Therefore, these results prove that renewable energy sources do not completely generate strong employment (Barros et al., 2017).

Comparisons made for regions and countries are also meaningful in this context. Hartono et al. (2020) stated that coal-fired power plants, one of the non-renewable energy resources in Indonesia, require relatively less investment volume than hydroelectric power plants. Additionally, it appears to have a significant impact on employment. This means lower investment volume and higher GDP and employment growth. However, it is pointed out that although the solar power plant requires a smaller investment volume, its economic and social impacts may be limited.

Ultimately, a foreseen change in energy policies can have an additional employment effect in renewable energy systems, as well as create an employment reduction effect in systems where non-renewable energy is used. More broadly, it may need to need more specifically applied evidence to determine impacts on labor markets. Thus, when the human capital factor, financing power, investment vision and possible effects are evaluated in a broad context, it is vital to measure the net outputs of both renewable energy investments and non-renewable energy sources, which currently represent the largest energy supply (Pestel, 2019).

21.3 Practice on Relationship Between Energy Investments and Employment

In this part of the study, the existence of the relationship between energy investments and employment will be tried to be revealed by econometric method. It was emphasized in the literature that energy investments are important for employment. In this context, it is thought that this situation should be examined empirically.

21.3.1 Literature Review

Employment is among the important macroeconomic targets in the economy and is seen as a factor affecting production and growth in the economy. For this reason, it is important to examine the sectors that will increase employment and affect employment. The effect of the energy sector, which has come to the fore among these sectors in recent years, on employment has been discussed by many studies in the literature. Studies on the relationship between energy and employment generally focus on renewable energy investments. Karaca et al. (2017) analyzed the employment-increasing effect of renewable energy investments using the COPRAS method. As a result of the study, it has been determined that energy investments increase employment. Similarly, Erdal (2012) discussed the employment creation effect of energy investments and it was predicted that investments to be made in the renewable energy sector will pave the way for the development of green jobs in sectors such as transportation, construction, energy and agriculture in the medium and long term. In the study conducted by Dinçer and Karakuş (2020a, 2020b), the relationship between renewable energy investments and employment was examined for G-7 countries. In this study, Kao and Pedroni Cointegration tests were used, and the relationship was determined as a result of the test.

Brown et al. (2020) made an employment forecast for energy investments for the 2020–2050 period in their study. In this study, input-output analysis has been made on the goods bills of the energy sector companies in the USA and it has been estimated that energy investments have an employment-increasing effect. Elder (2020) examined the impact of uncertainty on employment in industries that produce goods and services. In the study using the GARCH approach, it has been determined that the increasing uncertainty measured from energy prices has strong negative effects on employment. Nasirov et al. (2021) examined the effects of renewable energy on employment. In this study, the effects of renewable energy and non-renewable energy investments on employment are compared. Three energy scenarios were produced until 2026 in the study in which the SWITCH method was used on the country of Chile. According to the results of the study, it was estimated that renewable energy investments create more employment than non-renewable energy investments.

21.3.2 Data Set

The impact of energy investments on employment is examined for the OECD region. Three variables are used in the study and the period 1990–2019 is analyzed annually. The first variable of the study is energy production representing energy investments. The most important indicator of energy investments is energy production. The energy production data of the OECD region were obtained from the Global Energy Statistical Yearbook.[1] In addition, logarithmic data of energy production were used in the study. Another variable is the OECD region total employment data. Employment data were obtained from OECD statistics[2] and logarithmic data were used. The last variable is economic growth data. Economic growth data were obtained from OECD statistics[3] and handled proportionally. The variables discussed in the study were analyzed econometrically in EViews 9 package program.

21.4 Methods and Results

21.4.1 Unit Root Test

In order for a time series to be stationary, its mean and variance should not change over time, and the covariance between two periods should only depend on the distance between the two periods, not the period in which this covariance is calculated (Gujarati, 1999). The problem of bogus regression is encountered in models estimated with non-stationary time series and the results obtained do not reflect the real relationship. In such a case, the t and F test results are not valid. Therefore, the stability of variables is important. Stationary tests in time series are traditionally analyzed by Augmented Dickey—Fuller (ADF), Phillips-Perron (PP) and Kwiatkowski et al. (1992) tests. However, traditional unit root tests do not take into account structural breakage and therefore unit root results may be wrong. There may be a break in time series in some periods due to effects such as crisis. This has led to a revision of traditional unit root tests and the birth of unit root tests that take structural breaks into account. One of the unit root tests that take structural breaks into account was carried out by Perron (1990). Perron (1990) showed that the ADF unit root test, which is applied when a one-time break occurs in the trend function, fails to reject the null hypothesis, which is actually false. Perron claimed that in the case of a break in the trend function of the series, the series may

[1]Global Energy Statistical Yearbook 2020, https://yearbook.enerdata.net/ Date of Access: 19.03.2021.

[2]OECD statistic, https://stats.oecd.org/ Date of Access: 19.03.2021.

[3]OECD statistic, https://stats.oecd.org/ Date of Access: 19.03.2021.

Table 21.1 Breakpoint unit root test results

	Trend and intercept		intercept		
Variable name	t-stat	p-sign	t-stat	p-sign	Break date
Emp	−7.6405*	0.00	–	–	2009
Energy	−2.8000	0.9989	–	–	2007
$\Delta Energy$			−5.5747*	0.00	2016
Grow	−7.5934*	0.00	–	–	2009

Note: Δ symbol represents the first difference. In the table, the * symbol represents the 1% significance level

show stable fluctuations around a deterministic trend and developed an alternative test to the ADF unit root test (Emirmahmutoğlu et al., 2005; Atgur & Yigit, 2017).

The stationarity tests of the variables were analyzed using the ADF-based single break unit root test. Model C was preferred in the study. According to the results of the analysis, the energy variable was stationary at the first difference and the breakage date was determined as 2016. OECD overall growth in 2016 was 1.84%. In 2015, this rate was at the level of 2.63%. It is estimated that this break in energy in 2016 is due to growth. Among the variables, employment (emp) and growth (grow) were stable at the level and the breaking periods of both variables were determined as 2009. The reason for this break is estimated as the 2007 global crisis effect. The effects of the developments experienced in a period may sometimes affect the next one or two periods. The analysis results are indicated in Table 21.1.

21.4.2 VAR Analysis

VAR models are preferred for time series as they do not impose any restrictions on the structural model, and they do not require the intrinsic-extrinsic separation of variables. In addition, since VAR models include lagged values of dependent variables, it is possible to make strong predictions for the future. The VAR model gives information about whether there is a long-term relationship between two or more variables and the correlation analysis of the variables is a matter of question. The VAR model, which does not bring any restrictions to the structural model, is frequently considered in time series in terms of revealing dynamic relationships (Eroğlu & Kara, 2017; Mucuk & Alptekin, 2008).

In the process of performing the VAR analysis, the first step is to subject the variables to the unit root test. After ensuring the stationarities of the variables, information criteria are checked for the appropriate delay length. Likelihood Ratio (LR), Final Prediction Error (FPE), Akaike (AIC), Schwarz (SC) and Hannan Quinn (HQ) methods are considered for information criteria in the literature (Kalkavan et al., 2020). In our study, these information criteria were used and the length of the delay was found to be 2 according to the LR, FPE and HQ information criteria, 1 according to the SC information criterion and 4 according to the AIC information criterion. Since the lag length is accepted as 2 in three of these information criteria, the LR, FPE and HQ information criteria were also taken into account in our study.

Whether the VAR model is stationary in general is understood by looking at the positions of the inverse roots of the AR characteristic polynomials in the unit circle or the table modulus values. When the AR characteristic polynomial values of the variables are examined, the modulus values are between 0 and 1. This situation shows that the VAR model is stationary. After looking at the stationarity of the model, the autocorrelation state between the error terms is checked in order to test that the model does not contain any structural problems. In the autocorrelation test in which the inverse hypothesis works, the probability values must be above 0.05. In this study, the probability values are above 0.05 in all 3 delays and there is no autocorrelation problem between the error terms. Another test to examine the structural problem of the model is the normality test. For the normal distribution results of the error terms, the probability values of Jarque-Bera test statistics are checked. When analyzing these test statistics, the reverse hypothesis is taken into account and probability values are expected to be higher than 0.05. According to the results obtained, the probability values of Jarque-Bera test statistics are above 0.05. This situation shows that the error terms are normally distributed. Another prerequisite is to examine whether there is a problem of variance in the model. There is also an inverse hypothesis while analyzing White's variance results. Therefore, the probability value of the Chi-Square statistic should be higher than 0.05 in all observation values of the error term variance (Uysal, 2019; Gün, 2018; Yüksel & Canöz, 2017). In our study, the probability value was realized as 0.5693 and it was understood that there was no variance problem in the estimated model.

The model can be established after pre-tests in VAR analysis. The growth variable is also included in the model, examining the effects of energy investments on employment. It is accepted in the literature that there are at least three variables in model setup in VAR analysis. In this context, the effect of growth as well as energy investments on the employment variable is among the issues that are curious. According to the analysis results, it was determined that a lagged value of employment in the OECD region affects the current period of employment at a significance level of 1%, and that one and two lagged values of economic growth affect the employment variable at a 1% and 5% significance level, respectively. In other words, a one-unit increase in a lagged value of employment increases the current period employment by 0.748305 units, and a one-unit increase in the lagged values of economic growth increases employment by a coefficient close to 0. It seems that the effect of growth on employment is very small. Looking at energy investments, it is determined that energy investments do not affect employment for the OECD region. Energy is a value used in all areas of life. Considerable use of energy in industry, services and agriculture is thought to bring about employment. Considering that the share of energy in growth is important, it should be taken into account in terms of policy objectives that energy investments do not affect employment. The analysis results are summarized in Table 21.2.

Employment in the energy sector (renewable and non-renewable) needs to become more qualified. Considering the energy investments, increasing the diversity

Table 21.2 VAR analysis results

Model	Dependent variable	Independent variable	Symbols of independent variables	Coefficient	t-Statistic	Prob.
Model 1	EMP	EMP (−1)	C1	0.748305	3.619469	0.0006
		EMP (−2)	C2	0.268341	1.282618	0.2046
		GROW (−1)	C3	0.000831	4.084577	0.0001
		GROW (−2)	C4	0.000530	2.229737	0.0295
		DENER (−1)	C5	−0.011998	−0.344015	0.7320
		DENER (−2)	C6	0.041928	1.076807	0.2859
		Constant term	C7	−0.091981	−1.511074	0.1360
	R^2	0.997984				
	Adjusted R^2	0.997379			F statistic (p value)	0.00
	Durbin Watson statistic	2.192412				

of vocational education at the international level in this field is considered important in terms of the energy-employment relationship.

21.5 Conclusion

Economic growth is among the leading indicators in the economic performance of the world countries. Industry, agriculture and services sectors have an important place in achieving economic growth. With the development of technology, sub-sectors of the main sectors such as agriculture, industry and services started to develop and the energy sector came to the fore among these sub-sectors. The energy sector is considerably taken into account in the economy in terms of both consumption and production, and this sector also shows itself in economic growth.

The development of technology and industry in the economies of the country has also increased energy consumption. The increase in energy consumption has created a cost element in the economies of the country, and the failure to achieve energy production-balance in many countries has made countries dependent on foreign energy. For this reason, energy investments have come to the fore.

Apart from providing a balance between production and consumption, energy investments also contribute to employment. Besides being a leading macroeconomic policy, employment is a variable that contributes to economic growth. In this framework, it is deemed valuable to develop employment generating sectors in national economies. However, for energy investments, countries must either have sufficient energy resources or turn to renewable energy investments, which have been on the agenda recently. However, the high investment costs of renewable energy continue to import energy in many countries. In addition, since renewable energy investments require large costs, the private sector avoids investments and such investments are undertaken by the state. Naturally, this situation negatively affects the contribution of the energy sector to employment.

In our study, the contribution of energy investments to employment in the OECD region, which has an important share in the world economy, was tested with VAR analysis and the effect of energy investments on employment could not be determined. This result reveals that renewable and non-renewable energy investments are not employment generating. In this framework, it is recommended to diversify university profession departments in line with the field for qualified employment in energy investments and to implement policies that encourage employment in the field of energy.

In our study, the OECD region has been analyzed as time series within the framework of the relationship between energy and employment. Panel analysis for the OECD region is recommended to researchers who want to study on this subject.

References

Atgur, M., & Yigit, F. (2017). Causality analysis of the impact of monetary policy on stock markets: The case of Turkey 2006–2016. *Journal of Knowledge Management, Economics and Information Technology, 7*(4), 1–20.

Azam, M., & Haseeb, M. (2021). Determinants of foreign direct investment in BRICS-does renewable and non-renewable energy matter? *Energy Strategy Reviews, 35*, 100638.

Barros, J. J. C., Coira, M. L., de la Cruz López, M. P., & del Caño Gochi, A. (2017). Comparative analysis of direct employment generated by renewable and non-renewable power plants. *Energy, 139*, 542–554.

Böhringer, C., Keller, A., & Van der Werf, E. (2013). Are green hopes too rosy? Employment and welfare impacts of renewable energy promotion. *Energy Economics, 36*, 277–285.

Brown, M. A., Soni, A., & Li, Y. (2020). Estimating employment from energy-efficiency investments. *MethodsX, 7*, 100955.

Bulavskaya, T., & Reynès, F. (2018). Job creation and economic impact of renewable energy in the Netherlands. *Renewable Energy, 119*, 528–538.

Cayan, D. R., Maurer, E. P., Dettinger, M. D., Tyree, M., & Hayhoe, K. (2008). Climate change scenarios for the California region. *Climatic Change, 87*(1), 21–42.

Climate Change Science Program (US). (2014). *Climate change impacts in the United States, highlights: US national climate assessment*. US Global Change Research Program.

Cronin, J., Anandarajah, G., & Dessens, O. (2018). Climate change impacts on the energy system: A review of trends and gaps. *Climatic Change, 151*(2), 79–93.

Dinçer, H., & Karakuş, H. (2020a). Yenilenebilir Enerji Yatırımları ile İstihdam Arasındaki İlişkinin Belirlenmesi: G7 Ülkeleri Üzerine Ekonometrik Bir Analiz. *İstatistik ve Uygulamalı Bilimler Dergisi, 1*(1), 40–49.

Dinçer, H., & Karakuş, H. (2020b). Yenilenebilir Enerjinin Sürdürülebilir Ekonomik Kalkınma Üzerindeki Etkisi: BRICS ve MINT Ülkeleri Üzerine Karşılaştırmalı Bir Analiz. *Ekonomik ve Sosyal Araştırmalar Dergisi, 1*(1), 100–123.

Dincer, H., Yüksel, S., & Martinez, L. (2019). Balanced scorecard-based analysis about European energy investment policies: A hybrid hesitant fuzzy decision-making approach with quality function deployment. *Expert Systems with Applications, 115*, 152–171.

Du, L., Dinçer, H., Ersin, İ., & Yüksel, S. (2020). IT2 fuzzy-based multidimensional evaluation of coal energy for sustainable economic development. *Energies, 13*(10), 2453.

Elder, J. (2020). Employment and energy uncertainty. *The Journal of Economic Asymmetries, 21*, e00159.

Emirmahmutoğlu, F., Köse, N., & Yalçın, Y. (2005). Birim Kök Testlerinde Yapısal Kırılma Zamanının İçsel olarak Belirlenmesi Problemi: Alternatif Yaklaşımların Performansları. *VII. Uluslar arası Ekonometri ve İstatistik Sempozyumu Kitabı*, 1–16.

Erdal, L. (2012). Türkiye'de yenilenebilir enerji yatırımları ve istihdam yaratma potansiyeli. *Sosyal ve Beşeri Bilimler Dergisi, 4*(1), 171–181.

Eroğlu, N., & Kara, F. (2017). Türkiye'de makro ihtiyati para politikası araçlarının makroekonomik değişkenlere etkisinin var analizi ile incelenmesi. *İstanbul İktisat Dergisi, 67*(2), 59–88.

Eyraud, L., Clements, B., & Wane, A. (2013). Green investment: Trends and determinants. *Energy Policy, 60*, 852–865.

Fragkos, P., & Paroussos, L. (2018). Employment creation in EU related to renewables expansion. *Applied Energy, 230*, 935–945.

Füssel, H. M., & Klein, R. J. (2006). Climate change vulnerability assessments: an evolution of conceptual thinking. *Climatic Change, 75*(3), 301–329.

Garrett-Peltier, H. (2017). Green versus brown: Comparing the employment impacts of energy efficiency, renewable energy, and fossil fuels using an input-output model. *Economic Modelling, 61*, 439–447.

Gernaat, D. E., de Boer, H. S., Daioglou, V., Yalew, S. G., Müller, C., & van Vuuren, D. P. (2021). Climate change impacts on renewable energy supply. *Nature Climate Change, 11*(2), 119–125.

Gleditsch, N. P. (2021). This time is different! Or is it? NeoMalthusians and environmental optimists in the age of climate change. *Journal of Peace Research, 58*(1), 186–194. https://doi.org/10.1177/0022343320969785

Gujarati, D. N. (1999). *Basic econometrics* (3rd ed.). Mc Graw Hill, Literatür Yayıncılık.

Gün, M. (2018). The co-movement of Credıt default swaps and stock Markets in Emergıng Economıes. *Fınance & Econometrıcs, 55*.

Haiyun, C., Zhixiong, H., Yüksel, S., & Dinçer, H. (2021). Analysis of the innovation strategies for green supply chain management in the energy industry using the QFD-based hybrid interval valued intuitionistic fuzzy decision approach. *Renewable and Sustainable Energy Reviews, 143*, 110844.

Hartono, D., Hastuti, S. H., Halimatussadiah, A., Saraswati, A., Mita, A. F., & Indriani, V. (2020). Comparing the impacts of fossil and renewable energy investments in Indonesia: A simple general equilibrium analysis. *Heliyon, 6*(6), e04120.

Hillebrand, B., Buttermann, H. G., Behringer, J. M., & Bleuel, M. (2006). The expansion of renewable energies and employment effects in Germany. *Energy Policy, 34*(18), 3484–3494.

Kalkavan, H., Eti, S., & Yüksel, S. (2020). Türkiye'deki Bankacılık Sektörü, Sanayi Gelişimi ve Ekonomik Büyüme Arasındaki İlişkinin Var Analizi İle İncelenmesi. *Akademik Araştırmalar ve Çalışmalar Dergisi (AKAD), 12*(22), 56–74.

Karaca, C., Ulutaş, A., & Eşgünoğlu, M. (2017). Türkiye'de optimal yenilenebilir enerji kaynağının COPRAS yöntemiyle tespiti ve yenilenebilir enerji yatırımlarının istihdam artırıcı etkisi. *Maliye Dergisi, 172*, 111–132.

Karl, T. R., Melillo, J. M., Peterson, T. C., & Hassol, S. J. (Eds.). (2009). *Global climate change impacts in the United States*. Cambridge University Press.

Kečeк, D., Mikulić, D., & Lovrinčević, Ž. (2019). Deployment of renewable energy: Economic effects on the Croatian economy. *Energy Policy, 126*, 402–410.

Kwiatkowski, D., Phillips, P. C., Schmidt, P., & Shin, Y. (1992). Testing the null hypothesis of stationarity against the alternative of a unit root: How sure are we that economic time series have a unit root? *Journal of Econometrics, 54*(1–3), 159–178.

Lehr, U., Nitsch, J., Kratzat, M., Lutz, C., & Edler, D. (2008). Renewable energy and employment in Germany. *Energy Policy, 36*(1), 108–117.

Liao, X. (2018). Public appeal, environmental regulation and green investment: Evidence from China. *Energy Policy, 119*, 554–562.

Liu, J., Lv, J., Dinçer, H., Yüksel, S., & Karakuş, H. (2021). Selection of renewable energy alternatives for green Blockchain investments: A hybrid IT2-based fuzzy modelling. *Archives of Computational Methods in Engineering*, 1–15.

Mucuk, M., & Alptekin, V. (2008). Türkiye'de Vergi ve Ekonomik Büyüme İlişkisi: VAR Analizi (1975–2006). *Maliye Dergisi, 155*, 159–174.

Nagatomo, Y., Ozawa, A., Kudoh, Y., & Hondo, H. (2021). Impacts of employment in power generation on renewable-based energy systems in Japan—Analysis using an energy system model. *Energy, 226*, 120350.

Nasirov, S., Girard, A., Peña, C., Salazar, F., & Simon, F. (2021). Expansion of renewable energy in Chile: Analysis of the effects on employment. *Energy, 226*, 120410.

Perron, P. (1990). Testing for a unit root in a time series with a changing mean. *Journal of Business & Economic Statistics, 8*(2), 153–162.

Pestel, N. (2019). *Employment effects of green energy policies: Does a Switch in energy policy toward more renewable sources create or destroy jobs in an industrial country?* Forschungsinstitut zur Zukunft der Arbeit GmbH (IZA).

Qi, W., Huang, Z., Dinçer, H., Korsakienė, R., & Yüksel, S. (2020). Corporate governance-based strategic approach to sustainability in energy industry of emerging economies with a novel interval-valued intuitionistic fuzzy hybrid decision making model. *Sustainability, 12*(8), 3307.

Qiu, D., Dinçer, H., Yüksel, S., & Ubay, G. G. (2020). Multi-faceted analysis of systematic risk-based wind energy investment decisions in E7 economies using modified hybrid modeling with IT2 fuzzy sets. *Energies, 13*(6), 1423.

Ram, M., Aghahosseini, A., & Breyer, C. (2020). Job creation during the global energy transition towards 100% renewable power system by 2050. *Technological Forecasting and Social Change, 151*, 119682.

Sørensen, B. (2004). *Renewable energy: Its physics, engineering, environmental impacts. Economics & planning* (2nd ed., 2000). Academic Press.

Taconet, N., Méjean, A., & Guivarch, C. (2020). Influence of climate change impacts and mitigation costs on inequality between countries. *Climatic Change, 160*(1), 15–34.

Tezcan, N. (2015). OECD ve BRIC Ülkelerinin Enerji Göstergeleri Açısından Çok Boyutlu Ölçekleme Analizi İle Karşılaştırılması. *İstanbul Üniversitesi İşletme Fakültesi İşletme İktisadı Enstitüsü Yönetim Dergisi, 25*(77), 119–135.

Ucal, M., & Xydis, G. (2020). Multidirectional relationship between energy resources, climate changes and sustainable development: Technoeconomic analysis. *Sustainable Cities and Society, 60*, 102210.

Uysal, Ö. (2019). Hindistan'da Ekonomik Büyüme ve Enflasyon İlişkisi: VAR Analizi ve Nedensellik Analizi. *Sakarya İktisat Dergisi, 8*(3), 236–251.

Van den Hurk, B., Tank, A. K., Lenderink, G., van Ulden, A., van Oldenborgh, G. J., Katsman, C., ... Drijfhout, S. (2007). New climate change scenarios for the Netherlands. *Water Science and Technology, 56*(4), 27–33.

Wang, S., Liu, Q., Yuksel, S., & Dincer, H. (2019). Hesitant linguistic term sets-based hybrid analysis for renewable energy investments. *IEEE Access, 7*, 114223–114235.

Xie, Y., Zhou, Y., Peng, Y., Dinçer, H., Yüksel, S., & an Xiang, P. (2021). An extended Pythagorean fuzzy approach to group decision-making with incomplete preferences for analyzing balanced scorecard-based renewable energy investments. *IEEE Access, 9*, 43020–43035.

Yüksel, S., & Canöz, İ. (2017). Does Islamic banking contribute to economic growth and industrial development in Turkey. *Ikonomika, 2*(1), 93–102.

Yüksel, S., Dinçer, H., & Uluer, G. S. (2020). The role of technological development on renewable energy usage: An econometric analysis for G7 countries. In *Handbook of research on sustainable supply chain management for the global economy* (pp. 136–153). IGI Global.

Zahran, S., Brody, S. D., Grover, H., & Vedlitz, A. (2006). Climate change vulnerability and policy support. *Society and Natural Resources, 19*(9), 771–789.

Zhao, Y., Xu, Y., Yüksel, S., Dinçer, H., & Ubay, G. G. (2021). Hybrid IT2 fuzzy modelling with alpha cuts for hydrogen energy investments. *International Journal of Hydrogen Energy, 46*(13), 8835–8851.

The manufacturer's authorised representative in the EU is Springer Nature Customer Service Centre GmbH, Europaplatz 3, 69115 Heidelberg, Germany. If you have any concerns regarding our products, please contact ProductSafety@springernature.com

Printed and bound by CPI Group (UK) Ltd, Croydon, CR0 4YY
30/09/2026
02225690-0012